MW00760264

Fiber Optic Sensors and Applications

Fiber Optic Sensors and Applications

Editors

Swee Chuan Tjin
Lei Wei

MDPI • Basel • Beijing • Wuhan • Barcelona • Belgrade • Manchester • Tokyo • Cluj • Tianjin

Editors
Swee Chuan Tjin
Nanyang Technological University
Singapore

Lei Wei
Nanyang Technological University
Singapore

Editorial Office
MDPI
St. Alban-Anlage 66 4052 Basel,
Switzerland

This is a reprint of articles from the Special Issue published online in the open access journal *Sensors* (ISSN 1424-8220) (available at: https://www.mdpi.com/journal/sensors/special_issues/fiber_optic_sensors_applications).

For citation purposes, cite each article independently as indicated on the article page online and as indicated below:

LastName, A.A.; LastName, B.B.; LastName, C.C. Article Title. *Journal Name* **Year**, *Article Number*, Page Range.

ISBN 978-3-03936-696-5 (Hbk)
ISBN 978-3-03936-697-2 (PDF)

Contents

About the Editors

Swee Chuan Tjin is a Professor of Optics and Photonics at Nanyang Technological University in Singapore. Prof. Tjin's research interests are in fiber optic sensors, biomedical engineering, and biophotonics. Over the years, he has published more than 300 refereed journal papers and conference papers and has filed 40 patents in fiber optic sensors, biomedical engineering, and biophotonics. He is the Associate Provost (Graduate Education and Lifelong Learning) at Nanyang Technological University. He is also concurrently the Co-Director of The Photonics Institute (TPI) and the Founding Chair of the National Research Foundation (NRF)-funded LUX Photonics Consortium, which is a national consortium involving both industry and academia. Prof. Tjin has been in various leadership roles, including as Associate Provost (Graduate Education), Executive Director for the Office of Research & Technology in Defence and Security, Director (Projects) in the President's Office, Director of Research Support Office, Associate Chair (Research) for the School of Electrical and Electronic Engineering, Assistant Director of Research, Deputy Director of the CNRS-International-NTU-Thales Research Alliance (CINTRA), Co-Directors of the Thales@NTU Joint Research Laboratory, Co-Director of the Singapore-University of Washington Alliance in Bioengineering Programme (SUWA), and Founding Director of the Photonics Research Centre.

Lei Wei is an Associate Professor at Nanyang Technological University in Singapore. He received the B.E. degree in Electrical Engineering from Wuhan University of Technology (China) in 2005 and the Ph.D. degree in Photonics Engineering from the Technical University of Denmark (Denmark) in 2011. Then, he joined the Research Laboratory of Electronics at the Massachusetts Institute of Technology (USA) as a postdoctoral associate. In 2014, he joined the School of Electrical and Electronic Engineering at Nanyang Technological University in Singapore as a Nanyang Assistant Professor. In 2019, he was promoted to Associate Professor with tenure. His main research interests are fiber-based devices, multi-functional fibers, bio-fiber interfaces, and in-fiber energy generation and storage. He has served as a TPC chair or TPC member in many conferences. He currently serves as the Director of Centre for Optical Fibre Technology (COFT) at the Nanyang Technological University. He also serves as the Chair of The Optical Society (OSA) Singapore Section and an Executive Committee Member of IEEE Photonics Society Singapore Chapter.

Editorial

Special Issue "Fiber Optic Sensors and Applications": An Overview

Lei Wei * and Swee Chuan Tjin

School of Electrical and Electronic Engineering and The Photonics Institute, Nanyang Technological University, 50 Nanyang Avenue, Singapore 639798, Singapore; esctjin@ntu.edu.sg
* Correspondence: wei.lei@ntu.edu.sg

Received: 12 June 2020; Accepted: 15 June 2020; Published: 16 June 2020

Abstract: We present here the recent advance in exploring new detection mechanisms, materials, processes, and applications of fiber optic sensors.

Keywords: fiber optic sensors; detection mechanisms; materials; applications

1. Introduction

In this Special Issue, we aim to focus on all aspects of the recent research and development related to fiber optic sensors. Recent advances in fiber-based sensing technologies have enabled both fundamental studies and a wide spectrum of applications [1–3]. This Special Issue seeks to bring attention to the most recent results in the field of fiber optic sensors offered by their unique features and advantages, including new detection mechanisms, materials, processes, and applications.

2. Summary of Special Issue Papers

Below is a brief summary of all the articles covered in this Issue.

In "High-Sensitivity, Large Dynamic Range Refractive Index Measurement Using an Optical Microfiber Coupler" [4], a sensing strategy was proposed by utilizing the unique property of the dispersion turning point in an optical microfiber coupler mode interferometer. As a result, high sensitivity of larger than 5327.3 nm/RIU was achieved in the whole refractive index range of 1.333–1.4186. This sensor offered good performance in narrow refractive index ranges with high resolution and high linearity.

In "An Intra-Oral Optical Sensor for the Real-Time Identification and Assessment of Wine Intake" [5], an intra-oral optical fiber sensor was developed around an optical coupler topology and exemplified on the detection and assessment of wine intake. Its implementation exploited the advantages of fiber-optics sensing, and facilitated the integration into a mouthguard, holding considerable potential for real-time biomedical applications for the evaluation of risk factors in diet-related diseases.

In "Laser-Induced Deposition of Carbon Nanotubes in Fiber Optic Tips of MMI Devices" [6], a laser-induced technique was developed to obtain the deposition of carbon nanotubes (CNTs) onto the fiber optics tips of multimode interference (MMI) devices. The laser-induced deposition of CNTs performed in water-based solutions generated nonuniform deposits, while the laser-induced deposition performed with methanol solutions generated uniform deposits over the fiber tip, indicating the crucial role of the solvent on the spatial features of the laser-induced deposition process.

In "Fiber Link Health Detection and Self-Healing Algorithm for Two-Ring-Based RoF Transport Systems" [7], a two-ring-based radio over fiber (RoF) transport system with a two-step fiber link failure detection and self-healing algorithm was proposed to ensure quality of service (QoS) by automatically monitoring the health of each fiber link in the transport system and by resourcefully detecting, locating, and bypassing the blocked fiber links. Moreover, the proposed algorithm was able to find the blocked fiber links in the RoF transport system and animatedly adjust the status of preinstalled optical switches to restore all blocked network connections.

In "Design and Implementation of a Novel Measuring Scheme for Fiber Interferometer Based Sensors" [8], a measuring scheme for fiber interferometer based sensors was proposed to detect the equivalent changes of optical power corresponding to the variation in measuring parameters, and a signal processing system was used to analyze the optical power changes and to determine the spectrum shifts. A sensing device on polymer microcavity fiber interferometer was taken as an example for constructing a measuring system capable of long-distance monitoring of the temperature and relative humidity.

In "Nonlinearity Correction in OFDR System Using a Zero-Crossing Detection-Based Clock and Self-Reference" [9], a method for tuning nonlinearity correction in an optical frequency-domain reflectometry (OFDR) system was developed from the aspect of data acquisition and post-processing. The spatial resolution test and the distributed strain measurement test were both performed based on this nonlinearity correction method, which reduced the hardware and data burden for the system and has potential value for system integration and miniaturization.

In "Dynamic Deformation Reconstruction of Variable Section WING with Fiber Bragg Grating Sensors" [10], a dynamic reconstruction algorithm based on the inverse finite element method and fuzzy network was proposed to sense the deformation of the variable-section beam structure. Considering the installation error of the fiber Bragg grating (FBG) sensor and the dynamic un-modeled error caused by the difference between the static model and dynamic model, the real-time measured strain was corrected using a solidified fuzzy network.

In "Metal Forming Tool Monitoring Based on a 3D Measuring Endoscope Using CAD Assisted Registration" [11], a setup by combining a 3D measuring endoscope with a two-stage kinematic was demonstrated based on the projection of structured light, allowing time-effective measurements of larger areas. By the use of computer-aided design (CAD) data, registration was improved, allowing a detailed examination of local features like gear geometries while reducing the sensitivity to detect shape deviations.

In "A High Sensitivity Temperature Sensing Probe Based on Microfiber Fabry–Perot Interference" [12], a miniature Fabry–Perot temperature probe was designed by using polydimethylsiloxane to encapsulate a microfiber in one cut of hollow core fiber. The temperature sensing performance was experimentally demonstrated with a sensitivity of 11.86 nm/°C and an excellent linear fitting in the range of 43–50 °C, making it a promising candidate for exploring the temperature monitor or controller with ultrahigh sensitivity and precision.

In "Hybrid Plasmonic Fiber-Optic Sensors" [13], the development of plasmonics-based fiber-optic sensors was reviewed to reveal and explore the frontiers of such hybrid plasmonic fiber-optic platforms in various sensing applications. Coupled with the new advances in functional nanomaterials as well as fiber structure design and fabrication in recent years, new solutions continue to emerge to further improve the fiber-optic plasmonic sensors' performances in terms of sensitivity, specificity, and biocompatibility.

In "Carbon Allotrope-Based Optical Fibers for Environmental and Biological Sensing: A Review" [14], the development of carbon allotropes-based optical fiber sensors was reviewed. The first section provided an overview of four different types of carbon allotropes, including carbon nanotubes, carbon dots, graphene, and nanodiamonds. The second section discussed the synthesis approaches used to prepare these carbon allotropes, followed by some deposition techniques to functionalize the surface of the optical fiber, and the associated sensing mechanisms. Finally, a concluding section highlighting the technological deficiencies, challenges, and suggestions to overcome them was presented.

In "Relative Humidity Sensors Based on Microfiber Knot Resonators-A Review" [15], the recent research and development progress of relative humidity sensors using microfiber knot resonators (MKRs) were reviewed by considering the physical parameters of the MKR and coating materials sensitive to improve the relative humidity sensitivity. There are many advantages of the MKR, such as strong evanescent field, a high Q-factor, compact size, and high sensitivity provided a great diversity of sensing applications. The sensing performance of the MKR-based relative humidity sensors was also discussed, including sensitivity, resolution, and response time.

In "Dual-Polarized Fiber Laser Sensor for Photoacoustic Microscopy" [16], the recent progress in fiber-laser-based ultrasound sensors for photoacoustic microscopy was reviewed, especially the dual-polarized fiber laser sensor with high sensitivity. The principle, characterization, and sensitivity optimization of this type of sensor were presented. In vivo experiments demonstrated its excellent performance in the detection of photoacoustic signals in optical resolution photoacoustic microscopy (OR-PAM). This review also summarized the representative applications of fiber laser sensors in OR-PAM and discussed their further improvements.

Funding: This research received no external funding.

Acknowledgments: The Guest Editors would like to thank the authors' contribution to this Special Issue and all reviewers for providing valuable and constructive recommendations. In particular, we would like to thank in-house editor of *Sensors* journal for the administrative support.

Conflicts of Interest: The author declares no conflicts of interest.

References

1. Yan, W.; Page, A.; Nguyen, T.; Qu, Y.; Sordo, F.; Wei, L.; Sorin, F. Advanced Multi-Material Electronic and Optoelectronic Fibers and Textiles. *Adv. Mater.* **2019**, *31*, 1802348. [CrossRef] [PubMed]
2. Loke, G.; Alain, J.; Yan, W.; Khudiyev, T.; Noel, G.; Yuan, R.; Missakian, A.; Fink, Y. Computing Fabrics. *Matter* **2020**, *2*, 786–788. [CrossRef]
3. Yan, W.; Dong, C.; Xiang, Y.; Jiang, S.; Andreas, L.; Loke, G.; Xu, W.; Hou, C.; Zhou, S.; Chen, M.; et al. Thermally Drawn Advanced Functional Fibers: New Frontier of Flexible Electronics. *Mater. Today* **2020**, *35*, 168–194. [CrossRef]
4. Wang, J.; Li, X.; Fu, J.; Li, K. High-Sensitivity, Large Dynamic Range Refractive Index Measurement Using an Optical Microfiber Coupler. *Sensors* **2019**, *19*, 5078. [CrossRef] [PubMed]
5. Faragó, P.; Gălătuş, R.; Hintea, S.; Boşca, A.B.; Feurdean, C.N.; Ilea, A. An Intra-Oral Optical Sensor for the Real-Time Identification and Assessment of Wine Intake. *Sensors* **2019**, *19*, 4719. [CrossRef] [PubMed]
6. Cuando-Espitia, N.; Bernal-Martínez, J.; Torres-Cisneros, M.; May-Arrioja, D. Laser-Induced Deposition of Carbon Nanotubes in Fiber Optic Tips of MMI Devices. *Sensors* **2019**, *19*, 4512. [CrossRef] [PubMed]
7. Tsai, W.-S.; Chang, C.-H.; Lin, Z.-G.; Lu, D.-Y.; Yang, T.-Y. Fiber Link Health Detection and Self-Healing Algorithm for Two-Ring-Based RoF Transport Systems. *Sensors* **2019**, *19*, 4201. [CrossRef] [PubMed]
8. Ma, C.-T.; Lee, C.-L.; You, Y.-W. Design and Implementation of a Novel Measuring Scheme for Fiber Interferometer Based Sensors. *Sensors* **2019**, *19*, 4080. [CrossRef] [PubMed]
9. Zhao, S.; Cui, J.; Tan, J. Nonlinearity Correction in OFDR System Using a Zero-Crossing Detection-Based Clock and Self-Reference. *Sensors* **2019**, *19*, 3660. [CrossRef] [PubMed]
10. Fu, Z.; Zhao, Y.; Bao, H.; Zhao, F. Dynamic Deformation Reconstruction of Variable Section Wing with Fiber Bragg Grating Sensors. *Sensors* **2019**, *19*, 3350. [CrossRef] [PubMed]
11. Hinz, L.; Kästner, M.; Reithmeier, E. Metal Forming Tool Monitoring Based on a 3D Measuring Endoscope Using CAD Assisted Registration. *Sensors* **2019**, *19*, 2084. [CrossRef] [PubMed]
12. Li, Z.; Zhang, Y.; Ren, C.; Sui, Z.; Li, J. A High Sensitivity Temperature Sensing Probe Based on Microfiber Fabry-Perot Interference. *Sensors* **2019**, *19*, 1819. [CrossRef] [PubMed]
13. Qi, M.; Zhang, N.M.Y.; Li, K.; Tjin, S.C.; Wei, L. Hybrid Plasmonic Fiber-Optic Sensors. *Sensors* **2020**, *20*, 3266. [CrossRef] [PubMed]
14. Yap, S.H.K.; Chan, K.K.; Tjin, S.C.; Yong, K.-T. Carbon Allotrope-Based Optical Fibers for Environmental and Biological Sensing: A Review. *Sensors* **2020**, *20*, 2046. [CrossRef] [PubMed]

15. Han, Y.-G. Relative Humidity Sensors Based on Microfiber Knot Resonators-A Review. *Sensors* **2019**, *19*, 5196. [CrossRef] [PubMed]

16. Lin, X.; Liang, Y.; Jin, L.; Wang, L. Dual-Polarized Fiber Laser Sensor for Photoacoustic Microscopy. *Sensors* **2019**, *19*, 4632. [CrossRef]

Article

High-Sensitivity, Large Dynamic Range Refractive Index Measurement Using an Optical Microfiber Coupler

Jiajia Wang [1,†], Xiong Li [2,†], Jun Fu [3,*] and Kaiwei Li [4,*]

1 College of Agricultural Equipment Engineering, Henan University of Science and Technology, Luoyang 471003, China; johnnyjiajia@163.com
2 Tencent Robotics X, Shenzhen 518000, China; henricli@tencent.com
3 Key Laboratory of Bionic Engineering (Ministry of Education), Jilin University, Changchun 130025, China
4 Institute of Photonics Technology, Jinan University, Guangzhou 510632, China
* Correspondence: fu_jun@jlu.edu.cn (J.F.); likaiwei11@163.com (K.L.); Tel.: +86-1856-885-1373 (K.L.)
† These authors contributed equally to this work.

Received: 30 October 2019; Accepted: 19 November 2019; Published: 21 November 2019

Abstract: Wavelength tracking methods are widely employed in fiber-optic interferometers, but they suffer from the problem of fringe order ambiguity, which limits the dynamic range within half of the free spectral range. Here, we propose a new sensing strategy utilizing the unique property of the dispersion turning point in an optical microfiber coupler mode interferometer. Numerical calculations show that the position of the dispersion turning point is sensitive to the ambient refractive index, and its position can be approximated by the dual peaks/dips that lay symmetrically on both sides. In this study, we demonstrate the potential of this sensing strategy, achieving high sensitivities of larger than 5327.3 nm/RIU (refractive index unit) in the whole refractive index (RI) range of 1.333–1.4186. This sensor also shows good performance in narrow RI ranges with high resolution and high linearity. The resolution can be improved by increasing the length of the coupler.

Keywords: optical microfiber coupler; dispersion turning point; wavelength tacking method

---

1. Introduction

The reliable and accurate measurement of refractive index (RI) plays a crucial role in various fields, including biochemical analysis, environmental monitoring, food safety, and physical oceanography. Optical fiber-based RI sensors are one of the most important types of RI sensors, having merits such as high sensitivity, fast response time, anti-electromagnetic interference, remote monitoring ability, small footprint, and low cost [1]. Therefore, in recent years, researchers have carried out extensive and detailed research on optical fiber-based RI sensors.

Generally, these well-explored optical fiber RI sensors can be divided into several categories depending on their sensing mechanisms, including interference-based sensors [1–5], fiber Bragg-grating sensors [6,7], surface plasmonic resonance sensors [8–10], and microfiber resonators [11–13]. Compared with the other aforementioned sensor types, the interference-based fiber sensors offer the advantages of high sensitivity, simple structure, and ease of fabrication.

Therein, optical microfibers, including optical microfiber couplers (OMC), which rely on mode interference and evanescent field sensing mechanisms, are most explored [2,3,14,15]. By carefully optimizing the parameters of such sensors, we can obtain incredibly high sensitivities of tens of thousands of nanometers per refractive index unit (RIU) [3,4]. Another representative sensing scheme—the Fabry–Pérot interferometer—is also widely investigated for RI sensing [1,16]. Such sensors possess open Fabry–Pérot cavities near or on the end facet of the optical fiber, via which the liquid under test can enter the cavity and modify the effective optical pass of the probing light. Their sensing performance can be improved by optimizing the length of the cavity or incorporating the Vernier effect [16]. Photonic crystal fiber sensors (PCF), with air holes running along the fiber axial, are also employed to realize compact RI sensing platforms [17,18]. PCF can offer excellent performance in RI measurement, and their inherent air holes can work as microfluidic channels and facilitate the manipulation of fluids.

In most interference-based fiber sensors, the precise measurement of the physical parameter is normally realized by tracking the wavelength shift of the interference peaks with high resolution through spectrometers and complicated data processing methods. However, these interference-based sensors suffer from the well-known problem of fringe order ambiguity [19–21]. Generally, as the interference pattern of the optical spectrum is typically sine wave-like, the tracking of peaks can become problematic when the spectrum shifts exceed half of the free spectral range (FSR). It is difficult to figure out the direction of wavelength shift and the number of periods it has shifted, which significantly limits the dynamic range to only one FSR. Hence, it contains the practical application of the RI sensor. By increasing the FSR via adjusting the geometric parameters, we can improve the dynamic range to some extent; however, this would lead to a decrease in the resolution of peak recognition. In order to overcome this limitation, researchers developed several signal processing methods, including fast Fourier transform (FFT) [22], the two-peak method [23], and other advanced signal processing methods that can accurately recognize the peak order [19,21].

Previously, we developed an optical microfiber coupler based sensor enhanced by the dispersion turning point (DTP), which can achieve ultrahigh RI sensitivity in both the aqueous environment and gaseous environment [4,5,24]. Similar DTPs have also been discovered in optical-microfiber-based sensors, and ultrahigh sensitivities have been demonstrated by tracking the wavelength shift of the interference dips/peaks [25,26]. However, these sensors also suffer from the problem of fringe order ambiguity, which limits the applications of such sensors to a very narrow refractive index range.

In this paper, we propose a new measurement strategy to broaden the dynamic response range of the optical fiber coupler sensor by tracking the position of the DTP. Our study shows that the position of the DTP is quite sensitive to the surrounding refractive index (SRI), and it tends to redshift as the SRI increases. More importantly, the interference dual peaks/dips on both sides of the DTP are symmetrically distributed, making it possible to approximate the position of the DTP using the middle point of the dual peaks/dips. These peaks/dips show good fringe shape and possess narrow bandwidths, which are much easier to track than the real DTP. Using this simple detection method, we can realize a highly accurate measurement of SRI in a larger dynamic range.

2. Working Principle and Numerical Analysis

The typical structure of an OMC is shown in Figure 1a. It is normally fabricated by thermal tapering two closely packed single-mode telecommunication fibers. Thus, it contains two input ports, two output ports, two tapered transition regions, and a waist region, where the two standard optical fibers are reduced to two microfibers.

The basic working principle is as follows: when we guide light into one of the input ports, the light in fundamental mode travels along the down taper transition region and couples to the fundamental even mode and odd mode of the fiber coupler along the transition region. Then, the two new modes enter the uniform waist region, and interference between them occurs. Subsequently, the two modes enter the up-taper region and couple back to the fundamental mode of single-mode optical fiber. Assuming the initial input power is P_0, we can then get the output power at the through output port [4].

$$P_{out} = P_0 \cos^2\left(\frac{1}{2}\phi\right) \tag{1}$$

Here, ϕ represents the phase difference between the fundamental even mode and odd mode accumulated along the waist region with a length of L. The wavelengths of the dips in the output spectrum meet the criteria.

The wavelength of the Nth dip λ_N in the output spectrum satisfies [4]

$$\phi_N = \frac{2\pi L \left(n_{\text{eff}}^{\text{even}} - n_{\text{eff}}^{\text{odd}}\right)}{\lambda_N} = (2N - 1)\pi \tag{2}$$

where $n_{\text{eff}}^{\text{even}}$ and $n_{\text{eff}}^{\text{odd}}$ are the effective refractive index of the even mode and odd mode, respectively, which can be obtained through numerical simulations. λ_N denotes the wavelength of the Nth dip in the spectrum. In the sensing measurement, the change of SRI can lead to the corresponding change in $n_{\text{eff}}^{\text{even}}$ and $n_{\text{eff}}^{\text{odd}}$ and then cause a wavelength to the interference dip. The sensitivity can be calculated through [5].

$$S_{RI} = \frac{\partial \lambda_N}{\partial n} = \frac{\lambda_N}{n_g^{\text{even}} - n_g^{\text{odd}}} \frac{\partial \left(n_{\text{eff}}^{\text{even}} - n_{\text{eff}}^{\text{odd}}\right)}{\partial n} \tag{3}$$

where n_g^{even} and n_g^{odd} denote the group effective indexes of the fundamental even mode and odd mode, respectively. It can be calculated as $n_g = n_{\text{eff}} - \lambda_N \partial n_{\text{eff}}/\partial \lambda$. With Equation (3), we can numerically calculate the sensitivity of an OMC with given parameters or optimize the parameters of an OMC.

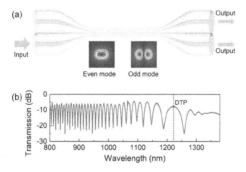

Figure 1. (a) Schematics of the OMC and modal field patterns of the two fundamental modes. (b) The typical transmission spectrum of an optical microfiber coupler with a DTP at around 1220 nm.

Our previous research shows that when the parameters of the OMC satisfy $n_g^{\text{even}} - n_g^{\text{odd}} = 0$, meaning the group effective index of the even mode and the odd mode are equal, the OMC shows a dispersion turning point in its transmission spectrum [4,5]. This DTP features a broad peak/dip, with concomitant dual peaks/dips symmetrically lying on both sides. The spectral responses around the DTP are unique. For example, in RI measurements, when the SRI increases, the dual peaks/dips tend to shift toward the DTP, along with increasing bandwidth. The dual peaks/dips meet at the DTP and vanish simultaneously. Then, the adjacent dual peaks/dips again sift toward the DTP, and so forth. The DTP acts similar to a slider in a zipper that can swallow the elements when we zip up. When the SRI drops steadily, the spectrum acts conversely; the DTP becomes a source that can spit out dual peaks/dips, similar to unzipping a zipper.

Based on this unique property, we have developed ultrasensitive RI sensors by tracking the shift of the dual peaks/dips around the DTP. However, such sensors suffer from a narrow dynamic range.

In fact, the DTP shifts towards short wavelengths as the SRI increases. Through numerical calculations, we found that the DTP blueshifts when the SRI increases from 1.333 to 1.4100 (shown in Figure 2), suggesting that the position of the DTP can be tracked to measure the value of the SRI. For example, for a fiber coupler with a waist width of 3.0 μm, when the SRI increases from

1.3329 to 1.4100, the position of the DTP shifts from 1529.5 to 906.2 nm, showing a high sensitivity of 7196.2 nm/RIU around 1.333. The sensitivity of the OMC, found by tracking the position of the DTP, gradually decreases as the width of the coupler shrinks. Even for a fiber coupler with a width of 2.0 μm, the sensitivity can reach as high as 4774.1 nm/RIU. This phenomenon reveals that the position of the DTP is a very sensitive indicator of the SRI.

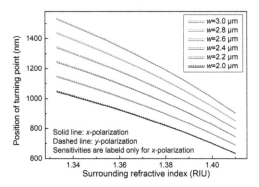

Figure 2. Positions of dispersion turning points (DTPs) for OMCs (w = 2.0–3.0 μm) as the surrounding refractive index increases from 1.3329 to 1.4100.

Here, we discuss the possibility of tracking the position of the DTP with high accuracy. Figure 3a displays a simulated spectrum with a DTP, where the period of the interference spectrum gets large, and the interference fringes narrow on both sides. It is easy to recognize the position of the DTP in this theoretical spectrum. However, the peaks or dips right at the position of the DTP are shallow, and the resolution is low. Moreover, in reality, the position of the DTP is always close to the cut-off point of the odd mode, and the attenuation becomes pronounced, making the precise measurement of its position difficult. A practical approach to obtain the position of the DTP is using the middle point of the dual peaks/dips that symmetrically lies adjacent to the DTP on both sides. These dual peaks/dips are quite distinct and show very narrow bands, which make them easy to be tracked. More importantly, the position of the DTP is determined by the waist width and is independent of the coupler's length. Thus, we can greatly improve the figure of merits of these dual peaks/dips simply by increasing the length of the OMC and, hence, increasing the detection accuracy of the position of the DTP (Figure 3b,c).

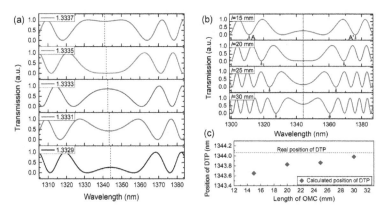

Figure 3. (a) The calculated spectral response of an OMC to SIR. (b) The calculated spectra of OMCs with different waist length. (c) Comparison of the calculated position of the DTP with the real position of the DTP. (Dotted lines marked the position of the DTP).

3. Results and Discussion

In order to demonstrate our sensing strategy, we fabricated an OMC with a width of about 2.6 μm and length of about 6 mm, by tapering two twisted, standard, single-mode fibers (SMF-28, Corning, NY, USA) that were heated using an oxyhydrogen flame. The microscopic photo of the OMC is shown in Figure 4a. Then, we fixed the as-fabricated coupler in the microchannel of a specially designed alumina sensor chip (Figure 4b,c) to keep it stable for IR sensing.

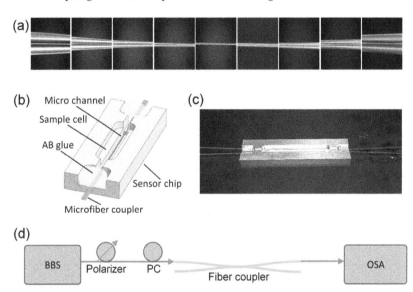

Figure 4. (**a**) Microscopic graph of an OMC. (**b**) Schematic diagram of the sensing chip with integrated microchannels at the bottom of the sample cell. (**c**) Photograph of the sensing chip with an OMC fixed inside the fluidic channel. (**d**) Diagram of the sensing system. PC: polarization controllers; OSA: optical spectrum analyzer.

The experimental setup for the SRI measurements is depicted in Figure 4d. We used a broadband source (BBS, SuperK, NKT photonics) as the input light source, which was connected to one input port of the coupler. We tuned the polarization state of the input light by using polarizers and polarization controllers (PC). The output spectrum of the coupler is analyzed by an optical spectrum analyzer (OSA, AQ6370C, YOKOGAWA). In this study, we used *x*-polarization.

First, we tested the sensing performance of our proposed sensing strategy using the as-fabricated coupler. We measured the response spectrum of the sensor to aqueous glycerol solutions with different SRIs in the wide RI range of 1.3330–1.4186, with an increment of around 0.85. The output spectra are recorded in Figure 5a. We can clearly recognize the broad dip/peak near the cut-off point in the interference spectrum, which is regarded as the region where the DTP exists. However, it is difficult to locate the position of the DTP directly. Using our new sensing strategy, we can calculate the position of the DTP as the middle point of the dual dips that are closely distributed on both sides. The positions of the unique dual dips/peaks are easy to follow, as they show good fringe shape and narrow bandwidth. As a result, the coupler shows a DTP at approximately 1310 nm for distilled water (RI: 1.3330), which is close to our simulated result of approximately 1344 nm as shown in Figure 2. Figure 5a also shows that the DTP continues to red-shift as the SRI increases steadily. Finally, when the SRI reaches a value of 1.4186, the DTP shifts to approximately 756 nm, causing a total wavelength shift of 554 nm. It is easy and convenient to calculate the position of the DTP through the neighboring dual peaks. We summarize the positions of the DTP for different SRIs in Figure 5b. It clearly shows a high RI sensitivity of 5237.3 nm/RIU at 1.3330. The sensitivity also increases as the SRI raises, reaching a sensitivity of 8445.2 nm/RIU at 1.4186. Although the sensitivity of our sensing strategy could be lower than that achieved by tracking the interference dips/peaks near the dispersion turning points, it is still much higher than conventional fiber optic refractive index sensors. These achievements demonstrate the feasibility of our strategy to broaden the dynamic response range of the OMC sensor from a very narrow range to 1.333–1.4186 while keeping a relatively high sensitivity.

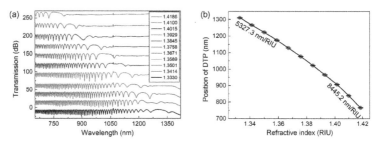

Figure 5. (**a**) Spectral response of an OMC to increasing SIR of 1.3330–1.4186. (**b**). Position of the DTP using our methodology as a function of the surrounding refractive index.

Then, we further tested the RI sensing performance in narrow RI ranges. Two SRI ranges of 1.3330–1.3344 and 1.3486–1.3502 were investigated. The response spectra are shown in Figure 6a,c, respectively. The positions of the DTP are depicted in Figure 6b,d, respectively. The DTP shows a linear response to the SRI with a sensitivity of 5327.3 nm/RIU and R2 of 0.9755 in the RI range of 1.3330–1.3344 and sensitivity of 5878.2 nm/RIU and R2 0.9898 in the RI range of 1.3486–1.3502, respectively. These results again demonstrate the capability of our sensing strategy. Moreover, the resolution of the sensor can be improved by increasing the length of the OMC.

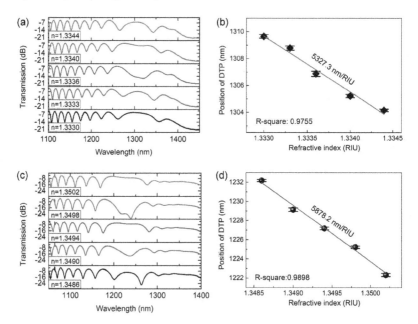

Figure 6. Spectral response of an OMC to small variation of SRI. (**a**) Spectral response of the OMC in the SRI range of 1.3330–1.3347. (**b**) A linear fit of the results. (**c**) Spectral response of the OMC in the SRI range of 1.3486–1.3502. (**d**) A linear fit of the results.

4. Conclusions

In summary, we have proposed and demonstrated a new strategy to tackle the problem of fringe order ambiguity in RI measuring by using the unique spectral characteristics of the dispersion turning point in an OMC. We first analyzed the reliability of using the position of the DTP for RI measurement. Our numerical calculations show that the position of the DTP is quite sensitive to the SRI. Moreover, the position of the DTP can be well approximated as the midpoint of the dual peaks/dips that lay nearly symmetrically on both sides of the DTP. Then, we experimentally tested this sensing strategy by using an optical fiber coupler with a width of 2.6 μm and a length of about 6 mm. A sensitivity of larger than 5327.3 nm/RIU was achieved in the whole RI range of 1.333–1.4186. This sensor also showed good performance in narrow RI ranges with high resolution and high linearity. Moreover, the resolution can be further improved by increasing the length of the coupler.

Author Contributions: J.W., J.F., and X.L. conceived and designed the work. J.W. and K.L. fabricated the devices and performed the experiments. X.L. and K.L. performed the simulations. All authors analyzed the data and discussed the results. J.W. and K.L. wrote the manuscript. J.F., J.W. and K.L. edited and revised the manuscript.

Funding: This research was funded by Tencent Robotics X Focused Research Program, grant number JR201983.

Conflicts of Interest: The authors declare no conflicts of interest.

References

1. Tian, J.; Lu, Z.; Quan, M.; Jiao, Y.; Yao, Y. Fast response Fabry–Perot interferometer microfluidic refractive index fiber sensor based on concave-core photonic crystal fiber. *Opt. Express* **2016**, *24*, 20132–20142. [CrossRef]
2. Liu, G.; Li, K.; Hao, P.; Zhou, W.; Wu, Y.; Xuan, M. Bent optical fiber taper for refractive index detection with a high sensitivity. *Sens. Actuators A Phys.* **2013**, *201*, 352–356. [CrossRef]
3. Luo, H.; Sun, Q.; Li, X.; Yan, Z.; Li, Y.; Liu, D.; Zhang, L. Refractive index sensitivity characteristics near the dispersion turning point of the multimode microfiber-based Mach-Zehnder interferometer. *Opt. Lett.* **2015**, *40*, 5042–5045. [CrossRef] [PubMed]
4. Li, K.; Zhang, T.; Liu, G.; Zhang, N.; Zhang, M.; Wei, L. Ultrasensitive optical microfiber coupler based sensors operating near the turning point of effective group index difference. *Appl. Phys. Lett.* **2016**, *109*, 101101. [CrossRef]
5. Li, K.; Zhang, N.M.Y.; Zheng, N.; Zhang, T.; Liu, G.; Wei, L. Spectral Characteristics and Ultrahigh Sensitivities Near the Dispersion Turning Point of Optical Microfiber Couplers. *J. Light. Technol.* **2018**, *36*, 2409–2415. [CrossRef]
6. Chong, J.H.; Shum, P.; Haryono, H.; Yohana, A.; Rao, M.K.; Lu, C.; Zhu, Y. Measurements of refractive index sensitivity using long-period grating refractometer. *Opt. Commun.* **2004**, *229*, 65–69. [CrossRef]
7. Shu, X.; Zhang, L.; Bennion, I. Sensitivity characteristics near the dispersion turning points of long-period fiber gratings in B/Ge codoped fiber. *Opt. Lett.* **2001**, *26*, 1755–1757. [CrossRef]
8. Li, K.; Zhou, W.; Zeng, S. Optical micro/nanofiber-based localized surface plasmon resonance biosensors: Fiber diameter dependence. *Sensors* **2018**, *18*, 3295. [CrossRef]
9. Monzón-Hernández, D.; Villatoro, J.; Talavera, D.; Luna-Moreno, D. Optical-fiber surface-plasmon resonance sensor with multiple resonance peaks. *Appl. Opt.* **2004**, *43*, 1216–1220. [CrossRef]
10. Zhang, N.M.Y.; Hu, D.J.J.; Shum, P.P.; Wu, Z.; Li, K.; Huang, T.; Wei, L. Design and analysis of surface plasmon resonance sensor based on high-birefringent microstructured optical fiber. *J. Opt.* **2016**, *18*, 065005. [CrossRef]
11. Xu, Z.; Luo, Y.; Liu, D.; Shum, P.P.; Sun, Q. Sensitivity-controllable refractive index sensor based on reflective θ-shaped microfiber resonator cooperated with Vernier effect. *Sci. Rep.* **2017**, *7*, 9620. [CrossRef] [PubMed]
12. Xu, Z.; Sun, Q.; Li, B.; Luo, Y.; Lu, W.; Liu, D.; Shum, P.P.; Zhang, L. Highly sensitive refractive index sensor based on cascaded microfiber knots with Vernier effect. *Opt. Express* **2015**, *23*, 6662–6672. [CrossRef]
13. Shi, L.; Xu, Y.; Tan, W.; Chen, X. Simulation of optical microfiber loop resonators for ambient refractive index sensing. *Sensors* **2007**, *7*, 689–696. [CrossRef]
14. Tian, Z.; Yam, S.S.-H.; Loock, H.-P. Refractive index sensor based on an abrupt taper Michelson interferometer in a single-mode fiber. *Opt. Lett.* **2008**, *33*, 1105–1107. [CrossRef] [PubMed]

15. Zhao, P.; Shi, L.; Liu, Y.; Wang, Z.; Zhang, X. Compact in-line optical notch filter based on an asymmetric microfiber coupler. *Appl. Opt.* **2013**, *52*, 8834–8839. [CrossRef] [PubMed]
16. Zhu, H.; He, J.J.; Shao, L.; Li, M. Ultra-high sensitivity optical sensors based on cascaded two Fabry-Perot interferometers. *Sens. Actuators B Chem.* **2018**, *277*, 152–156. [CrossRef]
17. Zhang, N.; Humbert, G.; Wu, Z.; Li, K.; Shum, P.P.; Zhang, N.M.Y.; Cui, Y.; Auguste, J.-L.; Dinh, X.Q.; Wei, L. In-line optofluidic refractive index sensing in a side-channel photonic crystal fiber. *Opt. Express* **2016**, *24*, 419–424. [CrossRef]
18. Juan Hu, D.J.; Lim, J.L.; Jiang, M.; Wang, Y.; Luan, F.; Ping Shum, P.; Wei, H.; Tong, W. Long period grating cascaded to photonic crystal fiber modal interferometer for simultaneous measurement of temperature and refractive index. *Opt. Lett.* **2012**, *37*, 2283–2285. [CrossRef]
19. Liu, G.; Hou, W.; Han, M. Unambiguous Peak Recognition for a Silicon Fabry-Pérot Interferometric Temperature Sensor. *J. Light. Technol.* **2018**, *36*, 1970–1978. [CrossRef]
20. Liu, G.; Sheng, Q.; Hou, W.; Han, M. High-resolution, large dynamic range fiber-optic thermometer with cascaded Fabry–Perot cavities. *Opt. Lett.* **2016**, *41*, 5134–5137. [CrossRef]
21. Liu, Z.; Liu, G.; Zhu, Y.; Sheng, Q.; Wang, X.; Liu, Y.; Jing, Z.; Peng, W.; Han, M. Unambiguous Peak Identification of a Silicon Fabry-Perot Temperature Sensor Assisted with an In-Line Fiber Bragg Grating. *J. Light. Technol.* **2019**, *37*, 4210–4215. [CrossRef]
22. Jiang, Y. Fourier transform white-light interferometry for the measurement of fiber-optic extrinsic Fabry-Pérot interferometric sensors. *IEEE Photonics Technol. Lett.* **2008**, *20*, 75–77. [CrossRef]
23. Jiang, Y. High-resolution interrogation technique for fiber optic extrinsic Fabry-Perot interferometric sensors by the peak-to-peak method. *Appl. Opt.* **2008**, *47*, 925–932. [CrossRef] [PubMed]
24. Li, K.; Zhang, N.; Zhang, N.M.Y.; Liu, G.; Zhang, T.; Wei, L. Ultrasensitive measurement of gas refractive index using an optical nanofiber coupler. *Opt. Lett.* **2018**, *43*, 679–682. [CrossRef] [PubMed]
25. Sun, L.-P.; Huang, T.; Yuan, Z.; Lin, W.; Xiao, P.; Yang, M.; Ma, J.; Ran, Y.; Jin, L.; Li, J.; et al. Ultra-high sensitivity of dual dispersion turning point taper-based Mach-Zehnder interferometer. *Opt. Express* **2019**, *27*, 23103–23111. [CrossRef]
26. Zhang, N.M.Y.; Li, K.; Zhang, N.; Zheng, Y.; Zhang, T.; Qi, M.; Shum, P.; Wei, L. Highly sensitive gas refractometers based on optical microfiber modal interferometers operating at dispersion turning point. *Opt. Express* **2018**, *26*, 29148–29158. [CrossRef] [PubMed]

Article

An Intra-Oral Optical Sensor for the Real-Time Identification and Assessment of Wine Intake

Paul Faragó [1,*], Ramona Gălătuș [1], Sorin Hintea [1], Adina Bianca Boșca [2],
Claudia Nicoleta Feurdean [3] and Aranka Ilea [3]

[1]   Bases of Electronics Department, Electronics, Faculty of Telecommunications and Information Technology,
      Technical University of Cluj-Napoca, 400027 Cluj-Napoca, Romania; Ramona.Galatus@bel.utcluj.ro (R.G.);
      Sorin.HINTEA@bel.utcluj.ro (S.H.)
[2]   Department of Histology, Faculty of Medicine, University of Medicine and Pharmacy "Iuliu Hatieganu"
      Cluj-Napoca, 400012 Cluj-Napoca, Romania; bianca.bosca@umfcluj.ro
[3]   Department of Oral Rehabilitation, Oral Health and Dental Office Management, Faculty of Dentistry,
      University of Medicine and Pharmacy "Iuliu Hațieganu" Cluj-Napoca, 400012 Cluj-Napoca, Romania;
      feurdeanclaudia@gmail.com (C.N.F.); aranka.ilea@umfcluj.ro (A.I.)
*   Correspondence: paul.farago@bel.utcluj.ro

Received: 20 September 2019; Accepted: 28 October 2019; Published: 30 October 2019

Abstract: Saliva has gained considerable attention as a diagnostics alternative to blood analyses. A wide spectrum of salivary compounds is correlated to blood concentrations of biomarkers, providing informative and discriminative data regarding the state of health. Intra-oral detection and assessment of food and beverage intake can be correlated and provides valuable information to forecast the formation and modification of salivary biomarkers. In this context, the present work proposes a novel intra-oral optical fiber sensor, developed around an optical coupler topology, and exemplified on the detection and assessment of wine intake, which is accounted for example for the formation of N^{ε}-carboxymethyllysine Advanced Glycation End-products. A laboratory proof of concept validates the proposed solution on four white and four red wine samples. The novel optical sensor geometry shows good spectral properties, accounting for selectivity with respect to grape-based soft drinks. This enables intra-oral detection and objective quality assessment of wine. Moreover, its implementation exploits the advantages of fiber-optics sensing and facilitates integration into a mouthguard, holding considerable potential for real-time biomedical applications to investigate Advanced Glycation End-products in the saliva and their connection with consumption of wine, for the evaluation of risk factors in diet-related diseases.

Keywords: optical fiber sensors; wearable sensors; intra-oral sensors; optical coupler; side-emitting fiber; wine spectroscopy; wine color analysis

1. Introduction

Clinical diagnosis is commonly based on invasive procedures for the determination of disease-signaling blood biomarkers. Although blood biomarkers are considered to be the most relevant in the diagnostics procedure [1], the research community targets to investigate the collection of biomarkers from alternative body fluids: sweat, saliva, tears, etc. [2–5].

Human saliva is indeed an attractive oral fluid, and its preference over other body fluids mainly consists of the ease of collection and relative simplicity of analysis, while exhibiting a reduced risk of cross-contamination and infection spread [6]. Indeed, a wide spectrum of compounds present in the salivary fluid has been proven as highly informative and discriminatory, and could be considered as targeted analytes for intra-oral sensors in order to investigate the oral and general health status of an individual [7]. At the present time, several electrochemical sensors have been developed for the analysis of various salivary components such as: glucose, urea, cytokines, mucins, and Advanced Glycation End-products. Moreover, it was determined that salivary compounds are well correlated to the blood concentrations of numerous analytes [8]. Thus, saliva analysis enables a painless diagnostic alternative to accurately reflect the healthy vs. diseased state conditions in humans, particularly useful for people with nervousness concerning the collection of blood samples or for those who require frequent clinical monitoring with multiple sampling in a relatively narrow time interval, e.g., every hour, multiple times per day, etc.

Point-of-care biosensensors are developed to help in the early diagnosis, periodic monitoring, and treatment of disease. Wearable sensors have recently received considerable interest for real-time monitoring of different parameters, specific for the wearer's health, in a wide range of biomedical point-of-care biosensors [5], sport [9], diet-related diseases [10], and military scenarios [11]. Most of the activity on wearable sensors has focused either on the non-invasive monitoring of vital signs via electrophysiological signals, as, for example, electrocardiography, electromyography, photopletismography, pulse oximetry, etc. [12–14], or on chemical sensing for chemical biomarker detection in body fluids [15–18].

The intra-oral concept has gained considerable interest as a wearable biosensing platform with real-time monitoring. For exemplifications, denture-based sensors have been reported for pH and temperature monitoring in [19]. Another example consists of a mouthguard biosensor for continuous salivary-based monitoring of metabolites, as reported in [20].

Integration of electrochemical sensor into the oral cavity is strongly limited by the necessity of a power supply, e.g., external supply, battery or energy harvesting, and corresponding electronics. As such, another direction of research targets in vivo monitoring with battery-less operation of the wearable sensor [21]. A solution for bacteria monitoring in the saliva, reported in [22], resembles a dental tattoo consisting of a resistor, inductor and capacitor (RLC) resonant circuit which has the resistance value, and, consequently, the resonant frequency, varied in the presence of the salivary compounds of interest. Supplying and reading the sensor are performed over the resonant coil, thus eliminating the need for onboard power and external connectivity.

Alternatively to electrophysiological monitoring and electrochemical sensing, optical sensing, in the shape of wearable and autonomous equipment, is enabled by 25 years of tremendous growth and development in the field of fiber optics. Fiber optic sensors offer a wide range of advantages over traditional sensing systems: small size for integration, ease of manufacturing and adaptation for patient-dedicated components, electrical passivity and immunity to electromagnetic interferences, electrical isolation, safety for interaction with the human body as well as environmentally friendly in the sense that fiber optics do not contaminate their surroundings [23,24]. From the signal point of view, fiber optic sensors exhibit a wide dynamic range and high sensitivity, thus having the ability to monitor a wide array of physical and chemical parameters.

Various technologies, including infrared spectroscopy, fluorescence spectroscopy, Raman spectroscopy, liquid chromatography–mass spectrometry (LC-MS), and gas chromatography–mass spectrometry (GC-MS), were proven capable of sensitive detection of different sample compounds, but require expensive equipment and complicated operation for home-based care applications. The optical fiber-based platform offers a suitable, sensitive alternative for point-of-care monitoring of different chemical parameters [25].

Advanced Glycation End-products (AGEs), familiarly known as Maillard products, have recently gained interest as novel biomarkers in the evaluation of risk factors in diet-related diseases. AGEs are inherently formed in foods, and are hypothesized for their endocrine disrupting properties considered as a significant concern for public health [26]. Endocrine disruptors (ED) are either natural or synthetic chemicals (such as pesticides, metals, additives or contaminants in food, and personal care products), which, at certain dosage, interfere with hormones of the living system. Thus, they have been suspected to be associated with health defects such as abnormal growth patterns and neurodevelopmental delays in children, as well as changes in the immune function, altered reproductive function in males and females, and increased incidence of breast cancer [27]. It is estimated that about one thousand out of one hundred thousand manufactured chemicals have endocrine disrupting properties [28]. As such, a new field of research aims towards the development of new salivary biosensor platforms for AGE identification [29–31].

Human exposure to EDs occurs via ingestion of food, water, alcohol, and dust, as well as inhalation of gases and particles in the air and through the skin. For instance, MG (methylglyoxal), GO (glyoxal), and 3-DG (3-deoxyglucosone) derived AGEs are highly formed during manufacturing of bakery products, carbonated soft drinks sweetened with high fructose corn syrup and fermented foods such as yoghurt, wine, or beer. Thus, the assessment of food and beverage intake provides results very well correlated to the formation of AGEs [32]. For a concrete exemplification, the intake of wine is correlated to the formation of N^ε-carboxymethyllysine (CML) and accounts for the intake of 11.20–32.80 AGEs kU/100 mL (estimated for Pinot Noir and Pinot Gris, respectively) [33], and is far more elevated in the cases of chronic alcohol misuse [34]. While salivary biomarker sensing is difficult to be performed with purely optical means, an alternative approach assumes to rather perform identification and assessment of wine in order to forecast the formation and/or modification of targeted biomarkers.

In this paper, we propose a novel intra-oral fiber-optics-based sensor geometry, aiming for the detection and assessment of wine intake. We have previously demonstrated fiber-optics-based salivary optical sensing in [35]. On the other hand, we have illustrated that intra-oral label-free sensing with a D-shaped side-polished optical fiber sensor is limited. Nevertheless, the implementation of intra-oral sensors with plastic optical fibers (POF) accounts for a series of advantages such as flexibility and ease of integration into an intra-oral device. Moreover, fiber-optics sensors can be straightforwardly extended with the deposition of a labeling layer, as well as a noble metal layer for surface plasmon resonance (SPR) [36].

The proposed sensor geometry drops the traditional transmission configuration [37] in favor of an optical coupler topology. The proof-of-concept intra-oral optical sensor is developed around a POF deployed in a T-optical coupler topology, having optical radiation applied laterally into a D-shaped side-polished optical fiber. Unlike the previously reported salivary sensor in [35], which is implemented around a fluorescent optical fiber and performs analyte detection via the modification of the fluorescence spectrum, the intra-oral sensor proposed in this work performs analyte identification via its spectral signature. Depending on the spectral analysis specifications for analyte identification, this can be performed on either the radiation spectrum, the transmission spectrum, or the absorption spectrum, respectively. In the present exemplification, which targets wine detection and assessment, sensing will be performed by investigating the absorption spectrum, and only the visible range will be considered. Two types of radiation sources are investigated: a broad-band white LED source to implement point sensing, and a side-emitting fiber to extend the sensing area. To replicate intra-oral sensing conditions, the proposed optical sensor is integrated into a mouthguard support. Tests are then carried out in a laboratory environment to validate the proposed sensor laboratory proof of concept.

2. Materials and Methods

2.1. Point Intra-Oral Sensor

The proposed intra-oral optical sensor implementation concept is illustrated in the diagram from Figure 1, resembling point sensing. The proposed optical sensor is developed around a D-shaped side-polished POF. Rather than having the fiber propagate optical radiation applied axially on one end, which is the case for side-polished evanescent-wave optical fiber sensors [36–39], the proposed solution assumes having optical radiation applied vertically into the polished fiber surface, implementing an optical coupler topology.

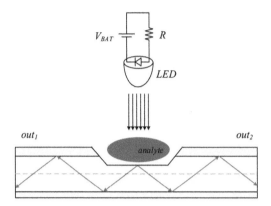

Figure 1. Implementation concept of the proposed intra-oral optical sensor, consisting of an LED source and a D-shaped side-polished optical fiber deployed in a vertical coupler topology, implementing point sensing.

Operation of the proposed intra-oral sensor is described as follows. A white LED source (supplied from battery voltage V_{BAT} over bias resistance R) applies wide-band radiation onto the polished surface of the optical fiber, i.e., the sensing area. This radiation is coupled into the optical fiber and is transmitted to the fiber receiving end. An analyte, present on the sensing area of the fiber, filters the incident radiation and consequently modifies the parameters of the propagated radiation. A spectrometer, deployed on the fiber receiving end, can then be employed to assess the spectrum of the propagated radiation, enabling analyte identification based on its spectral signature.

As illustrated, the proposed intra-oral sensor concept implements vertical coupling of the optical radiation into the fiber, resembling the topology of a T coupler [40,41]. Accordingly, the T-coupler spectral attenuation measures, expressed in terms of insertion loss, coupling ratio, and excess loss, are considered for spectral characterization of the proposed intra-oral sensor.

Insertion loss (*IL*) is expressed as

$$IL_i(\lambda) = 10\lg\frac{P_{in}(\lambda)}{P_{out,i}(\lambda)}, \tag{1}$$

where $P_{in}(\lambda)$ is the power of the incident light source and $P_{out,i}(\lambda)$, $i = \{1, 2\}$, is the power measured at both optical fiber receiving ends.

Coupling ratio (*CR*) is expressed as the difference between the insertion loss measurements in dB, at the two fiber ends, as

$$CR(\lambda) = IL_1(\lambda) - IL_2(\lambda). \tag{2}$$

Excess loss (*EL*) is expressed as

$$EL_i(\lambda) = 10\lg\frac{P_{in}(\lambda)}{P_{out,1}(\lambda) - P_{out,2}(\lambda)}. \tag{3}$$

As expressed in Equations (1)–(3), the insertion loss, coupling loss, and excess loss are wavelength dependent. Thus, the spectral attenuation measures are employed to characterize the intra-oral sensor in the wavelength domain.

The proposed intra-oral sensor exhibits several advantages in comparison to a classical evanescent-wave side-polished optical fiber sensor. On one hand, lateral application of the incident radiation through the polished fiber surface eliminates the limiting effects of sensing area length and polishing depth, which is the case with evanescent-wave optical sensors [38]. On the other hand, lateral illumination of the optical fiber accounts for increased flexibility and is more convenient for applying light into the fiber core, while relaxing the specifications for expensive coupling optics, which is the case for axial illumination [42–44]. In addition, the proposed sensor topology can be straightforwardly applied to extend the sensing area.

2.2. Extended Sensing Area Intra-Oral Sensor

For the realization of extended area sensing, the sensing area on the optical fiber is extended by having polished a longer fiber section in contrast to the point sensor. Then, the incident radiation must also be distributed to feed light into the optical fiber along the entire sensing area. For this purpose, a side-emitting fiber was deployed as illustrated in the diagram of the extended sensing area intra-oral sensor from Figure 2.

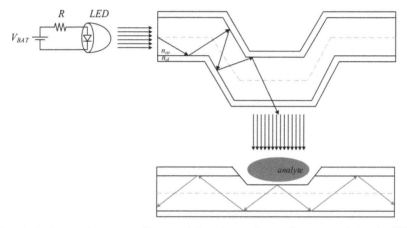

Figure 2. Implementation concept of the extended sensing area intra-oral sensor, consisting of an LED source, a side-emitting fiber, and an end-emitting fiber. Sensing is achieved by having the end-emitting fiber incident illumination applied via a side-emitting fiber, which distributes the incident LED source onto the entire sensing surface.

As illustrated, the side-emitting fiber and the end-emitting optical fiber are deployed in parallel, in an emitter–receiver structure, implementing vertical coupling of the optical radiation [40]. The sensing procedure is the same as for the point sensor. The analyte present on the sensing area will filter the incident illumination, thus modifying the parameters of the propagated radiation. However, in contrast to the point sensor, the extended sensing area sensor enables intra-oral detection along the whole polished fiber length.

Some considerations regarding the deployment of the side-emitting fiber must be specified at this stage. Propagation of light in an optical fiber is performed via total internal reflection (TIR), provided that the core and the cladding refraction indices, n_{co} and n_{cl} respectively, follow

$$n_{co} > n_{cl},$$ (4)

determining a critical angle θ_c expressed by

$$\theta_c = 90^\circ - \arcsin\left(\frac{n_{cl}}{n_{co}}\right) = \arccos\left(\frac{n_{cl}}{c_{co}}\right).$$ (5)

In contrast to the end-emitting optical fiber, the core and cladding refraction indices of side-emitting fibers are chosen such as to violate the TIR condition expressed in Equation (4) and have the refraction indices be approximately equal [45]. As a consequence, radiation applied axially into the side-emitting fiber fiber will gradually leave the fiber via diffraction during propagation. Typically, the side glow intensity I_S decreases exponentially with length,

$$I_s(d) = (4\pi)^{-1} \cdot I_{in} \cdot (1 - \exp(-kd)),$$ (6)

where k is the scattering coefficient, I_{in} is the input source intensity, and d is the distance variable along the fiber length [46]. Considering, however, that the targeted fiber length for integration into the intra-oral device is of only a few centimeters, it is sensible to consider the side-emission intensity constant.

With respect to integration into the intra-oral device, bending the side-emitting fiber to fit the curvature of the mouthguard produces a macrobending. By definition, a macrobending of the fiber is a curvature with a radius larger than the core diameter [47,48], as depicted in Figure 3.

The macrobending accounts for a geometric asymmetry in between the fiber core and cladding [45], which alters the incidence angle as follows. The macrobending produces a local change of the core refractive index profile expressed by [44,47–49]:

$$n_{co}'(r, \theta) = \sqrt{n_{co}^2(r) + \frac{2n_{cl}^2}{R}r\cos\theta},$$ (7)

where R is the bending radius, θ is the bending angle, and r is the radial distance from the fiber axis. The critical bending radius R_c is derived as

$$R_C = \frac{2n_{cl}\lambda}{4\pi\sqrt{(n_{co} - n_{cl})^3}}.$$ (8)

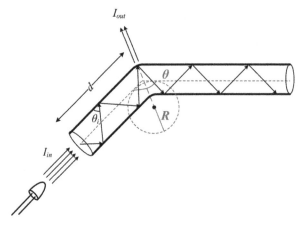

Figure 3. A macrobending of an optical fiber, with the illustration of the input source intensity I_{in}, intensity of the light, which leaves the fiber through the bending I_{out}, the propagation parameters: incidence angle θ_i, and length variable d, and the bending parameters: bending radius R and bending angle θ.

Accordingly, the core refractive index n_{co}' can be decreased, as prescribed in Equation (7), by increasing the bending radius R above R_c:

$$R\uparrow (>> R_C) \Rightarrow n_{co}'\downarrow,$$ (9)

and increasing the bending angle θ around or above the critical angle θ_c, yet below 90° [48]:

$$\theta \uparrow (\geq \theta_C) \Rightarrow n_{co}' \downarrow . \tag{10}$$

This violates the TIR condition in Equation (4) even further, translating to a larger amount of light leaving the fiber. This determines an increased side-glow intensity, while, on the other hand, it determines a propagation loss a_{MB} expressed as

$$a_{MB}(d, \lambda, r, \theta)[dB] = 10 \lg \exp(2 \cdot d \cdot \alpha_{MB}(\lambda, n_{co}'(r, \theta))), \tag{11}$$

where $\alpha_{MB}(\lambda, n_{co}')$ is the macrobending loss coefficient. The phenomenon of propagation loss due to fiber macrobending is undesirable for fiber optics data transmission. In this work, however, it contributes to the increase of the side-glow intensity of the side-emitting fiber, and, consequently, the optical sensor incident light source intensity.

As was the case with the point intra-oral sensor, the extended sensing area sensor is assimilated to an optical coupler, with the side-emitting fiber as the emitter and the end-emitting fiber as the receiver. The sensor is then characterized in the spectral domain as an optical coupler. However, considering the nature of the sensor, assimilation to an optical coupler assumes a continuum of emitters and receivers along the sensing area, making it impractical to assess individual loss components (for example, in terms of insertion or excess loss, respectively) as was the case with the point sensor.

On the other hand, the extended sensing area intra-oral sensor is also affected by additional sources of loss. The side-emitting fiber exhibits a wavelength dependency of the side glow, given by the scattering coefficient k in Equation (6), as well as by the macrobending parameters R_c and a_{MB} defined in Equations (8) and (11), respectively. The scattering coefficient of side-glow fibers exhibits an inverse proportionality vs. wavelength [50]. With respect to propagation, this accounts for a strong attenuation of short wavelengths (blue to green) for short distances and the transmission of long wavelengths (yellow to red) towards longer distances. This phenomenon was confirmed by our laboratory measurements. On the other hand, the bending parameters R_c and a_{MB} exhibit a direct proportionality vs. wavelength [51]. The scattering coefficient k and the macrobending parameters R_c and a_{MB} follow opposite wavelength dependency trends, with different magnitudes.

As such, an input/output spectral attenuation measure would be better suited to assess the wavelength dependency of the superimposed losses. For this purpose, we have considered the wavelength dependent coupling loss [40,52] as a spectral measure of the optical coupler for the characterization of the proposed extended sensing area intra-oral optical sensor. The wavelength dependent coupling loss (*CL*) is defined as an input/output ratio:

$$CL(\lambda) = 10 \lg \frac{P_{in}(\lambda)}{P_{out}(\lambda)} \tag{12}$$

and resembles a superposition of individual loss components, rather than providing an individual estimate for each individual loss component which occurs during propagation and coupling, as was the case with the point sensor. Thus, the benefit of employing this measure is that, besides insertion loss and excess loss, the wavelength dependent coupling loss of the extended sensing area intra-oral sensor accounts for the wavelength dependency of the side-emitting fiber input/side-glow relationship, as well as the wavelength dependency of the bending loss.

2.3. Sensor Integration Into a Mouthguard

Two laboratory proofs of concept have been realized for the proposed intra-oral optical sensor: one for the point sensor and one for the extended sensing area sensor. The parameters of the plastic optical fibers are as follows: a 1 mm diameter end-emitting optical fiber with a 1.49 core refractive index and a 0.5 numerical aperture, and a 1 mm side-emitting fiber with a 1.525 core refractive index and a 0.6 numerical aperture.

The point intra-oral optical sensor consists of an LED source that applies light laterally into a side-polished optical fiber. A Chibitronix SMD form factor white LED (Chibitronics, Lewes, DE, USA), mounted on a flexible triangular sticker substrate alongside the bias resistance and pads for a 3V external power supply, was employed for this purpose. The D-shaped side-polishing of the end-emitting optical fiber was realized using the Industrial Fiber Optics IF CPK 2000 grit polishing paper (Industrial Fiber Optics, Inc., Tempe, AZ, USA). Polishing was realized by describing an 8-shaped motion with the bended fiber on the polishing film. The footprint of the LED and the fiber has been drilled into the mouthguard. The LED was deployed underneath the side-polished optical fiber, with a 0.5 mm spacing in between the LED and the fiber, such as to have the light applied over the polishing into the fiber core, as illustrated in Figure 4.

Figure 4. Practical realization of the point intra-oral sensor laboratory proof of concept, which has the LED source and the optical fiber from the sensor structure integrated into a lateral mouthguard.

The extended sensing area intra-oral optical sensor consists of a side-emitting fiber and an end-emitting optical fiber, deployed in parallel in a vertical coupler topology. Integration into the mouthguard follows the diagram illustrated in Figure 5 and is described as follows.

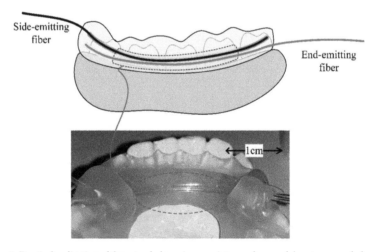

Figure 5. Practical realization of the extended sensing area intra-oral sensor laboratory proof of concept, which has the side emitting fiber and the end-emitting fiber deployed in parallel and integrated into a frontal mouthguard. The source, which feeds light axially into the side-emitting fiber, is external to the sensor.

Two parallel canals were drilled into the mouthguard, such as to have the fibers deployed in parallel with a 0.5 mm spacing. This deployment is then parallel to the lingual slope and perpendicular to the dental arch. The advantage of this deployment is that the sensing surface is inherently exposed to the saliva.

Particular care was paid to having the curvature of the side-emitting fiber envisioned at this stage of the integration process. The side-emitting fiber and the side-polished end-emitting fiber are next fitted into the designated canals, such that the polished surface of the end-emitting optical fiber is oriented towards the curvature of the side-emitting fiber. To counter the inherent elasticity of the plastic fiber, a thin layer of acrylic resin was used to fix the fibers, thus preventing them to jump out from the designated canals. The sensing surface, however, was left exposed to the intra-oral environment in the shape of a window.

2.4. Spectral Analysis of Wine Color

Wine color analysis is performed in this work, with the proposed intra-oral optical sensor, by acquiring and assessing the wine absorption spectrum in the visible range, via the proposed intra-oral optical sensor. The measurement setup is depicted in Figure 6. A white LED applies wide-band radiation onto the analyte, either directly or via a side-emitting fiber, depending on whether the point sensor or the extended sensing area sensor is being tested. A KMAC SV2100 spectrometer (KMAC, Daejeon, Korea) is employed for the acquisition of the sensor output spectrum.

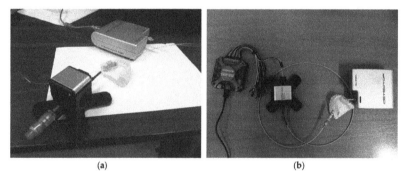

(a)   (b)

Figure 6. The intra-oral optical sensor test setup for wine color analysis: (**a**) point sensor test setup, and (**b**) extended sensing area sensor test setup.

The typical absorption spectra of white and red wines, respectively, are illustrated in the qualitative plots from Figure 7. There are specific particularities in the shape of the wine absorption spectra that are exploited for wine color analysis. White wines exhibit a peak of the absorption spectra in the 400–480 nm wavelength range. Red wines, on the other hand, exhibit an absorption maximum at 520 nm, i.e., absorption of red—standing for the representative absorption of anthocyanins and their flavylium combinations, and a trough at 420 nm, i.e., absorption of yellow—standing for the absorption of tanins and flavonols [53].

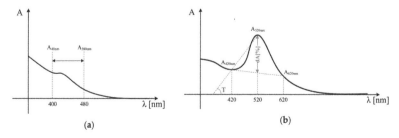

(a)   (b)

Figure 7. Qualitative plot of a typical absorption spectrum for: (**a**) white wines, and (**b**) red wines.

Wine color is mainly defined in terms of intensity, chromaticity, and brightness [53–56]. Each of these measures is expressed, based on the wine absorption spectrum. It should be noticed, however, that these measures are employed in practice only for the assessment of red wines.

Intensity (I) provides an objective quantitative assessment of wine color. The Sudraud method [57] defines wine intensity as the sum of the absorbance measurements at the 420 nm (yellow) and 520 nm (red) wavelengths, respectively,

$$I = A_{420nm} + A_{520nm},$$ (13)

where A_λ is the absorbance measurement at wavelength λ. Wine intensity in Equation (13) is well correlated to the average n of sensory test scores [58], as expressed by

$$n = 23 \cdot \lg I.$$ (14)

Although wine intensity in Equation (13) is a good estimate for the sensorial ranking of wines, as prescribed by Equation (14), it is noteworthy that Equation (14) only holds for red and rose wines, whereas, for white wines, n results in negative values and the equation become inapplicable.

In addition to the yellow and red absorption components, the Glories method [59] also considers the absorbance at the 620 nm, i.e., absorption of blue, standing for the absorbance of quinonic forms of free and combined anthocyanins. Accordingly, the intensity expression is redefined as

$$I' = A_{420nm} + A_{520nm} + A_{620nm}.$$ (15)

Intensity further on enables the assessment of the wine color composition [53–56], defined by the contribution of each absorption component to the overall intensity, particularly useful in the assessment of wine color saturation, as follows:

$$A_{420nm}[\%] = \frac{A_{420nm}}{I'} \cdot 100,$$ (16)

$$A_{520nm}[\%] = \frac{A_{520nm}}{I'} \cdot 100,$$ (17)

$$A_{620nm}[\%] = \frac{A_{620nm}}{I'} \cdot 100.$$ (18)

Hue (*T*) provides an objective assessment of wine chromaticity and is defined as the angle of the line that connects absorbance components at the 420 nm and 520 nm wavelengths, respectively, on the absorbance spectrum [53], as illustrated in Figure 7b. An estimation of this angle is given by the definition of hue as the ratio of the absorbance measures at the 420 nm and 520 nm wavelengths, respectively [57],

$$T = \frac{A_{420nm}}{A_{520nm}}.$$ (19)

With respect to wine aging, red wine assumes a shift from strong red towards orange-red, due to the transition from monomeric to polymeric anthocyanins, respectively. In the spectral domain, this translates to a decrease of A_{520nm} and an increase of A_{420nm}, which influences the hue accordingly [60].

Brightness (*dA*[%]) measures the clearness of the wine and is linked to the shape of the absorption spectrum, as prescribed by equation

$$dA[\%] = (1 - \frac{A_{420nm} + A_{620nm}}{2 \cdot A_{520nm}}) \cdot 100,$$ (20)

and corresponds to the A_{420nm}, A_{520nm} and A_{620nm} triangle median length, as illustrated in Figure 7b. Brightness measures the contribution of red, i.e., the flavilium cations of the free and combined anthocyanins [53,56], to the wine color. Accordingly, a larger brightness value stands for a dominance of the red color of the wine.

3. Results

The laboratory proof of concept of the proposed intra-oral optical sensor is tested in laboratory environment for the intra-oral detection and assessment of wine. Wavelength selective attenuation measures are first assessed in order to have a spectral characterization of the intra-oral optical sensor. Next, measurements on wine samples are carried out. The test results are presented as follows to validate the proposed intra-oral optical sensor.

3.1. Spectral Characterization of the Intra-Oral Optic Sensor

The spectral attenuation of the point intra-oral sensor is expressed in terms of the T-coupler insertion loss, coupling ratio, and excess loss. The test setup is illustrated in Figure 8a.

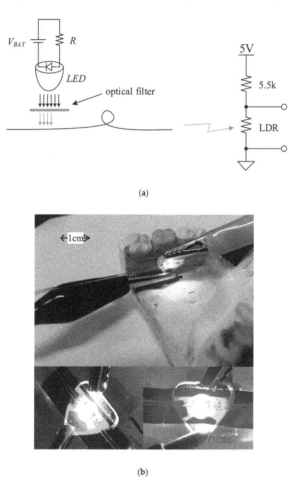

(a)

(b)

Figure 8. Test setup for the assessment of the point intra-oral sensor spectral attenuation measures: (a) schematic diagram, (b) practical realization with external LED supply and optical filter foils.

The optical power from Equations (1)–(3) was replaced with the electrical power measured on a light dependent resistor (LDR). In the measurement setup, the optical fiber output end was focused onto the LDR, connected to a 5 V DC supply in a resistive divider topology with an additional 5.5 K resistance. The output electrical power $P_{out,i}(\lambda)$, $i = \{1, 2\}$, was measured at each of the coupler outputs by having each fiber end illuminate the LDR successively. The reference power $P_{in}(\lambda)$ was measured by having the white LED source (measured broadband spectrum with a 457 nm spectral peak and 8.03 µW emission power) illuminate the LDR over a 1 m long POF optical fiber.

In the absence of a monochromator, as was described in the procedure from [41], wavelength scanning was performed with the use of optical filter foils deployed in between the LED source and the side-emitting fiber, as illustrated in Figure 8b. This allowed for the acquisition of the output radiation power estimate for a discrete set of wavelengths, as listed in Table 1 alongside the spectral attenuation measures. The corresponding spectral attenuation measures were then interconnected with cubic spline interpolation.

Table 1. Resistance values of light dependent resistor measured for different wavelengths of the sensor input, and the corresponding spectral attenuation measures: insertion loss (IL), coupling ration (CR) and excess loss (EL).

Wavelength	$R_{LDR}@end_1$	$R_{LDR}@end_2$	IL_1	IL_2	CR	EL
455 nm	1.7 kΩ	1.6 kΩ	0.74	1.09	0.35	6.02
456 nm	2.2 kΩ	2 kΩ	0.4	0.6	0.18	6.43
470 nm	1.7 kΩ	1.4 kΩ	1.03	1.7	0.66	5.57
480 nm	1.9 kΩ	1.7 kΩ	0.7	0.95	0.25	6.1
533 nm	2.2 kΩ	2 kΩ	0.42	0.55	0.13	6.44
538 nm	2.9 kΩ	2.8 kΩ	−0.16	−0.13	0.03	7.07
556 nm	2.8 kΩ	2.6 kΩ	−0.12	−0.04	0.08	7.02
562 nm	1.9 kΩ	1.7 kΩ	0.7	0.95	0.25	6.1
607 nm	2.5 kΩ	2.5 kΩ	0.64	−0.61	0.02	7.56
623 nm	2.1 kΩ	2 kΩ	0.4	0.52	0.12	6.46
630 nm	2.3 kΩ	1.9 kΩ	0.35	0.68	0.32	6.41

The point intra-oral sensor spectral attenuation measures are plotted in Figure 9, with dashed line for 1st order interpolation (straight-line interconnection of the points listed in Table 1) and with continuous lines for cubic spline interpolation.

As illustrated by the insertion loss measures plotted in Figure 9a, as well as by the coupling ratio plotted in Figure 9b, the proposed intra-oral optical sensor can indeed be assimilated to a T coupler with a split ratio close to 50–50%. Indeed, the insertion loss measures at the two fiber ends are approximately equal, and the wavelength dependency of the coupling ratio exhibits a rather small fluctuation—less than 1dB peak-to-peak. Similarly, the excess loss plotted in Figure 9c exhibits a peak to peak variation of maximum 2 dB, and can therefore be assimilated to be relatively flat. Accordingly, the point intra-oral sensor is to be considered transparent with respect to color.

A similar procedure was adopted to characterize the spectral behavior of the extended sensing area intra-oral sensor, expressed in terms of wavelength dependent coupling loss. The test setup is illustrated in Figure 10. An addition to the measurement setup is that an external CLM1C-WKW cool white high brightness LED (HBL) (measured broadband spectrum with 445 nm and 567 nm spectral peaks and 174 µW emission power) was used as white radiation source, which applies wideband radiation axially into the side-emitting fiber over a PMMA fiber. This latter PMMA fiber was used as transmission medium to connect the HBL to the intra-oral sensor in order to counter the effect of wavelength selective attenuation of the side-emitting fiber propagated radiation.

The wavelength dependent coupling loss of the extended sensing area intra-oral sensor is plotted in Figure 11, illustrating as expected a rather strong loss for short wavelengths, i.e., the blue spectrum. It should be noticed that, besides the superposition of individual loss components, e.g., insertion loss, excess loss, bending loss, and side-emitting fiber input/side-glow relationship which are impractical to be assessed individually for the extended sensing area intra-oral sensor, the wavelength dependent coupling loss measured with the setup from Figure 10 also accounts for the double connector that joins the PMMA fiber to the side-emitting fiber for application of the incident radiation [61].

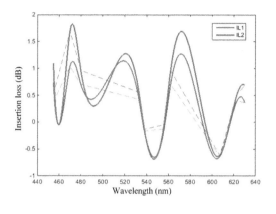

(a)

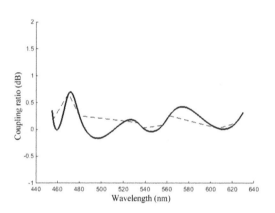

(b)

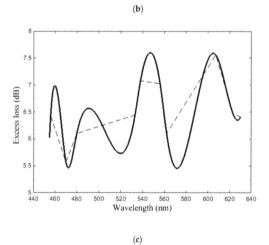

(c)

Figure 9. Spectral attenuation measures of the point intra-oral optical sensor expressed in terms of: (a) insertion loss, (b) coupling ratio, (c) excess loss.

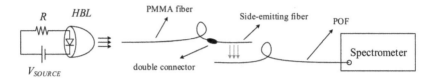

Figure 10. Test setup for the assessment of the wavelength dependent coupling loss of the extended sensing area intra-oral sensor.

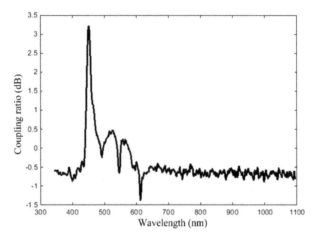

Figure 11. Wavelength dependent coupling loss of the extended sensing area intra-oral sensor.

3.2. Wine Color Analysis Results

The proposed intra-oral optical sensor implementations were tested to acquire the wine absorption spectrum with eight types of wine: four white wines and four red wines, respectively. A set of reference measurements were first carried out following the standard spectrometric method for wine color determination. The test setup is illustrated in Figure 12 and consists of a KMAC TH 2100 broadband source, i.e., 400–800 nm Tungsten Halogen, and a KMAC SV2100 spectrometer. The samples of wine to be tested for the reference absorption spectra were applied into plastic cuvettes. It should be noticed that plastic cuvettes were employed instead of quartz, which is actually the standard, in order to attenuate the incident radiation on the spectrometer sensor and prevent saturation. The reference absorption spectra will be plotted alongside the wine absorption spectra acquired with the proposed sensors for comparison.

Tests of the proposed intra-oral sensor aim to replicate the intra-oral sensing environment in a wine tasting scenario. The sensor, integrated into a mouthguard and supplied from an external 3 V source, was mounted onto a custom jaw mold as illustrated in Figure 13. Accordingly, the sensing plane is parallel to the lingual slope and perpendicular to the dental arch, thus the analyte will pour through, rather than stagnate, on the sensing surface. The sensor was initially moistened with Ringer solution to replicate saliva in the oral cavity. Wine samples were then applied onto the sensing area with a pipette, and the KMAC SV2100 spectrometer, deployed on the sensor receiving end, acquired the wine spectrum. The sensing area was rinsed with water in between successive tests.

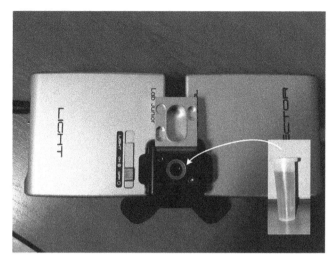

Figure 12. Test setup, consisting of a TH 2100 Tungsten Halogen source and SV 2100 spectrometer, for the acquisition of the reference absorption spectra of the wine samples in plastic cuvettes.

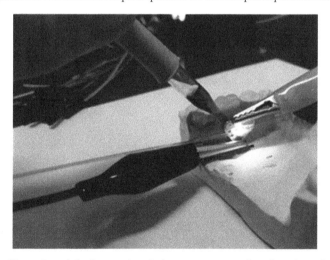

Figure 13. Illustration of the intra-oral optical sensor test procedure for wine color analysis, which assumes having the sensor integrated into a mouthguard and mounted onto a custom jaw mold.

The spectrum of the blank system, i.e., in the absence of the analyte, is first plotted for reference in Figure 14, to illustrate that spectra acquired in the following tests are indeed a consequence of the analyte, i.e., wine. Further on, our preliminary tests illustrate that water, Ringer solution, and human saliva don't inflict any modifications to the spectra plotted in Figure 14, and can be considered transparent in the visible range for the proposed sensing systems.

First, the experimental results obtained with white wine are presented. Four types of white wine have been considered in our tests: Sauvignon Blanc, Yellow of Transylvania (i.e., Fetească Regală, a variety of wine mainly cultivated and produced in Romania), blended dry wine, and blended medium dry wine. The absorption spectra of the white wines considered in our tests, acquired via the proposed intra-oral sensors, are plotted in Figure 15. The results obtained with the point and the extended sensing area sensor respectively are plotted alongside the reference for comparison.

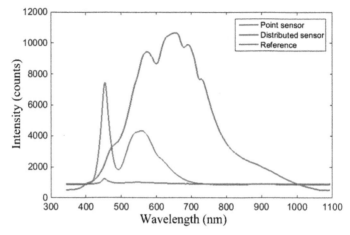

Figure 14. Spectrum of the blank system, i.e., in the absence of the analyte, with a blue line for the point sensor (with white LED source), orange for the extended sensing area sensor (with white LED source), and green line for the reference system (with Tungsten Halogen source).

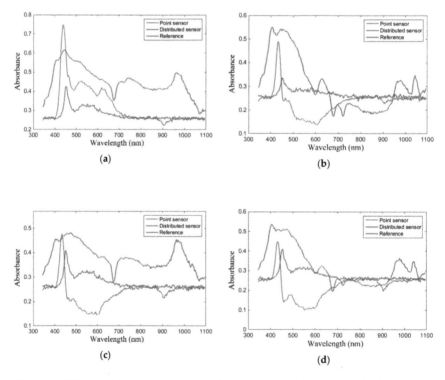

Figure 15. White wine absorption spectra, acquired via the proposed intra-oral sensor implementations, with a blue line for the point sensor, orange line for the extended sensing area sensor, and green line for standard spectroscopic method, respectively: (**a**) blended dry, (**b**) blended medium dry, (**c**) Yellow of Transylvania, (**d**) Sauvignon Blanc.

As illustrated in Figure 15, the white wine absorption spectra follow the specifications of having a spectral peak in the 420–480 nm range, followed by a monotonous roll-off towards longer wavelengths. It should be noticed, however, that the maxima of the absorption spectra acquired with the point and extended sensing area sensors respectively, as listed in Table 2, exhibit a wavelength shift of about 20 nm. This is attributed to the side-emitting fiber wavelength-selective behavior to attenuate short wavelengths inherent to side-emitting fiber operation, this phenomenon being visible on the spectrum of the blank system. Accordingly, wavelengths rejected by the side-emitting fiber don't even reach being applied onto the analyte to be absorbed by the latter, translating into the 20 nm shift.

It should be noticed that literature doesn't provide a definition for white wine color characteristics based on the absorption spectra, as is the case for the red wine color characteristics defined in Equations (13)–(19). The CIE Lab color scheme is rather used for the objective assessment of white wine color. Nevertheless, the authors in [62] provide an illustration of the employment of the red wine color characteristics for the assessment of white wines, and, for that purpose, the absorption spectra plotted in Figure 15 allow for the reading of absorption magnitudes at specific wavelengths of interest.

Second, the experimental results obtained with red wine are presented. Four types of red wine have been considered in our tests: Cabernet Sauvignon, Fetească Neagră (a variety of grapes cultivated in Romania), blended medium dry, and blended medium sweet. The absorption spectra of the red wines considered in our tests, acquired with the proposed intra-oral sensors, are plotted in Figure 16. The results obtained with the point and the extended sensing area sensor, respectively, are plotted alongside the reference for comparison.

Table 2. White wine absorption peaks, acquired with the point and extended sensing area intra-oral sensors, respectively.

Wine Type	Absorption Peak	
	Point Sensor	Extended Sensing Area Sensor
blended dry	438 nm	452 nm
blended medium dry	432 nm	451 nm
Yellow of Transylvania	434 nm	450 nm
Sauvignon Blanc	431 nm	453 nm

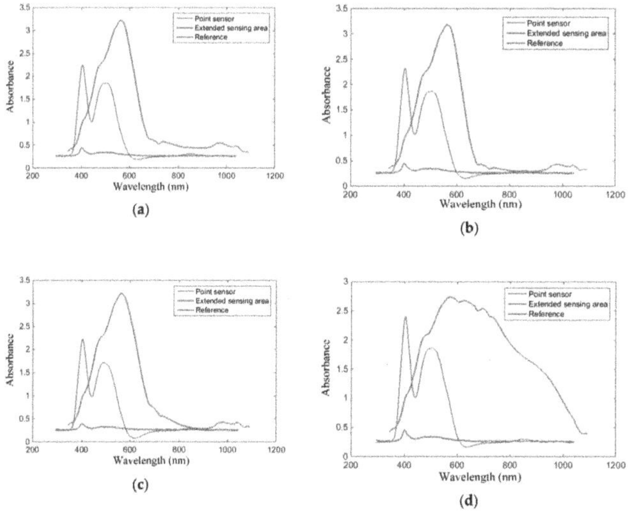

(a)

(b)

(c)

(d)

Figure 16. Red wine absorption spectra, acquired via the proposed intra-oral sensor implementations, with the blue line for the point sensor, orange line for the extended sensing area sensor, and green line for standard spectroscopic method, respectively: (**a**) blended medium dry, (**b**) blended medium sweet, (**c**) Fetească Neagră, (**d**) Cabernet Sauvignon.

As plotted in Figure 16, the red wine absorption spectra follow the specifications of having a peak around 520 nm and a trough around 420 nm. Again, the spectral troughs exhibit a 10–20 nm right shift, as was the case for white wine.

The magnitude of the absorption spectrum at 420 nm, 520 nm and 620 nm, acquired with the point intra-oral sensor, are listed in Table 3 alongside the intensity I, estimated sensory test score n, modified intensity I', hue T and brightness dA[%]. To be notices is that the spectral shift is very strong for Fetească Neagră which, according to our measurement results, accounts for a hue value larger than unity.

Table 3. Red wine absorption spectrum magnitudes and the estimated wine color measures, determined with the point intra-oral sensor.

Wine Type	A_{420nm}	A_{520nm}	A_{620nm}	I	n	I'	T	dA[%]
blended medium dry	1.5	1.7	0.19	3.2	26.7	3.39	0.88	50.3
blended medium sweet	1.2	1.8	0.17	3	25.2	3.17	0.67	61.9
Fetească Neagră	1.6	1.5	0.07	3.1	26	3.17	n.a.	44.3
Cabernet Sauvignon	1.6	1.8	0.19	3.4	28.1	3.59	0.88	50.2

The magnitude of the absorption wavelength at 420 nm, 520 nm and 620 nm, acquired with the extended sensing area intra-oral sensor, are listed in Table 4 alongside the intensity I, estimated sensory test score n, modified intensity I', hue T and brightness dA[%]. Additionally, a multiplicative correction factor of 5 was added to each absorption value to compensate for the strong attenuation inherent to the side-emitting fiber.

Table 4. Red wine absorption spectrum magnitudes and the estimated wine color measures, determined with the extended sensing area intra-oral sensor.

Wine Type	A_{420nm}	A_{520nm}	A_{620nm}	I	n	I'	T	dA[%]
blended medium dry	1.5	1.65	1.3	3.15	26.4	4.5	0.9	13.6
blended medium sweet	1.45	1.7	1.3	3.15	26.4	4.5	0.85	17.6
Fetească Neagră	1.45	1.55	1.3	3	25.2	4.35	0.93	9.6
Cabernet Sauvignon	1.5	1.65	1.4	3.15	26.4	4.55	0.9	12.1

As listed in Tables 3 and 4, the measurement results are comparable with respect to intensity and estimated sensory test score. The hue measure is also comparable for the blended medium dry and Cabernet Sauvignon wine types. On the other hand, the measures for modified intensity and brightness change due to the A_{620nm} magnitude which is different for the two sensors. Indeed, A_{620nm} is considerably smaller than A_{520nm} and A_{420nm} for the point sensor, whereas they are comparable for the extended sensing area sensor.

The difference between the results obtained with the point and extended sensing area sensors respectively consists mainly of the magnitude of the absorption spectral components and the 10–20 nm shift in between the spectral peaks. As specified earlier, the spectral shift is attributed to the side-emitting fiber, and, since this phenomenon is known and expected, it can be accounted for during operation. Thus, both sensors are applicable for the intra-oral detection and assessment of wine. Considering however the nature of the application, which assumes real-time assessment of wine intake, a sip of wine would fill the oral cavity and won't require localizing the presence of wine. From this point of view, we continue our tests with the point sensor.

In the following test scenario, we aimed to detect and assess red wine in the saliva. The normal flow for unstimulated saliva is approximately 0.1 mL/min and for the stimulated saliva is 0.2 mL/min [63]. This amount of saliva does not induce important dilution of the wine; therefore, the wine color is not significantly affected. Rather, saliva contains several peptide families, including the proline-rich proteins that interact with the phenolic compounds in wine, and form complexes that modify the perception of astringency and bitterness [64]. The Fetească Neagră red wine was chosen for this test, as it exhibits the strongest color tonality. Tests were carried out using the point sensor. The absorption spectra for different dilutions of wine in saliva are plotted in Figure 17.

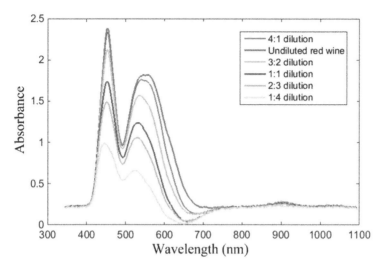

Figure 17. Fetească Neagră red wine absorption spectra, acquired with the point sensor, for different dilutions of wine in saliva.

As illustrated, the red wine absorption spectrum follows the specifications depicted in the qualitative plot from Figure 7b. Accordingly, the proposed sensor is applicable for wine identification in the oral cavity. On the other hand, the larger the dilution of wine, the smaller the magnitudes of the absorption spectrum samples. Thus, while it is still possible to detect and assess wine even in cases of rather large dilutions, applicability of the proposed sensor becomes limited for the detection of traces of wine in the saliva.

The optical sensor we developed is also suitable for the assessment of non-alcoholic wines, based on the wine chemistry. The plyphenolic content of wines is responsible for the wine color. Depending on the method used for dealcoholization, the reduced alcohol wines have different physical and chemical properties: higher concentration in polyphenols and organic acids, and increased color intensity [65].

With regard to sensor selectivity, we have performed the same spectrometric analyses as for wines on grape-based drinks, using the point sensor. Three types of commercially available drinks have been considered: white grape (grapes 5%) and aloe vera drink, white grape (grapes 11%) and raspberry, red grape fizzy drink (grapes 0.1%). The absorption spectra of the grape-based drinks considered in our tests, acquired with the proposed point sensors, are plotted in Figure 18.

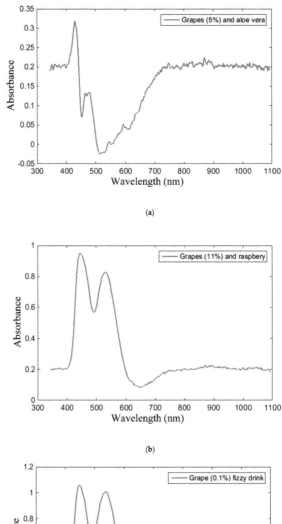

Figure 18. Grape-based drink absorption spectra, acquired via the proposed intra-oral point sensor: (**a**) white grape (5%) and aloe vera, (**b**) white grape (11%) and raspberry, and (**c**) red grape (0.1%) fizzy drink.

As illustrated, the absorption spectrum of the white grape and aloe vera drink plotted in Figure 18a are clearly distinguished from that of white wines. On the other hand, the shape of the absorption spectra plotted in Figure 18b,c resembles that of the red wines. This is attributed to the red color of the raspberry and the red grape. The magnitudes of the spectral peaks, however, are considerably smaller than those determined for red wines, and, on the other hand, the difference between the two peak magnitudes is considerably smaller than for the red wines. As such, the absorption spectra of the soft drinks is clearly distinguished from the red wine spectra as well.

Another aspect to be considered regards the implementation of the proposed sensors using POFs. Alcohol will dissolve plastic in time and, in the absence of a protective foil, the lifetime of the proposed sensor becomes limited. We have performed five repetitions of each test and have obtained similar results. Then, we have inspected the POFs after each repetition and didn't find any degradation caused by alcohol.

4. Discussion

Test results obtained after testing the proposed intra-oral optical sensors for the detection and assessment of wine in the oral cavity lead to a series of valuable conclusions, enumerated in the discussion which follows.

The wine absorption spectra acquired with the proposed intra-oral sensor and a KMAK SV2100 spectrometer resemble the typical wine absorption spectrum, exhibiting the typical absorption peak, trough and roll-off range. Moreover, the acquired spectra exhibit specific differences with respect to the type of wine being analyzed. Thus, the acquired wine spectra allow for wine discrimination based on the spectral signature.

The wine absorption spectra acquired with the proposed sensor exhibit a resolution fine enough, in both wavelength and amplitude domains, to express the measures of wine colors defined in terms of intensity, chromaticity, and brightness. Accordingly, the proposed intra-oral sensor enables characterization of wine color, which provides valuable objective information with respect to the assessment of wine quality.

As illustrated by all of the test results, the proposed intra-oral optical sensor, with either point or extended area sensing, is validated in the laboratory environment and proves to be applicable for real-time intra-oral sensing.

At this stage, a discussion concerns which of the point or extended sensing area sensor should be chosen for a specific application. Considering the nature of wine testing, which is the scenario targeted in this work, a sip of wine fills the oral cavity. This makes the deployment of an extended sensing area sensor for wine detection unjustified. Moreover, the large attenuation introduced by the side-emitting fiber is too high a price for the target of detecting the same type of analyte all along the dental arch. On the other hand, the target of detecting specific compounds that are uniquely localized inside the oral cavity, e.g., blood in the saliva as was the case in [35], which requires a thorough cartography of the oral cavity, renders the extended sensing area intra-oral sensor well argued.

Further experiments aimed to test the applicability of the proposed sensor for the identification of wine in contact with saliva. Several dilutions of wine with saliva were tested. The acquired absorption spectra follow the typical specifications for wine; therefore, indentification can be performed. The magnitudes of the absorption spectrum samples, however, change with wine dilution, and, although the shape of the spectrum is preserved, assessment of wine must account for this phenomenon.

Sensor selectivity was illustrated with a series of tests carried out with grape-based soft and fizzy drinks. In each situation, the acquired absorption spectrum differs from that of wine. Discrimination of wine from other drinks can thus be performed.

Our tests also illustrate the repeatability of the results. While alcohol clearly alters the parameters of the POF in time, as it dissolves plastic, we were able to confirm that we received similar results for five repetitions of the tests. From this point of view, the 10%–15% alcoholic content of wine won't limit the sensor lifetime as far as making it inapplicable for in-vivo monitoring.

Furthermore, the proposed sensor performs label-free intra-oral sensing of wine, replicating a real-life sensing scenario. Accordingly, we target the employment of the proposed sensor for in-vivo intra-oral monitoring. It should be noticed that the proposed optical sensor does not detect the alcohol in the wine, but the distinction between the alcoholic and non-alcoholic wines resides in their health effects. Even though the wine consumption could be associated with some health risks due to the alcohol, the polyphenols present in both alcoholic and non-alcoholic wines exert multiple beneficial effects. Moreover, taking into consideration that the alcohol intake is associated with the production of oxidative stress and induces the indirect formation of AGEs, our further research will focus on the inter-relation between alcoholic wine consumption and the salivary levels of AGEs, such as CML.

5. Conclusions

This paper presented an intra-oral optical sensor for the intra-oral detection and assessment of wine. The proposed sensor was developed around an optical coupler topology, having an emitter apply lateral illumination onto a D-shaped side-polished plastic optical fiber. Two types of emitters have been employed: a white LED—thus implementing a point sensing, and a side-emitting fiber—thus implementing extended area sensing.

The proposed intra-oral optical sensor resembles the topology of an optical coupler and is characterized in the spectral domain by means of insertion loss, coupling ratio, and excess loss for the point sensor, and wavelength selective coupling loss for the extended sensing area sensor, respectively.

Spectral characterization results illustrate a rather flat variation of the spectral attenuation measures vs. wavelength. Accordingly, the proposed intra-oral sensor is considered to be transparent with respect to color, and is therefore proven to be applicable for spectroscopic sensing applications. However, the proposed sensor illustrates a rather large attenuation on the output/input relationship, and the explanation for this phenomenon is threefold. On one hand, lateral illumination is not focused on the polished surface of the end-emitting optical fiber, thus only a small fraction of the LED emitted light is coupled into the fiber. In addition, having saliva as the optical coupler transmission medium contributes some attenuation. On the other hand, the side-emitting optical fiber from the extended sensing area sensor structure introduces an attenuation inherent to operation. To counter the effect of the attenuation inherently introduced by the side-emitting fiber, the extended sensing area sensor employs a high brightness LED as radiation source.

Based on the spectral characterization results, we have considered the proposed intra-oral optical sensor suitable for the detection and assessment of beverages in the oral cavity, provided they exhibit color. In this respect, we have exemplified the proposed sensor on wine samples, and have tested the proposed sensor for the identification and assessment of wine. To replicate the intra-oral sensing environment in a wine tasting scenario, we have applied wine samples onto the sensing area previously moistened with a Ringer solution. Next, we have tested several dilutions of wine in saliva to demonstrate the applicability of the sensor for intra-oral wine detection and assessment. Finally, selectivity of the proposed sensor was illustrated with an experiment which tests several grape-based drinks.

The proposed intra-oral sensor geometry exploits the advantages of fiber-optics sensing and facilitates integration into a mouthguard. Based on the test results and discussions, the proposed solution holds considerable potential for real-time biomedical applications regarding in-vivo monitoring of food and beverage intake, aiming to forecast the formation of disease-signaling salivary biomarkers. In the context of hybrid patient monitoring, the correlation and combination of data collected by electrochemical sensors and the proposed optical sensor could provide more reliable findings for biochemical and immunological profiling of saliva.

Author Contributions: Conceptualization, P.F. and A.I.; Optical sensor design, P.F. and R.G.; Development of mouthguard, C.N.F.; Optical sensor integration into the mouthguard, P.F., C.N.F., and A.I.; Formal analysis, P.F.; Methodology, P.F., R.G. and A.B.B.; Project administration, A.I.; Supervision, S.H.; Validation, S.H., A.B.B., C.N.F., and A.I.; Writing—original draft, P.F. and R.G.; Writing—review and editing, S.H., A.B.B., C.N.F., and A.I.

Sensors **2019**, *19*, 4719

Funding: This research was partially supported by the project 21 PFE in the frame of the program PDI-PFE-CDI 2018 and partially supported by the COFUND-ERA-HDHL ERANET Project, European and International Cooperation—Subprogram 3.2—Horizon 2020, PNCDI III Program—Biomarkers for Nutrition and Health—"Innovative technological approaches for validation of salivary AGEs as novel biomarkers in evaluation of risk factors in diet-related diseases", No. 25/1.09.2017.

Conflicts of Interest: The authors declare no conflict of interest.

References

1. Malon, R.S.P.; Sadir, S.; Balakrishnan, M.; Córcoles, E.P. Saliva-Based Biosensors: Noninvasive Monitoring Tool for Clinical Diagnostics. *BioMed Res. Int.* **2014**, *2014*, 962903. [CrossRef] [PubMed]
2. Chiappin, S.; Antonelli, G.; Gatti, R.; de Palo, E.F. Saliva specimen: A new laboratory tool for diagnostic and basic investigation. *Clin. Chim. Acta* **2007**, *383*, 30–40. [CrossRef] [PubMed]
3. Kaufman, E.; Lamster, I.B. The diagnostic applications of saliva—A review. *Crit. Rev. Oral Biol. Med.* **2002**, *13*, 197–212. [CrossRef] [PubMed]
4. Lee, Y.H.; Wong, D.T. Saliva: An emerging biofluid for early detection of diseases. *Am. J. Dentistry.* **2009**, *22*, 241–248.
5. Tricoli, A.; Nasiri, N.; De, S. Wearable and Miniaturized Sensor Technologies for Personalized and Preventive Medicine. *Adv. Fun. Mater.* **2017**, *27*, 1605271. [CrossRef]
6. Jung, D.G.; Jung, D.; Kong, S.H. A Lab-on-a-Chip-Based Non-Invasive Optical Sensor for Measuring Glucose in Saliva. *Sensors* **2017**, *17*, 2607. [CrossRef] [PubMed]
7. Ilea, A.; Andrei, V.; Feurdean, C.N.; Băbțan, A.M.; Petrescu, N.B.; Câmpian, R.S.; Boșca, A.B.; Ciui, B.; Tertiș, M.; Săndulescu, R.; et al. Saliva, a Magic Biofluid Available for Multilevel Assessment and a Mirror of General Health—A Systematic Review. *Biosensors* **2019**, *9*, 27. [CrossRef]
8. Aguirre, A.; Testa-Weintraub, L.A.; Banderas, J.A.; Haraszthy, G.G.; Reddy, M.S.; Levine, M.J. Sialochemistry: A Diagnostic Tool? *Crit. Rev. Oral Biol. Med.* **1993**, *4*, 343–350. [CrossRef]
9. Li, R.T.; Kling, S.R.; Salata, M.J.; Cupp, S.A.; Sheehan, J.; Voos, J.E. Wearable Performance Devices in Sports Medicine. *Sports Health* **2016**, *8*, 74–78. [CrossRef]
10. Nasiri, N.; Tricoli, A. Advances in Wearable Sensing Technologies and Their Impact for Personalized and Preventive Medicine. In *Wearable Technologies*; Ortiz, J.H., Ed.; IntechOpen: London, UK, 2018.
11. Wyss, T.; Roos, L.; Beeler, N.; Veenstra, B.; Delves, S.; Buller, M.; Friedl, K. The comfort, acceptability and accuracy of energy expenditure estimation from wearable ambulatory physical activity monitoring systems in soldiers. *J. Sci. Med. Sport* **2017**, *20*, S133–S134. [CrossRef]
12. Faragó, P.; Groza, R.; Ivanciu, L.; Hintea, S. A Correlation-based Biometric Identification Technique for ECG, PPG and EMG. In Proceedings of the 2019 42nd International Conference on Telecommunications and Signal Processing (TSP), Budapest, Hungary, 1–3 July 2019; pp. 716–719.
13. Faragó, P.; Groza, R.; Hintea, S. High Precision Activity Tracker Based on the Correlation of Accelerometer and EMG Data. In Proceedings of the 2019 42nd International Conference on Telecommunications and Signal Processing (TSP), Budapest, Hungary, 1–3 July 2019; 2019; pp. 428–431.
14. Faragó, P.; Groza, R.; Ignat, C.; Cirlugea, M.; Hintea, S. A Fuzzy Expert System for Infection Screening Based on Vital Signs and Activity Data. In Proceedings of the 2018 41st International Conference on Telecommunications and Signal Processing (TSP), Athens, Greece, 4–6 July 2018; pp. 1–5.
15. Jia, W.; Bandodkar, A.J.; Valdes-Ramirez, G.; Windmiller, J.R.; Yang, Z.; Ramirez, J.; Chan, G.; Wang, J. Wearable salivary uric acid mouthguard biosensor with integrated wireless electronics. *Anal. Chem.* **2013**, *85*, 6553–6560. [CrossRef] [PubMed]
16. Thomas, N.; Lähdesmäki, I.; Parviz, B.A. A contact lens with an integrated lactate sensor. *Sens. Actuators B* **2012**, *162*, 128–134. [CrossRef]
17. Schabmueller, C.G.J.; Loppow, D.; Piechotta, G.; Schütze, B.; Albers, J.; Hintsche, R. Micromachined sensor for lactate monitoring in saliva. *Biosens. Bioelectron.* **2006**, *21*, 1770–1776. [CrossRef] [PubMed]
18. Claver, J.B.; Valencia Mirón, M.C.; Capitán-Vallvey, L.F. Disposable electrochemiluminescent biosensor for lactate determination in saliva. *Analyst* **2009**, *134*, 1423–1432. [CrossRef]
19. Minamitani, H.; Suzuki, Y.; Iijima, A.; Nagao, T. A Denture Base Type of Sensor System for Simultaneous Monitoring of Hydrogen Ion Concentration pH and Tissue Temperature in the Oral Cavity. *Ieice Trans. Inf. Sys.* **2002**, *85*, 22–29.

20. Kim, J.; Valdés-Ramírez, G.; Bandodkar., A.J.; Jia, W.; Martinez, A.G.; Ramírez, J.; Mercier, P.; Wang, J. Non-invasive mouthguard biosensor for continuous salivary monitoring of metabolites. *Analyst* **2014**, *139*, 1632–1636. [CrossRef]

21. Magnelli, L.; Amoroso, F.A.; Crupi, F.; Cappuccino, G.; Iannaccone, G. Design of a 75-nW, 0.5-V subthreshold complementary metal–oxide–semiconductor operational amplifier. *Int. J. Circuit Theory Appl.* **2014**, *42*, 967–977. [CrossRef]

22. Mannoor, M.S.; Tao, H.; Clayton, J.D.; Sengupta, A.; Kaplan, D.L.; Naik, R.R.; Verma, N.; Omenetto, F.; McAlpine, M.C. Graphene-based wireless bacteria detection on tooth enamel. *Nat. Comm.* **2012**, *3*, 763. [CrossRef]

23. Culshaw, B.; Kersey, A. Fiber-optic sensing: A historical perspective. *J. Light Technol.* **2008**, *26*, 1064–1078. [CrossRef]

24. Bogue, R. Fibre optic sensors: A review of today's applications. *Sens. Rev.* **2011**, *31*, 304–309. [CrossRef]

25. Bosch, M.E.; Sánchez, A.J.R.; Rojas, F.S.; Ojeda, C.B. Recent Development in Optical Fiber Biosensors. *Sensors* **2007**, *7*, 797–859. [CrossRef]

26. Ravichandran, G.; Lakshmanan, D.K.; Raju, K.; Elangovan, A.; Nambirajan, G.; Devanesan, A.A.; Thilagar, S. Food advanced glycation end products as potential endocrine disruptors: An emerging threat to contemporary and future generation. *Environ. Int.* **2019**, *123*, 486–500. [CrossRef] [PubMed]

27. Street, M.E.; Angelini, S.; Bernasconi, S.; Burgio, E.; Cassio, A.; Catellani, C.; Amarri, S. Current Knowledge on Endocrine Disrupting Chemicals (EDCs) from Animal Biology to Humans, from Pregnancy to Adulthood: Highlights from a National Italian Meeting. *Int. J. Mol. Sci.* **2018**, *19*, 1647. [CrossRef] [PubMed]

28. Schug, T.T.; Johnson, A.F.; Birnbaum, L.S.; Colborn, T.; Guillette, L.J., Jr.; Crews, D.P.; Heindel, J.J. Minireview: Endocrine Disruptors: Past Lessons and Future Directions. *Mol. Endocrinol. (Balt. Md.)* **2016**, *30*, 833–847. [CrossRef]

29. Faragó, P.; Gălătuș, R.; Ilea, A.; Cîrlugea, C.; Faragó, C.; Hintea, S. Smart sensor interface in biomedical monitoring systems. In *Workshop on Integrated Nanodevices for Environmental Analysis*; UTCN: Cluj-Napoca, Romania, 2017.

30. Ilea, A.; Bianca, A.B.; Crișan, M.; Băbțan, A.; Petrescu, N.; Câmpian, R.S. Advanced Glycation End Products (AGEs) in oral pathology–Quo vadis. In *Napoca Biodent Symposium*, 7th ed.; Clujul Medical: Cluj-Napoca, Romania, 2017; Supplement 1, p. S15. 2017; p-ISSN 1222-2119; e-ISSN 2066-8872.

31. Ciui, B.; Tertis, M.; Feurdean, C.N.; Ilea, A.; Sandulescu, R.; Wang, J.; Cristea, C. Cavitas electrochemical sensor toward detection of N-epsilon (Carboxymethyl)lysine in oral cavity. *Sens. Actuators B. Chem.* **2019**, *281*, 399–407. [CrossRef]

32. Boșca, A.B.; Parvu, A.E.; Ana, U.; Tăulescu, M.; Negru, M.; Băbțan, A.M.; Petrescu, N.B.; Bondor, C.; Ilea, A. Effect of the Hyperlipidic Diet on The Systemic Oxidative Stress and The Accumulation of Ages in an Animal Model. In Proceedings of the 8th National and 1st International Congress Nutrition–Medicine of the Future, Cluj-Napoca, Romania, 15–18 May 2019.

33. Uribarri, J.; Woodruff, S.; Goodman, S.; Cai, W.; Chen, X.; Pyzik, R.; Yong, A.; Striker, G.E.; Vlassara, H. Advanced glycation end products in foods and a practical guide to their reduction in the diet. *J. Am. Diet. Assoc.* **2010**, *110*, 911–916. [CrossRef]

34. Kalousová, M.; Zima, T.; Popov, P.; Spacek, P.; Braun, M.; Soukupová, J.; Pelinkova, K.; Kientsch-Engel, R. Advanced glycation end-products in patients with chronic alcohol misuse. *Alcohol. Alcohol.* **2004**, *39*, 316–320. [CrossRef]

35. Farago, P.; Băbțan, A.M.; Galatus, R.; Groza, R.; Roman, N.M.; Feurdean, C.N.; Ilea, A. A Side-Polished Fluorescent Fiber Sensor for the Detection of Blood in the Saliva. In *IFMBE Proceedings, Proceedings of the 6th International Conference on Advancements of Medicine and Health Care through Technology, Cluj-Napoca, Romania, 17–20 October 2018*; Vlad, S., Roman, N., Eds.; Springer: Singapore, 2019; Volume 71.

36. Gaston, A.; Lozano, I.; Perez, F.; Auza, F.; Sevilla, J. Evanescent Wave Optical-Fiber Sensing (Temperature, Relative Humidity, and pH Sensors). *IEEE Sens. J.* **2003**, *3*, 806–8011. [CrossRef]

37. Bilro, L.; Alberto, N.; Pinto, J.L.; Nogueira, R. Optical sensors based on plastic fibers. *Sensors* **2012**, *12*, 12184–12207. [CrossRef]

38. Melo, A.A.; Santiago, M.F.S.; Silva, T.B.; Moreira, C.S.; Cruz, R.M.S. Investigation of a D-Shaped Optical Fiber Sensor with Graphene Overlay. *IFAC-PapersOnLine* **2018**, *51*, 309–314. [CrossRef]

39. Sequeira, F.; Duarte, D.; Bilro, L.; Rudnitskaya, A.; Pesavento, M.; Zeni, L.; Cennamo, N. Refractive Index Sensing with D-Shaped Plastic Optical Fibers for Chemical and Biochemical Applications. *Sensors* **2016**, *16*, 2119. [CrossRef] [PubMed]

40. Pathak, S. Photonics Integrated Circuits. In *Nanoelectronics*; Kaushik, B.K., Ed.; Elsevier: Amsterdam, The Netherlands, 2019; pp. 219–270.

41. López, A.; Losada, M.Á.; Mateo, J.; Antoniades, N.; Jiang, X.; Richards, D. Characterization of a Y-Coupler and Its Impact on the Performance of Plastic Optical Fiber Links. *Fibers* **2018**, *6*, 96. [CrossRef]

42. Gălătuș, R.; Faragó, P.; Miluski, P.; Valles, J.A. Distributed fluorescent optical fiber proximity sensor: Towards a proof of concept. *Spectrochim. Acta A Mol. Biomol. Spectrosc.* **2018**, *198*, 7–18. [CrossRef] [PubMed]

43. Egalon, C.O. Multipoint side illuminated absorption based optical fiber sensor for relative humidity. *Proc. SPIE Int. Soc. Opt. Eng.* **2013**, *8847*, 88471H.

44. Farago, P.; Galatus, R.; Fărcaș, C.; Oltean, G.; Tosa, N. Low-cost Quasi-distributed position sensing platform based on blue fluorescent optical fiber. In Proceedings of the 2017 IEEE 23rd International Symposium for Design and Technology in Electronic Packaging (SIITME), Constanta, Romania, 26–29 October 2017.

45. Křemenáková, D.; Militký, J.; Meryová, B.; Lédl, V. Characterization of side emitting plastic optical fibers light intensity loss. *World J. Eng.* **2013**, *10*, 223–228. [CrossRef]

46. Spigulis, J. Side-Emitting Fibers Brighten Our World. *Op. Photonics News* **2005**, *16*, 34–39. [CrossRef]

47. Amanu, A.A. Macro Bending Losses in Single Mode Step Index Fiber. *Adv. Appl. Sci.* **2016**, *1*, 1–6.

48. Faragó, P.; Gălătuș, R.; Cîrlugea, M.; Hintea, S. Fluorescent Fiber Implementation of an Angle Sensor. In Proceedings of the 20th International Conference on Transparent Optical Networks (ICTON), Bucharest, Romania, 1–5 July 2018; pp. 1–4.

49. Jay, J.A. An Overview of Macrobending and Microbending of Optical Fibers. *White Paper Corning* **2010**, *WP1211*, 1–10.

50. Shabahang, S.; Forward, S.; Yun, S. Polyethersulfone optical fibers with thermally induced microbubbles for custom side-scattering profiles. *Opt. Express* **2019**, *27*, 7560–7567. [CrossRef]

51. Morgan, R.; Barton, J.; Harper, P.; Jones, J. Wavelength dependence of bending loss in monomode optical fibers: Effect of the fiber buffer coating. *Opt. Lett.* **1990**, *15*, 947–949. [CrossRef]

52. Yang, B.; Duan, J.; Xie, Z.; Xiao, H.F. Wavelength dependent loss of splice of single-mode fibers. *J. Cent. South Univ.* **2013**, *20*, 1832–1837. [CrossRef]

53. Heredla, F.J.; Guzman-Chozas, M. The Color of Wine: A Historical Perspective. I. Spectal Evaluations. *J. Food Qual.* **1993**, *16*, 429–437. [CrossRef]

54. White paper. Determination of Wine Colour with UV-VIS Spectroscopy Following Sudraud Method. Available online: https://solutions.shimadzu.co.jp/an/n/en/uv/appl_uv_wine-colour_06d_en.pdf (accessed on 13 September 2019).

55. Babincev, L.M.; Gurešić, D.M.; Simonović, R.M. Spectrophotometric Characterization of Red Wine Color from the Vineyard Region of Metohia. *J. Agric. Sci.* **2016**, *61*, 281–290. [CrossRef]

56. Ribereau-Gayon, P.; Dubourdieu, D.; Doneche, B. *Handbook of Enology*, 2nd ed.; John Wiley & Sons Ltd.: Hoboken, NJ, USA, 2006; p. 178.

57. Sudraud, P. Interpretation des courbes d'absorption des vins rouges. *Ann. Technol. Agric.* **1958**, *7*, 203–208.

58. Kerenyi, A.; Kampis, A. Comparison between the sensorially established and instrumentally measured colour of red wine. *Acta Aliment.* **1984**, *13*, 325–342.

59. Glories, Y. La couleur des vins rouges. 2e partie. Mesure, origine et interpretation. *Connaiss. Vigne Vin* **1984**, *18*, 253–271.

60. Leinders, J. Determination of Wine Colour by UV-VIS Spectroscopy Following Sudraud Method. Available online: https://www.shimadzu.eu/sites/shimadzu.seg/files/3_wine_color_uv.pdf (accessed on 13 September 2019).

61. Richards, D.H.; Losada, M.A.; Antoniades, N.; López, A.; Mateo, J.; Jiang, X.; Madamopoulos, N. Methodology for Engineering SI-POF and Connectors in an Avionics System. *J. Lightw. Technol.* **2013**, *31*, 468–475. [CrossRef]

62. Sen, I.; Tokatil, F. Differentiation of wines with the use of combined data of UV–visible spectra and color characteristics. *J. Food Compos. Anal.* **2016**, *45*, 101–107. [CrossRef]

63. Humphrey, S.P.; Williamson, R.T. A review of saliva: Normal composition, flow, and function. *J. Prosthet. Dent.* **2001**, *85*, 162–169. [CrossRef]

64. Ferrer-Gallego, R.; Soares, S.; Mateus, N.; Rivas-Gonzalo, J.; Escribano-Bailón, M.T.; de Freitas, V. New Anthocyanin-Human Salivary Protein Complexes. *Langmuir* **2015**, *31*, 8392–8401. [CrossRef]
65. Pham, D.T.; Stockdale, V.J.; Wollan, D.; Jeffery, D.W.; Wilkinson, K.L. Compositional Consequences of Partial Dealcoholization of Red Wine by Reverse Osmosis-Evaporative Perstraction. *Molecules* **2019**, *24*, 1404. [CrossRef] [PubMed]

Article

Laser-Induced Deposition of Carbon Nanotubes in Fiber Optic Tips of MMI Devices

Natanael Cuando-Espitia [1,*], Juan Bernal-Martínez [2], Miguel Torres-Cisneros [3] and Daniel May-Arrioja [4]

1   CONACyT, Applied Physics Group, DICIS, University of Guanajuato, Salamanca, Guanajuato 368850, Mexico
2   Unidad de Investigación Biomédica y Nanotecnología, Calle Cañada Honda 129, Ojocaliente 1 Aguascalientes, Ags. C.P. 20190, Mexico; drjuanbernal@gmail.com
3   Applied Physics Group, DICIS, University of Guanajuato, Salamanca, Guanajuato 368850, Mexico; torres.cisneros@ugto.mx
4   Centro de Investigaciones en Óptica, Prol. Constitución 607, Fracc. Reserva Loma Bonita, Aguascalientes 20200, Mexico; darrioja@cio.mx
*   Correspondence: natanael.cuando@ugto.mx

Received: 11 September 2019; Accepted: 11 October 2019; Published: 17 October 2019

Abstract: The integration of carbon nanotubes (CNTs) into optical fibers allows the application of their unique properties in robust and versatile devices. Here, we present a laser-induced technique to obtain the deposition of CNTs onto the fiber optics tips of multimode interference (MMI) devices. An MMI device is constructed by splicing a section of no-core fiber (NCF) to a single-mode fiber (SMF). The tip of the MMI device is immersed into a liquid solution of CNTs and laser light is launched into the MMI device. CNTs solutions using water and methanol as solvents were tested. In addition, the use of a polymer dispersant polyvinylpyrrolidone (PVP) in the CNTs solutions was also studied. We found that the laser-induced deposition of CNTs performed in water-based solutions generates non-uniform deposits. On the other hand, the laser-induced deposition performed with methanol solutions generates uniform deposits over the fiber tip when no PVP is used and deposition at the center of the fiber when PVP is present in the CNTs solution. The results show the crucial role of the solvent on the spatial features of the laser-induced deposition process. Finally, we register and study the reflection spectra of the as-fabricated CNTs deposited MMI devices.

Keywords: optical fibers; carbon nanotubes; multimode interference; laser-induced deposition

---

1. Introduction

Engineered devices based on carbon nanotubes (CNTs) represent one of the most promising and active areas of nanotechnology development. Since its discovery, CNTs have attracted the attention of the scientific community due to the extraordinary properties exhibited by these materials. In particular, their electrical, chemical and optical properties have been exploited to develop electrochemical devices [1–5], photonic and laser applications [6–10] and optical biosensors [11–13] to name a few. Moreover, CNTs have been used as sensing material in pressure [14], flow [15,16] strain [17] and protein sensors [18]. The deposition of CNTs on the surface of optical fiber tips allows for the integration of the unique properties of CNTs into robust and versatile photonic devices. Chemical vapor deposition (CVD) [19], drop-casting [20,21], and Langmuir-Blodgett deposition [22,23] are some of the conventional techniques reported to incorporate CNTs into fiber optic tips. More recently, light-induced deposition has been demonstrated to generate saturable absorbers for ultra-short pulsed lasers [24,25]. Light-induced deposition is an inexpensive and reliable technique which has been also used to develop micro heaters and microbubble generators [26,27]. In order to obtain a CNTs deposit onto the tip of a fiber, a near-infrared (NIR) optical source is used to launch an optical power of 30–100 mW to a fiber tip immersed into a CNTs solution. According to previous reports, a thermal gradient attracts the CNTs to the tip of the fiber generating a deposit of the nanostructures at the core of the fiber [27].

In contrast with conventional techniques, light-induced deposition can be controlled by means of few experimental parameters such as wavelength, power and irradiation time. Moreover, as this technique relies on thermal gradients at the fiber-liquid interface, it is reasonable to think that the specific characteristics of the liquid solution may affect the performance of the deposition technique. However, very limited information can be found on the effects of different solutions over the light-induced deposition of CNTs. Moreover, the laser-induced deposition of CNTs has been demonstrated mainly in single-mode fibers as the specific features of nanotubes are expected to occur at localized spatial regions [24,25]. Laser-induced deposition on different fibers and fiber structures are yet to be demonstrated which may lead to attractive features in the areas of sensing and opto-fluidics. For example, multimode interference (MMI) fiber sensors are low cost, easy to fabricate and highly stable devices with adequate capabilities to be used as pressure [28,29], temperature [30,31], refractive index [32,33] and liquid level sensors [34,35]. Although the incorporation of carbon nanostructures into an MMI fiber device has been addressed previously [36], the method used to achieve carbon nanostructures deposition onto the optical fiber was drop-casting; leaving laser-induced deposition method yet to be investigated in the framework of MMI structures.

In this study, we explore the laser-induced deposition method to obtain CNTs deposits onto the fiber tips of MMI devices. The aim of this work is to demonstrate laser-induced CNTs deposits into MMI structures and to study the influence of different CNTs solutions on the as-fabricated deposits. These two specific aims will help in the design and development of future MMI sensors in which the spectral and spatial features of the CNTs deposits will be of relevant importance. In particular, we investigate the feasibility of laser-inducing CNTs deposits onto the tip of an MMI structure comprised of a single-mode fiber (SMF) spliced to a section of no-core fiber (NCF). Solutions of single-wall carbon nanotubes (SWCNTs) and multiwall carbon nanotubes (MWCNTs) in deionized water as well as in methanol were tested. In addition, solutions with a widely used surfactant (polyvinylpyrrolidone, PVP, [37–40]) were also studied. Moreover, PVP has been previously used as a surfactant in solutions of carbon nanostructures [41–44]. The laser-induced deposits were studied by means of optical microscopy and the reflected spectra of a broadband source which was registered by means of an optical spectrum analyzer (OSA).

2. Materials and Methods

The MMI fiber structure was built by splicing a segment of a commercially available silica-based no-core fiber (NCF) with refractive index of 1.444 (at a wavelength of 1550 nm) and diameter of 125 μm (Prime Optical Fiber Co. Ltd., NCF125) to a commercial single-mode fiber (SMF-28) using a Fujikura splicer (FSM S70). Then, the end of the no-core fiber was cleaved to obtain a spliced segment of the no-core fiber of 58.8 mm. The spectral response of the MMI structure depends on the geometrical and optical properties of the device as follows [45,46]:

$$\lambda_0 = p\left(\frac{nD^2}{L}\right), \tag{1}$$

Here, λ_0 is the peak wavelength, n is the refractive index of the NCF, D is the diameter of the NCF, L is the length of the NCF segment and p is an integer. Briefly, in an MMI structure the initial single-mode launched from the SMF is coupled to all the available radial modes of the NCF and due to interference effects, the initial mode is refocused at a given length within the NCF. This phenomenon is known as self-image formation and occurs at a periodic length indexed by the integer parameter p in Equation (1). Moreover, Equation (1) relates the parameters of the NCF (refractive index n and diameter D) to the launched wavelength λ_0 and the length L at which the p-th image is formed. Further details of

MMI theory can be found elsewhere [32,47–49]. According to Equation (1) and for $p = 4$, the fabricated MMI device has a calculated peak wavelength of 1534.8 nm. Notice that the proposed MMI structure relies on the detection of reflected light at the end of the no-core fiber. After traveling a distance L over the no-core fiber, part of the light is reflected back and travels a distance L before coupling again to the SMF. In terms of spectral features, this is equivalent to a conventional transmission MMI structure with $p = 8$ and a length of $2L$. In other words, the peak wavelength of an MMI device with a length L and for $p = 4$ is the same using Equation (1) with a length of $2L$ and $p = 8$. Moreover, it has been shown that the maximum coupling factor in MMI devices occurs periodically at every 4th image (i.e., $p = 4$, $p = 8$, $p = 12$) [32,50]. For a different wavelength, the self-image is formed at a different length which in turns decreases the corresponding coupling factor and ultimately provides MMI with filtering capabilities.

In contrast with transmission-based MMI structures, reflection-based MMI structures are more suitable in sensing applications as the same fiber tip is used to send and collect light. After fabrication of the MMI structure, the fiber device was incorporated into an optical fiber setup which comprises a wavelength division multiplexer (WDM), an optical circulator, a laser diode and a superluminescent diode centered at 1550 nm. The experimental setup is shown schematically in Figure 1.

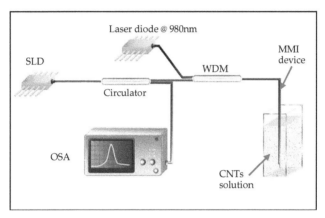

Figure 1. Schematic representation of the optical setup used in this study. SDL: superluminescent laser diode operating at 1550 nm with FWHM of 80 nm; WDM: wavelength division multiplexer; MMI device: multimode interference fiber device; CNTs solution: carbon nanotubes solution and OSA: optical spectrum analyzer. The depicted setup allows the fabrication of laser-induced deposition of CNTs and also the registration of the reflected spectrum of a broadband source.

As Figure 1 shows, the WDM allows for launching light from two different sources to the MMI device. In particular, we have chosen a laser diode emitting at 980 nm as pump source for laser-inducing the CNTs deposits while we have used an SLD centered at 1550 nm to probe the reflection features of the as-fabricated CNTs deposits. The circulator depicted in Figure 1 allows for collecting the reflected light from the tip of the MMI device and analyzing the spectra with an OSA. A v-groove engraved steel plate was mounted on a translation stage to secure the position of the MMI device with a small circular magnet. Finally, a quartz cuvette is used to allocate the CNTs solutions and the MMI device is immersed into the CNTs solutions using the translation stage.

For the CNTs solutions, SWCNTs and MWCNTs were purchased from Cheap Tubes Incorporated (https://www.cheaptubes.com/). The length of the SWCNTs and the MWCNTs was in both cases reported by the manufacturer to be 200–500 nm. Outer diameter was chosen to be also similar; outer diameter of the SWCNTs used in this set of experiments was 1–2 nm according to the manufacturer. The outer diameter of the MWCNTs used in this study was reported by the manufacturer to be less than 8 nm. Previous reports have achieved light-induced CNTs deposits using polar solvents such as ethanol and dimethylformamide (DMF) [24–27]. Methanol was selected as polar solvent due to its simple chemical structure and its similar ability to form hydrogen bonds compared with water [51]. In order to explore the feasibility of generating CNTs deposits using a biocompatible solvent, solutions based on deionized water were also studied. The concentration of CNTs in the liquid solutions was kept fixed at 2.5 mg/mL for all the samples studied here. As one of the key factors when using CNTs solutions is the effective dispersion of the nanostructures over the liquid volume, we have added a widely used dispersant in the study. Since PVP is a polymer which has proved to be effective in dispersing CNTs solutions [41–44], a commercial PVP purchased from Sigma-Aldrich was used in all the experiments described here. A PVP concentration of 50 mg/mL was selected since this concentration has shown to be effective in dispersing CNTs in liquid solutions [43]. The general idea of adding a dispersant in laser-induced CNTs deposition experiments is to allow more homogeneous solutions and thus assisting in the realization of more homogeneous CNTs deposits onto the optical fiber end-faces. A total of 8 CNTs solutions were prepared as shown in Table 1.

Table 1. Aqueous samples studied in this report. For all cases, the CNTs concentration was 2.5 mg/mL. The samples with PVP were prepared at a concentration of 50 mg/mL.

Sample Number	Description	Abbreviation
1	Deionized water + SWCNT	W-SW
2	Deionized water + SWCNT +PVP	W-SW-PVP
3	Deionized water + MWCNT	W-MW
4	Deionized water + MWCNT + PVP	W-MW-PVP
5	Methanol + SWCNT	M-SW
6	Methanol + SWCNT + PVP	M-SW-PVP
7	Methanol + MWCNT	M-MW
8	Methanol + MWCNT + PVP	M-SW-PVP

For the experiments, the optical fiber is attached to the v-groove plate with a small magnet. Care was taken to place the magnet over a section of the SMF avoiding stress-induced features on the multimode fiber section of the device. Then, the MMI device is immersed in a set CNTs solution using a translation stage. Each CNTs solution was sonicated for 60 min before conducting the corresponding experiment. Once the MMI device is immersed, the laser diode is turned on to deliver a preset power to the tip of the MMI device for five minutes. After this period of time, the laser diode is turned off and the MMI device is taken out from the CNTs solution. Once air-dried, the tip of the MMI device is studied by means of optical microscopy. Finally, the SLD is used as a broadband source to probe the spectral features of the as-fabricated device.

3. Results

3.1. Laser-Induced Deposition of CNTs in Water Samples

In general, water-based solutions of PVP/CNTs have been preferred as these solutions allow biocompatibility and stability for subsequent biomedical applications [42,43]. However, previous reports on the light-induced deposition of carbon nanostructures onto fiber tips have centered their attention on studying alcohol-based solutions of CNTs [24–27]. Figure 2 shows the tip of the MMI fiber device after the laser-induced deposition in water-based solutions of CNTs.

As Figure 2 shows, some accumulation of material is evident over the tip of the MMI device in all cases owing to some degree of CNTs deposition. However, the fiber tips exhibit non-uniform deposits when using water-based solutions of CNTs and the surface of the fiber tip is partially covered with CNTs material. According to MMI theory, the energy distribution of these fiber devices corresponds to radial-symmetric modes; which in turn concentrate the energy at the center of the fiber and in concentric ring-shaped structures [50]. The number of radial modes can be calculated as $\sim V/\pi$ for fibers with large V parameter [52]. In particular, for 980nm and the NCF used in this work the V parameter can be calculated as:

$$V = 2\pi \frac{a}{\lambda_0} \sqrt{n_{NCF}^2 - n_{Water}^2},\tag{2}$$

In Equation (2), a is the NCF radius (62.5 μm), λ_0 is the wavelength of the laser (980 nm), n_{NCF} represents the refractive index of the NCF at 980 nm (1.45) and n_{Water} is the refractive index of water (1.327). Equation (2) leads to a V parameter of 234.88 and thus a number of radial modes of ~75 which are enough to generate the characteristic ring structures of MMI effects. Nevertheless, the images of Figure 2 show no clear tendency of attached material around the center of the fiber or in ring-shaped structures. However, some differences can be extracted from the depositions using PVP in contrast to the depositions without PVP by analyzing the images shown in Figure 2. The corresponding deposits using water-based solutions with PVP (Figure 2b,d) show zones of attached material that exhibits smaller island-like features than their counterparts of deposits using water-based solutions without PVP (Figure 2a,c). These island-like structures on the deposits using PVP may suggest that the polymer is effectively reducing the agglomeration on the CNTs solutions during the deposition process.

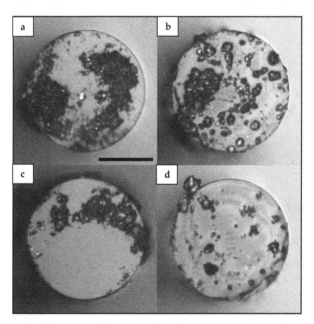

Figure 2. Laser-induced deposition of CNTs onto MMI fiber tip using water-based aqueous solutions: (**a**) Deposition performed with the W-SW sample; (**b**) Deposition performed with the W-SW-PVP sample; (**c**) Deposition performed with the W-MW sample; (**d**) Deposition performed with the W-MW-PVP sample. A 980 nm laser is used at 88.3 mW to obtain the laser-induced depositions in all cases shown in this figure. The images were taken using an optical microscope with a 20× microscope objective. The reference bar in panel (**a**) corresponds to 60 μm.

3.2. Laser-Induced Deposition of CNTs in Methanol Samples

The images of the deposits on MMI fiber tips obtained with methanol-based solutions are presented in Figure 3. It is clear from Figure 3 that the behavior of the obtained deposits using methanol as solvent is essentially different than the deposits obtained with water-based solutions.

Figure 3. Laser-induced deposition of CNTs onto MMI fiber tip using methanol-based aqueous solutions: (**a**) Deposition performed with the M-SW sample; (**b**) Deposition performed with the M-SW-PVP sample; (**c**) Deposition performed with the M-MW sample; (**d**) Deposition performed with the M-MW-PVP sample. A 980 nm laser is used at 88.3 mW to obtain the laser-induced depositions in all cases shown in this figure. The images were taken using an optical microscope with a 20× microscope objective. The reference bar in panel (**a**) corresponds to 60 μm.

Figure 3a,c shows uniform and homogeneous CNTs deposits covering the surface area of the fiber tip. On the other hand, Figure 3b,d show images of deposits in which the material is found mainly around the center of the fiber. This result is a remarkable finding as to the difference between the left column (Figure 3a,c) and the right column (Figure 3b,d) is the use of PVP on the CNTs solutions. Figure 3b,d show similar island-like features as the ones observed in Figure 2b,d. In particular, Figure 3b exhibits scattered material on the surface of the fiber tip that resembles a ring-shaped structure. However, and comparing the upper and lower rows of Figure 3, it can be said that in general, deposits with methanol-based solutions of SWCNTs and MWCNTs present very similar behavior. This is important as this result shows that the proposed laser-induced deposition in MMI fiber devices allows the deposition of both SWCNTs and MWCNTs. The results shown in Figure 3 suggest that the action of the dispersant used in these experiments is not only related to promoting less agglomeration of carbon nanostructures, but also to a decrease in the interaction of carbon nanostructures with the surface of the fiber tip. In other words, as the presence of PVP in the solution reduces the interaction of the CNTs with the surface of the fiber, the deposits are formed at the center of the fiber where more energy is concentrated.

3.3. Deposition Features Dependence on Laser Power

In order to study the effect of different laser powers on the proposed deposition technique, we performed laser-induced deposition experiments varying laser power. In particular, we ran laser-induced deposition as described previously with, 40.0, 88.3 and 139.1 mW. As expected, we found that as laser power increases the amount of material deposited on the fiber tip also increases. Figure 4 shows a series of images taken from depositions using M-SW and M-SW-PVP samples at different laser powers. Although Figure 4 shows images from depositions using M-SW and M-SW-PVP samples, we found that the increase of deposited material as laser power increases was present for all the samples studied here.

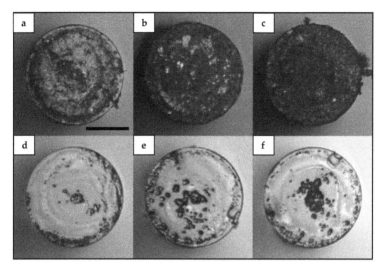

Figure 4. Laser-induced deposition of CNTs with different laser power: (**a**) M-SW sample at 40.0 mW; (**b**) M-SW sample at 88.3 mW; (**c**) M-SW sample at 139.1 mW; (**d**) M-SW-PVP sample at 40.0 mW; (**e**) M-SW-PVP sample at 88.3 mW; (**f**) M-SW-PVP sample at 139.1 mW. The images were taken using an optical microscope with a 20× microscope objective. The reference bar in panel (**a**) corresponds to 60 μm.

Figure 4a–c show homogeneous deposition on the fiber tip; similar to the depositions seen in Figure 3a,c. However, the images of Figure 4 show an increase of deposited material as laser power increases. Similarly, Figure 4d-f show similar features to the images depicted in Figure 3b,d with an increase of deposited material at the center of the fiber as laser power increases. It is important to note that although the amount of deposited material increases with laser power in both cases, the deposition features with and without PVP are essentially different. For the upper row in Figure 4 (M-SW) the deposited material is found all over the surface of the optical fiber tip, while the lower row (M-SW-PVP) of Figure 4 shows accumulation of material around the center of the fiber. In other words, the distribution of the attached material over the surface of the fiber tip differentiates the deposits using aqueous solutions with and without PVP. It is clear from Figure 4 that while a homogeneous deposition over the tip of the fiber is allowed using CNTs solution without PVP, a central deposition of material is promoted with solutions including PVP. This essential difference in deposition features may indicate that the fundamental phenomena interplaying in the deposition processes are substantially altered with the presence of a dispersant such as PVP. It is important to note that although the NCF allows part of the energy to interact with the surroundings; Figures 2–4 show that most of the CNTs material is attached to the fiber tips and very few CNTs material was observed to be attached along the outer surface of the fibers. This is related to the small amount of energy that interacts with the surroundings as the evanescent field compared to the energy leaving from the tip of the fiber. Notice also that the deposition of CNTs unto optical fibers has been associated with thermal gradients which are more difficult to induce only by means of evanescent fields propagating along the outer surface of the fiber. These interesting results can be useful in several areas such as fiber sensing and optofluidics where different devices can be engineered based on the specific distribution of the CNTs material onto the fiber tips. However, the exact mechanisms of the laser-induced deposition in optical fibers and the effect of dispersants on the processes governing the deposition of CNTs onto optical fiber tips are beyond this study.

3.4. Spectral Characterization of the CNTs Deposited MMI Fiber Probe

In order to evaluate the spectral changes of MMI devices after the laser-induced CNTs deposition process, we have recorded the reflection spectra of the proposed MMI fiber devices. As shown schematically in Figure 1, an SLD was used as a light source which provides a broadband spectrum of 80 nm of full width at half maximum (FWHM) centered on 1550 nm. An optical circulator and an OSA complete the fiber setup for spectral recording. Notice that the optical setup depicted in Figure 1 has been designed to take advantage of the fiber devices versatility to allow laser-induced deposition process as well as reflection sensing with a single optical setup.

Firstly, we registered the reflection spectrum in the air of a bare MMI device (*L* = 58.8 mm) with no CNTs deposition. The SLD was set at 100 mW for all the experiments. Then, the SLD was turned off, the MMI device was immersed in sample 1 solution and the 980 nm laser was then turned on to operate at 88.3 mW following the deposition process described previously. Notice that within this approach, the 980nm laser is used to induce the CNTs depositions while the SLD is used to probe the spectral features of the MMI device and both sources are not launched simultaneously. Upon drying, the corresponding spectrum is registered. Consequently, the MMI device is subjected to a multi-step cleaning/checking procedure. In the first step, the material attached to the outer surface of the fiber is removed by means of lens tissues and isopropyl alcohol. Then, the fiber is immersed in isopropyl alcohol and sonicated for 2 min. The isopropyl alcohol used in the previous step is then discarded and replaced with fresh isopropyl alcohol to sonicate again for 2 min. Once the fiber has been subjected to two sonication cycles, the tip of the fiber is evaluated by means of the optical microscope. If CNTs material or dust is found in the fiber, a two-cycle sonication is completed until the fiber is clean. Once no CNTs material or dust is found in the fiber, the reflection spectrum is measured again in order to corroborate the intactness of the MMI device. If the measured spectrum corresponds to the spectrum of the pristine MMI, the deposition using sample 2 is then performed and the test continues. However, if the spectrum has changed from the initial reference, the MMI is discarded and the fabrication process starts over. The test finished once the eight samples are effectively used to obtain CNTs depositions with the same MMI device. This procedure allows for a fair comparison of the spectra recorded.

Figure 5 summarizes the results of this set of experiments. The acquired reflection spectrum of the bare MMI device is shown in Figure 5 as a red solid curve. Figure 5 also shows a rescaled spectrum of the SLD source as a solid black curve. This spectral narrowing is one of the key features of MMI devices [28–31,34,35] with a peak wavelength that can be calculated according to Equation (1). The calculated peak wavelength for this device was 1534.8 nm which is very similar to the experimental peak wavelength of 1534.9 nm. As expected, a narrow spectrum (FWHM = 8 nm) was observed for all the depositions studied here. Moreover, a spectral shape with one dominant peak close to the wavelength of the bare MMI device was also found in all the reflection spectra recorded. In general, the intensity of the reflection spectra was found to be less for the case of the depositions using samples with PVP comparing to the depositions using samples without PVP. This is depicted in the left panel of Figure 5, where the maximum intensity (i.e., the intensity at the peak wavelength) is plotted as a function of the sample number. The dashed line in the left panel of Figure 5 represents the maximum intensity of the reflection spectrum of the bare MMI device.

In addition to the reflection spectra of SLD and bare MMI device, Figure 5 depicts two representative spectra of this set of experiments. Namely, the corresponding spectra of deposits obtained with M-SW and M-SW-PVP samples are depicted as solid blue and solid green lines respectively. As Figure 5 shows, the corresponding spectrum of the M-SW sample exhibits a small shift to shorter wavelengths without considerable loss in intensity. On the other hand, the corresponding spectrum of the M-SW-PVP sample shows an important shift to longer wavelengths with a considerable loss in intensity compared to the initial reflection spectrum. A possible explanation for this reduction in intensity can be a PVP film covering the surface of the fiber tip. In general, a film of material with a refractive index higher than the refractive index of air would decrease the contrast index and thus decreasing the reflectivity of the fiber probe. This scenario is further supported by observing the images from Figures 2–4; in which rainbow-colored features are present in most of the depositions performed using solutions with PVP. However, this laser-induced PVP film has not been confirmed yet and more analysis is needed in order to clarify the nature of the rainbow-colored features and their relationship with a change in reflectivity in the tip of the fiber.

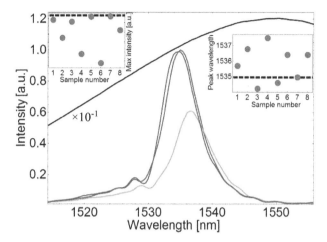

Figure 5. Spectral features of the as-fabricated MMI devices. The solid black line corresponds to the spectrum of the SLD source rescaled by a constant factor of 10^{-1} to emphasize spectral narrowing. The red solid line corresponds to the MMI device prior to CNTs deposition. Solid blue and green lines correspond to reflection spectra of CNTs deposited MMI device using M-SW and M-SW-PVP samples respectively. The left panel shows the maximum intensity as the function of the sample number used for the deposition experiments. The dashed line in the left panel corresponds to the maximum intensity of the spectrum of the bare MMI device. The right panel shows the peak wavelength as the function of the sample used in the laser-induced CNTs deposition experiments. The dashed line in the right panel corresponds to the peak wavelength of the bare MMI device.

The right panel of Figure 5 shows peak wavelength as a function of sample number for the as-fabricated deposits with an MMI device of 58.8 mm. The horizontal dashed line in the right panel of Figure 5 corresponds to the designed peak wavelength (i.e., the peak wavelength of the bare MMI device). In general, these results display a tendency in which the spectra from samples with PVP seem to shift to longer wavelengths while the corresponding spectra from samples without PVP seem to shift to shorter wavelengths. In contrast with a decrease in intensity in which the source can be modulated to compensate the losses, a shift in peak wavelength may drastically alter the resolution, accuracy and range of a fiber sensor based on this deposition technique. Therefore, more details of this particular behavior need to be addressed. In particular, the wavelength shift upon different designed peak wavelengths is important for future sensing applications. In order to explore this behavior, we have performed the proposed deposition technique on MMI fiber devices with designed peak wavelengths between 1529.6 and 1583.3 nm by allowing the no-core fiber length L to be 59–57 mm. In particular, MMI devices with $L = 57, 58$ and 59 mm were fabricated. Then, laser-induced deposition experiments using the samples shown in Table 1 were carried out as previously described for each of the MMI devices constructed. A fixed optical power of 88.3 mW was used for this set of experiments. Finally, the corresponding spectra were registered and the wavelength shift from the designed wavelength peak of the bare MMI device was obtained. The results of this set of experiments are shown in Figure 6.

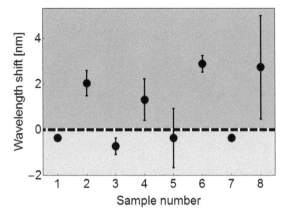

Figure 6. Wavelength shift as a function of sample number (see Table 1). Positive wavelength shifts represent a shift to longer wavelengths. Each symbol represents the mean value obtained from three MMI devices (see text). The bars represent the standard deviation.

In Figure 6, the wavelength shift has been defined as the difference between the experimental peak wavelength and the designed wavelength for the corresponding deposition experiment. Then, the horizontal dashed line in Figure 6 corresponds to the designed wavelength; a positive wavelength shift corresponds to a shift to longer wavelengths and a negative wavelength shift corresponds to a shift to shorter wavelengths. The results shown in Figure 6 indicate that the behavior of red/blue shifting with/without PVP is present in the deposition experiments regardless of the initially designed wavelength. These results provide important considerations for the proposed fiber devices in future sensing applications in which the peak wavelength of the reflection spectrum has to be carefully chosen.

4. Discussion

The systematic study of the laser-induced deposition of CNTs onto MMI fiber tips with different aqueous solutions has revealed interesting features that suggest the interplay of a variety of factors on the spatial and spectral features of the laser-induced CNTs deposits. One of the relevant findings has been the non-homogeneous deposits obtained with water-based solutions. Although the use of water solutions with PVP has been extensively reported as dispersant solutions for CNTs [37,42–44], we obtained deposits in which the attached material was found scattered over the fiber tip and partially covering the surface of the fiber tip as seen in Figure 2. These results indicate that the effect of the solvent on the CNTs is only one factor of the solvent in the complex process of laser-induced deposition. Moreover, these results indicate that the material properties of the solvent are critical for obtaining a homogenous deposition of the CNTs material onto the fiber tip. According to references [25,27], the deposition process of light-absorbing nanoparticles onto the fiber tip is attributed to thermal gradients and convective currents. As the light is absorbed by the CNTs, part of the energy is dissipated to the surroundings generating thermal gradients and convective currents over the solution volume. According to this description, a solvent with a higher density such as water would decrease current velocities as compared with a solvent with a lower density such as methanol. The irregular deposits found in water-based solutions may be related to irregular thermal gradients due to a higher density but also due to a less efficient thermal diffusion in the nanoparticle-water interface. The exact profile of the light-induced thermal gradient is not trivial as it depends on the material properties of the particles and solvent at nanometric scale. Nanofluidics has shown that at nanoscale, heat diffusion phenomena may greatly differ from their macroscopic counterparts [53–55]. Moreover, it has been shown that the thermal conductivity of water at nanoscale is significantly dependent on temperature [56–58]. Although more experimental evidence is needed to unveil the rich processes behind the laser-induced deposition of nanostructures, the experimental results presented here show the pertinence of a detailed theoretical study in which the material properties of the solvent in laser-induced deposition experiments are adequately addressed.

On the other hand, depositions obtained with methanol solutions showed a clear difference when PVP was added to the CNTs solution. In a counterintuitive result, the homogeneous deposits are obtained for solutions without PVP. This result can be explained taking into account the dispersion achieved by the intense sonication performed before the deposition and the fact that the deposition process takes a few minutes. However, the intriguing tendency of the CNTs material to attach to the center of the fiber for methanol-based solutions with PVP cannot be explained in terms of sonication or in terms of the initial dispersion conditions of the CNTs solution. As PVP has been used previously as the polymer matrix in optical-enhancing films on silica substrates [59–62], one possible scenario is a thin film of PVP covering the tip of the fiber preventing the direct interaction of the CNTs with the surface of the fiber. This film would also increase the optical energy needed to allow the deposition of CNTs material onto the fiber tip and thus enabling the deposition process mainly at the center of the fiber where the optical energy is concentrated. In addition, it has been shown that PVP promotes the dispersion of CNTs in aqueous solutions by adhering to the external surface of the CNTs structure [63–66]. This PVP wrapping may affect the amount of light absorbed by the CNTs and also may substantially alter the way the heat is diffused into the solvent volume. Again, the exact photothermal processes are beyond the aim of this experimental study as the objective of this work is to serve as the starting point in the design of MMI fiber sensors based on the laser-induced deposition of CNTs.

Finally, we have found that regardless of the solvent or CNTs structure used, the reflection spectra from depositions using solutions with PVP present a wavelength shift to longer wavelengths. This may reinforce the idea of a thin PVP film on the fiber tip as MMI fiber devices have been used to interrogate the refractive index of a liquid by analyzing the spectral response of the fiber device [32–35,67]. As the peak wavelength of any fiber sensor is one of the key sensing parameters, the shift induced by the deposition process must be taken into account when designing a fiber sensor based on the proposed deposition technique. In general, the results presented here demonstrate the feasibility of a fiber sensor based on laser-induced CNTs deposition. The use of a biocompatible optical fiber in combination with a reflection geometry represents attractive features for biomedical applications. Moreover, the use of functionalized CNTs within this approach will complete the reference frame to design and construct high-sensitivity fiber sensors with custom spatial and spectral deposition features.

5. Conclusions

We have demonstrated the laser-induced deposition of CNTs onto an MMI fiber device. Compared to conventional deposition methods, the laser-induced method represents a cost-effective, straightforward yet versatile and robust technique to obtain CNTs deposition onto optical fiber tips. The characteristics of the solvent in the CNTs solutions have shown to be crucial on the spatial and spectral features of the depositions studied here. The results presented here may lead to micro-patterned deposition of CNTs based on modal laser control. Depending on the spatial laser profile and the solvent used in this technique, a heterogeneous multistep CNTs deposition can be envisioned allowing deposition of CNTs with different features over the tip of the fiber. The structures presented here represent a robust platform to sense the change in the optical properties of the deposited CNTs. For example, a variation on the optical absorbance of the deposited CNTs would be tracked as an intensity variation on the reflected light at the fiber tip. As the peak intensity of the spectral features in these devices would be related to the change of CNTs optical properties, functionalized and conjugated CNTs [68,69] are ideal candidates to be incorporated into the presented methodology to develop high-sensitivity fiber sensors. In particular, pH-sensitive CNTs have been reported in which the absorbance of CNTs solutions is modulated through the pH of the solvent in which the CNTs are immersed [70]. Moreover, and due to the intrinsic spectral shift sensitivity of MMI devices on the surrounding refractive index, the structures presented here would help to develop multivariable/multiplexed sensors by relating a peak wavelength shift with the refractive index of the liquid in contact with the outer surface of the device. Finally, future work on the presented approach utilizing functionalized CNTs to detect specific targets will be particularly important for biomedical applications.

Author Contributions: Conceptualization, N.C.-E., J.B.-M., M.T.-C. and D.M.-A.; methodology, N.C.-E. and J.B.-M.; investigation, N.C.-E. and J.B.-M.; resources, M.T.-C. and D.M.-A.; writing—original draft preparation, N.C.-E. and J.B.-M.; writing—review and editing, M.T.-C. and D.M.-A.; supervision, M.T.-C. and D.M.-A.; funding acquisition, M.T.-C. and D.M.-A.

Funding: The authors acknowledge funding from the Consejo Nacional de Ciencia y Tecnología through projects CB2016-286368 and CB2016-286629 and from Universidad de Guanajuato through grant CIIC no. 310/2019.

Acknowledgments: The authors thank Guanajuato University-CONACYT National Laboratory for material analysis. NCE thanks CONACyT for economic support through Catedras CONACyT program, Project 379. Authors are also grateful to Claudio Frausto Reyes and Gustavo Acevedo Ramírez for their technical help with the optical microscope.

Conflicts of Interest: The authors declare no conflict of interest.

References

1. Gao, M.; Dai, L.; Wallace, G.G. Biosensors based on aligned carbon nanotubes coated with inherently conducting polymers. *Electroanal. Int. J. Devoted Fundam. Pract. Asp. Electroanal.* **2003**, *15*, 1089–1094. [CrossRef]

2. Hirsch, A. Functionalization of single-walled carbon nanotubes. *Angew. Chem. Int. Ed.* **2002**, *41*, 1853–1859. [CrossRef]

3. Hrapovic, S.; Liu, Y.; Male, K.B.; Luong, J.H. Electrochemical biosensing platforms using platinum nanoparticles and carbon nanotubes. *Anal. Chem.* **2004**, *76*, 1083–1088. [CrossRef] [PubMed]

4. Lin, Y.; Lu, F.; Wang, J. Disposable carbon nanotube modified screen-printed biosensor for amperometric detection of organophosphorus pesticides and nerve agents. *Electroanal. Int. J. Devoted Fundam. Pract. Asp. Electroanal.* **2004**, *16*, 145–149. [CrossRef]

5. Wang, J.; Musameh, M. Carbon nanotube/teflon composite electrochemical sensors and biosensors. *Anal. Chem.* **2003**, *75*, 2075–2079. [CrossRef] [PubMed]

6. Lefebvre, J.; Austing, D.G.; Bond, J.; Finnie, P. Photoluminescence imaging of suspended single-walled carbon nanotubes. *Nano Lett.* **2006**, *6*, 1603–1608. [CrossRef]

7. Wang, F.; Dukovic, G.; Brus, L.E.; Heinz, T.F. Time-resolved fluorescence of carbon nanotubes and its implication for radiative lifetimes. *Phys. Rev. Lett.* **2004**, *92*, 177401. [CrossRef]

8. Tatsuura, S.; Furuki, M.; Sato, Y.; Iwasa, I.; Tian, M.; Mitsu, H. Semiconductor carbon nanotubes as ultrafast switching materials for optical telecommunications. *Adv. Mater.* **2003**, *15*, 534–537. [CrossRef]

9. Rozhin, A.G.; Sakakibara, Y.; Kataura, H.; Matsuzaki, S.; Ishida, K.; Achiba, Y.; Tokumoto, M. Anisotropic saturable absorption of single wall carbon nanotubes aligned in polyvinyl alcohol. *MRS Online Proc. Libr. Arch.* **2004**, *858*. [CrossRef]

10. Della Valle, G.; Osellame, R.; Galzerano, G.; Chiodo, N.; Cerullo, G.; Laporta, P.; Svelto, O.; Morgner, U.; Rozhin, A.; Scardaci, V. Passive mode locking by carbon nanotubes in a femtosecond laser written waveguide laser. *Appl. Phys. Lett.* **2006**, *89*, 231115. [CrossRef]

11. Liu, Z.; Tabakman, S.; Welsher, K.; Dai, H. Carbon nanotubes in biology and medicine: In vitro and in vivo detection, imaging and drug delivery. *Nano Res.* **2009**, *2*, 85–120. [CrossRef] [PubMed]

12. Chen, Z.; Zhang, X.; Yang, R.; Zhu, Z.; Chen, Y.; Tan, W. Single-walled carbon nanotubes as optical materials for biosensing. *Nanoscale* **2011**, *3*, 1949–1956. [CrossRef] [PubMed]

13. Heller, D.A.; Jeng, E.S.; Yeung, T.-K.; Martinez, B.M.; Moll, A.E.; Gastala, J.B.; Strano, M.S. Optical detection of DNA conformational polymorphism on single-walled carbon nanotubes. *Science* **2006**, *311*, 508–511. [CrossRef] [PubMed]

14. Niranjana, S. *Characterization of Nanocarbon Thin Films*; Manipal Institute of Technology: Manipal, India, 2012.

15. Kawano, T.; Chiamori, H.C.; Suter, M.; Zhou, Q.; Sosnowchik, B.D.; Lin, L. An electrothermal carbon nanotube gas sensor. *Nano Lett.* **2007**, *7*, 3686–3690. [CrossRef]

16. Wang, S.; Zhang, Q.; Yang, D.; Sellin, P.; Zhong, G. Multi-walled carbon nanotube-based gas sensors for NH_3 detection. *Diam. Relat. Mater.* **2004**, *13*, 1327–1332. [CrossRef]
17. Dharap, P.; Li, Z.; Nagarajaiah, S.; Barrera, E. Nanotube film based on single-wall carbon nanotubes for strain sensing. *Nanotechnology* **2004**, *15*, 379. [CrossRef]
18. Jacobs, C.B.; Peairs, M.J.; Venton, B.J. Carbon nanotube based electrochemical sensors for biomolecules. *Anal. Chim. Acta* **2010**, *662*, 105–127. [CrossRef]
19. Yamashita, S.; Inoue, Y.; Maruyama, S.; Murakami, Y.; Yaguchi, H.; Jablonski, M.; Set, S. Saturable absorbers incorporating carbon nanotubes directly synthesized onto substrates and fibers and their application to mode-locked fiber lasers. *Opt. Lett.* **2004**, *29*, 1581–1583. [CrossRef]
20. Shabaneh, A.; Girei, S.; Arasu, P.; Mahdi, M.; Rashid, S.; Paiman, S.; Yaacob, M. Dynamic response of tapered optical multimode fiber coated with carbon nanotubes for ethanol sensing application. *Sensors* **2015**, *15*, 10452–10464. [CrossRef]
21. Shabaneh, A.; Girei, S.; Arasu, P.; Rashid, S.; Yunusa, Z.; Mahdi, M.; Paiman, S.; Ahmad, M.; Yaacob, M. Reflectance response of optical fiber coated with carbon nanotubes for aqueous ethanol sensing. *IEEE Photonics J.* **2014**, *6*, 1–10. [CrossRef]
22. Penza, M.; Cassano, G.; Aversa, P.; Antolini, F.; Cusano, A.; Cutolo, A.; Giordano, M.; Nicolais, L. Alcohol detection using carbon nanotubes acoustic and optical sensors. *Appl. Phys. Lett.* **2004**, *85*, 2379–2381. [CrossRef]
23. Consales, M.; Campopiano, S.; Cutolo, A.; Penza, M.; Aversa, P.; Cassano, G.; Giordano, M.; Cusano, A. Carbon nanotubes thin films fiber optic and acoustic VOCs sensors: Performances analysis. *Sens. Actuators B Chem.* **2006**, *118*, 232–242. [CrossRef]
24. Kashiwagi, K.; Yamashita, S.; Set, S.Y. Optically manipulated deposition of carbon nanotubes onto optical fiber end. *Jpn. J. Appl. Phys.* **2007**, *46*, L988. [CrossRef]
25. Nicholson, J.; Windeler, R.; DiGiovanni, D. Optically driven deposition of single-walled carbon-nanotube saturable absorbers on optical fiber end-faces. *Opt. Express* **2007**, *15*, 9176–9183. [CrossRef]
26. Pimentel-Domínguez, R.; Moreno-Álvarez, P.; Hautefeuille, M.; Chavarría, A.; Hernández-Cordero, J. Photothermal lesions in soft tissue induced by optical fiber microheaters. *Biomed. Opt. Express* **2016**, *7*, 1138–1148. [CrossRef]
27. Pimentel-Domínguez, R.; Hernández-Cordero, J.; Zenit, R. Microbubble generation using fiber optic tips coated with nanoparticles. *Opt. Express* **2012**, *20*, 8732–8740. [CrossRef]
28. Ruiz-Pérez, V.; Basurto-Pensado, M.; LiKamWa, P.; Sánchez-Mondragón, J.; May-Arrioja, D. Fiber optic pressure sensor using multimode interference. *J. Phys. Conf. Ser.* **2011**, *274*, 012025. [CrossRef]
29. May-Arrioja, D.A.; Ruiz-Perez, V.I.; Bustos-Terrones, Y.; Basurto-Pensado, M.A. Fiber optic pressure sensor using a conformal polymer on multimode interference device. *IEEE Sens. J.* **2015**, *16*, 1956–1961. [CrossRef]
30. Fuentes-Fuentes, M.; May-Arrioja, D.; Guzman-Sepulveda, J.; Torres-Cisneros, M.; Sánchez-Mondragón, J. Highly sensitive liquid core temperature sensor based on multimode interference effects. *Sensors* **2015**, *15*, 26929–26939. [CrossRef]
31. Irace, A.; Breglio, G. All-silicon optical temperature sensor based on Multi-Mode Interference. *Opt. Express* **2003**, *11*, 2807–2812. [CrossRef]
32. Wang, Q.; Farrell, G. All-fiber multimode-interference-based refractometer sensor: Proposal and design. *Opt. Lett.* **2006**, *31*, 317–319. [CrossRef] [PubMed]
33. Biazoli, C.R.; Silva, S.; Franco, M.A.; Frazão, O.; Cordeiro, C.M. Multimode interference tapered fiber refractive index sensors. *Appl. Opt.* **2012**, *51*, 5941–5945. [CrossRef] [PubMed]
34. Antonio-Lopez, J.E.; Sanchez-Mondragon, J.; LiKamWa, P.; May-Arrioja, D.A. Fiber-optic sensor for liquid level measurement. *Opt. Lett.* **2011**, *36*, 3425–3427. [CrossRef] [PubMed]
35. Antonio-Lopez, J.E.; May-Arrioja, D.A.; LiKamWa, P. Fiber-optic liquid level sensor. *IEEE Photonics Technol. Lett.* **2011**, *23*, 1826–1828. [CrossRef]
36. Yao, B.C.; Wu, Y.; Yu, C.B.; He, J.R.; Rao, Y.J.; Gong, Y.; Fu, F.; Chen, Y.F.; Li, Y.R. Partially reduced graphene oxide based FRET on fiber-optic interferometer for biochemical detection. *Sci. Rep.* **2016**, *6*, 23706. Available online: https://www.nature.com/articles/srep23706#supplementary-information (accessed on 9 August 2019). [CrossRef] [PubMed]

37. El Achaby, M.; Arrakhiz, F.-E.; Vaudreuil, S.; Essassi, E.M.; Qaiss, A.; Bousmina, M. Nanocomposite films of poly(vinylidene fluoride) filled with polyvinylpyrrolidone-coated multiwalled carbon nanotubes: Enhancement of β-polymorph formation and tensile properties. *Polym. Eng. Sci.* **2013**, *53*, 34–43. [CrossRef]

38. Van den Mooter, G.; Wuyts, M.; Blaton, N.; Busson, R.; Grobet, P.; Augustijns, P.; Kinget, R. Physical stabilisation of amorphous ketoconazole in solid dispersions with polyvinylpyrrolidone K25. *Eur. J. Pharm. Sci.* **2001**, *12*, 261–269. [CrossRef]

39. Li, Z.; Zhang, Y. Monodisperse Silica-Coated Polyvinylpyrrolidone/NaYF4 Nanocrystals with Multicolor Upconversion Fluorescence Emission. *Angew. Chem. Int. Ed.* **2006**, *45*, 7732–7735. [CrossRef]

40. Simonelli, A.P.; Mehta, S.C.; Higuchi, W.I. Dissolution Rates of High Energy Polyvinylpyrrolidone (PVP)-Sulfathiazole Coprecipitates. *J. Pharm. Sci.* **1969**, *58*, 538–549. [CrossRef]

41. Ntim, S.A.; Sae-Khow, O.; Witzmann, F.A.; Mitra, S. Effects of polymer wrapping and covalent functionalization on the stability of MWCNT in aqueous dispersions. *J. Colloid Interface Sci.* **2011**, *355*, 383–388. [CrossRef]

42. Bernal-Martínez, J.; Seseña-Rubfiaro, A.; Godínez-Fernández, R.; Aguilar-Elguezabal, A. Electrodes made of multi-wall carbon nanotubes on PVDF-filters have low electrical resistance and are able to record electrocardiograms in humans. *Microelectron. Eng.* **2016**, *166*, 10–14. [CrossRef]

43. bernal-martinez, J.; Godínez-Fernández, R.; Aguilar-Elguezabal, A. Suitability of the Composite Made of Multi Wall Carbon Nanotubes-Polyvinylpyrrolidone for Culturing Invertebrate Helix aspersa Neurons. *J. Mater. Sci. Chem. Eng.* **2017**, *5*, 41–50. [CrossRef]

44. Haghighat, F.; Mokhtary, M. Preparation and Characterization of Polyvinylpyrrolidone-functionalized Multiwalled Carbon Nanotube (PVP/f-MWNT) Nanocomposites. *Polym. Plast. Technol. Eng.* **2017**, *56*, 794–803. [CrossRef]

45. Socorro, A.B.; Del Villar, I.; Corres, J.M.; Arregui, F.J.; Matias, I.R. Mode transition in complex refractive index coated single-mode–multimode–single-mode structure. *Opt. Express* **2013**, *21*, 12668–12682. [CrossRef] [PubMed]

46. Walbaum, T.; Fallnich, C. Multimode interference filter for tuning of a mode-locked all-fiber erbium laser. *Opt. Lett.* **2011**, *36*, 2459–2461. [CrossRef] [PubMed]

47. Ruiz-Pérez, V.I.; Basurto-Pensado, M.A.; May-Arrioja, D.; Sánchez Mondragón, J.J.; LiKamWa, P. Intrinsic fiber optic pressure sensor based on multimode interference device as sensitive element. In Proceedings of the Frontiers in Optics 2010/Laser Science XXVI, Rochester, NY, USA, 24 October 2010; p. JWA36.

48. Mohammed, W.S.; Mehta, A.; Johnson, E.G. Wavelength tunable fiber lens based on multimode interference. *J. Lightwave Technol.* **2004**, *22*, 469–477. [CrossRef]

49. Wang, Q.; Farrell, G. Multimode-fiber-based edge filter for optical wavelength measurement application and its design. *Microw. Opt. Technol. Lett.* **2006**, *48*, 900–902. [CrossRef]

50. Wang, Q.; Farrell, G.; Yan, W. Investigation on Single-Mode–Multimode– Single-Mode Fiber Structure. *J. Lightwave Technol.* **2008**, *26*, 512–519. [CrossRef]

51. Ayala, R.; Martínez, J.M.; Pappalardo, R.R.; Sánchez Marcos, E. Theoretical Study of the Microsolvation of the Bromide Anion in Water, Methanol, and Acetonitrile: Ion–Solvent vs Solvent–Solvent Interactions. *J. Phys. Chem. A* **2000**, *104*, 2799–2807. [CrossRef]

52. Saleh, B.E.; Teich, M.C. *Fundamentals of Photonics*; John Wiley & Sons: Hoboken, NJ, USA, 2019.

53. Buongiorno, J. Convective Transport in Nanofluids. *J. Heat Transf.* **2005**, *128*, 240–250. [CrossRef]

54. Yu, W.; France, D.M.; Routbort, J.L.; Choi, S.U.S. Review and Comparison of Nanofluid Thermal Conductivity and Heat Transfer Enhancements. *Heat Transf. Eng.* **2008**, *29*, 432–460. [CrossRef]

55. Mahian, O.; Kolsi, L.; Amani, M.; Estellé, P.; Ahmadi, G.; Kleinstreuer, C.; Marshall, J.S.; Siavashi, M.; Taylor, R.A.; Niazmand, H.; et al. Recent advances in modeling and simulation of nanofluid flows-Part I: Fundamentals and theory. *Phys. Rep.* **2019**, *790*, 1–48. [CrossRef]

56. Das, S.K.; Putra, N.; Thiesen, P.; Roetzel, W. Temperature Dependence of Thermal Conductivity Enhancement for Nanofluids. *J. Heat Transf.* **2003**, *125*, 567–574. [CrossRef]

57. Xu, G.; Fu, J.; Dong, B.; Quan, Y.; Song, G. A novel method to measure thermal conductivity of nanofluids. *Int. J. Heat Mass Transf.* **2019**, *130*, 978–988. [CrossRef]

58. Mintsa, H.A.; Roy, G.; Nguyen, C.T.; Doucet, D. New temperature dependent thermal conductivity data for water-based nanofluids. *Int. J. Therm. Sci.* **2009**, *48*, 363–371. [CrossRef]

59. Bailey, J.; Sharp, J.S. Infrared dielectric mirrors based on thin film multilayers of polystyrene and polyvinylpyrrolidone. *J. Polym. Sci. Part B Polym. Phys.* **2011**, *49*, 732–739. [CrossRef]

60. Slistan-Grijalva, A.; Herrera-Urbina, R.; Rivas-Silva, J.F.; Ávalos-Borja, M.; Castillón-Barraza, F.F.; Posada-Amarillas, A. Synthesis of silver nanoparticles in a polyvinylpyrrolidone (PVP) paste, and their optical properties in a film and in ethylene glycol. *Mater. Res. Bull.* **2008**, *43*, 90–96. [CrossRef]

61. England, M.W.; Sato, T.; Urata, C.; Wang, L.; Hozumi, A. Transparent gel composite films with multiple functionalities: Long-lasting anti-fogging, underwater superoleophobicity and anti-bacterial activity. *J. Colloid Interface Sci.* **2017**, *505*, 566–576. [CrossRef]

62. Chen, Y.-Y.; Wei, W.-C.J. Formation of mullite thin film via a sol-gel process with polyvinylpyrrolidone additive. *J. Eur. Ceram. Soc.* **2001**, *21*, 2535–2540. [CrossRef]

63. Karpushkin, E.; Gvozdik, N.; Klimenko, M.; Filippov, S.K.; Angelov, B.; Bessonov, I.; Sergeyev, V. Structure and flow behavior of dilute dispersions of carbon nanotubes in polyacrylonitrile–dimethylsulfoxide solution. *Colloid Polym. Sci.* **2016**, *294*, 1187–1195. [CrossRef]

64. Karpushkin, E.; Berkovich, A.; Sergeyev, V. Stabilization of Multi-Walled Carbon Nanotubes Aqueous Dispersion with Poly-N-vinylpyrrolidone via Polymer-Wrapping. *Macromol. Symp.* **2015**, *348*, 63–67. [CrossRef]

65. Zhang, W.-B.; Xu, X.-L.; Yang, J.-H.; Huang, T.; Zhang, N.; Wang, Y.; Zhou, Z.-W. High thermal conductivity of poly(vinylidene fluoride)/carbon nanotubes nanocomposites achieved by adding polyvinylpyrrolidone. *Compos. Sci. Technol.* **2015**, *106*, 1–8. [CrossRef]

66. Boguslavsky, Y.; Fadida, T.; Talyosef, Y.; Lellouche, J.-P. Controlling the wettability properties of polyester fibers using grafted functional nanomaterials. *J. Mater. Chem.* **2011**, *21*, 10304–10310. [CrossRef]

67. Zhou, X.; Chen, K.; Mao, X.; Yu, Q. A reflective fiber-optic refractive index sensor based on multimode interference in a coreless silica fiber. *Opt. Commun.* **2015**, *340*, 50–55. [CrossRef]

68. Sun, Y.-P.; Fu, K.; Lin, Y.; Huang, W. Functionalized carbon nanotubes: Properties and applications. *Acc. Chem. Res.* **2002**, *35*, 1096–1104. [CrossRef] [PubMed]

69. Balasubramanian, K.; Burghard, M. Chemically functionalized carbon nanotubes. *Small* **2005**, *1*, 180–192. [CrossRef]

70. Zhao, W.; Song, C.; Pehrsson, P.E. Water-soluble and optically pH-sensitive single-walled carbon nanotubes from surface modification. *J. Am. Chem. Soc.* **2002**, *124*, 12418–12419. [CrossRef]

Article

Fiber Link Health Detection and Self-Healing Algorithm for Two-Ring-Based RoF Transport Systems

Wen-Shing Tsai [1], Ching-Hung Chang [2,*], Zhin-Guei Lin [2], Dong-Yi Lu [2] and Tsung-Ying Yang [2]

[1] Department of Electrical Engineering, Ming Chi University of Technology, New Taipei City 24301, Taiwan; wst@mail.mcut.edu.tw

[2] Department of Electrical Engineering, National Chiayi University, Chiayi City 60004, Taiwan; tklingary@gmail.com (Z.-G.L.); xzericzx222666@gmail.com (D.-Y.L.); joe487001@gmail.com (T.-Y.Y.)

* Correspondence: chchang@mail.ncyu.edu.tw

Received: 23 August 2019; Accepted: 25 September 2019; Published: 27 September 2019

Abstract: A two-ring-based radio over fiber (RoF) transport system with a two-step fiber link failure detection and self-healing algorithm is proposed to ensure quality of service (QoS) by automatically monitoring the health of each fiber link in the transport system and by resourcefully detecting, locating, and bypassing the blocked fiber links. With the assistance of the fiber Bragg grating remote sensing technique, preinstalled optical switches, and novel single-line bidirectional optical add/drop multiplexers, the optical routing pathways in the RoF transport system can be dynamically adjusted by the proposed algorithm when some fiber links are broken. Simulation results show that except in some extreme situations, the proposed algorithm can find the blocked fiber links in the RoF transport system and animatedly adjust the status of preinstalled optical switches to restore all blocked network connections, thereby ensuring QoS in the proposed RoF transport system.

Keywords: fiber Bragg grating; optical fiber transport system; optical add/drop multiplexer; self-healing

---

1. Introduction

Radio over fiber (RoF) transport systems have been developed to support broadband wireless communication technologies, such as 5G and WiGig. Typically, RoF transport systems in ring topology can utilize wavelength division multiplexing (WDM) technology to extend the network capacity [1,2] and to provide a better self-healing function compared with tree-topology architectures to overcome fiber line failures [3–6]. However, if the ring-topology comprises a traditional optical add/drop multiplexer (OADM) [7,8], the transport system may be unable to exercise its self-healing function when a fiber link failure is present because both the added and passed optical signals in an OADM are transmitted in the same direction of the fiber ring [9]. To overcome such drawback, several single-line bidirectional OADMs (SBOADMs) [7,10,11] are developed based on optical multiplexer/de-multiplexer, multiport optical circulators (OCs), optical switches (SW), fiber Bragg gratings (FBGs), or optical amplifiers. Based on the published SBOADMs, the downstream light waves from central office (CO) can be transmitted to each SBOADM either in the clockwise (CW) or counterclockwise (CCW) direction of the fiber ring to prevent the blockage of fiber link. Nevertheless, when upstream light waves are added from the SBOADM to the ring-based network, they are transmitted along the same transmission direction of the downstream light waves; in this case, the upstream light waves are unable to avoid the impact of the blocked point. A backup fiber ring or SW is needed to reconfigure the optical passway in order to transmit the optical signals in either the general or backup fiber ring [12]. However, in normal conditions, the backup fiber ring is redundant. Besides, the complex and expensive devices inside the published SBOADMs can significantly increase the deployment cost and gradually introduce complications to the management of the network.

To address these bottlenecks, a novel SBOADM was proposed in our previous research [13] for building a large-scale optical fiber sensor network. Based on two FBGs, two three-port OCs, and three four-port OCs, the novel SBOADM can drop dedicated light waves from the fiber ring regardless of whether the light waves are transmitted in either the CW or CCW direction of the ring and can add and route upstream light waves back to the CO along the reverse transmission direction of the downstream light waves. No backup fiber ring is required in the network, and no dedicated optical multiplexer or optical switch is employed inside the SBOADM.

To further extend the application of SBOADM and to enhance its flexible routing ability and self-healing function for ring-based optical fiber transport systems, a two-ring-based RoF transport system with a self-healing functionality is built based on SBOADM. A dedicated algorithm, named two-step fiber link failure detection and self-healing algorithm, is also proposed to accurately determine the location of fiber line failure based on FBG remote sensing technique and to adjust the optical routing passway accordingly. FBG sensor can reflect specific wavelengths of light signals and the signals of the other wavelengths will penetrate directly. Its applications have been applied in various fields, such as geological hydrology, aerospace industry, pipeline monitoring, and so on [14–19]. Unlike other embedded self-healing architectures [6,20–26] that lack any algorithm or method for determining the breakpoint location, the proposed algorithm can locate two breakpoints and bypass the failure fiber links by adjusting the status of the pre-installed SWs in the proposed network.

2. Materials and Methods

The proposed two-ring-based RoF transport system is shown in Figure 1, and its self-healing function was embedded by using SWs, FBGs, SBOADMs, and optical splitters. A total of 6 RoF groups located in 6 areas were connected to the CO by 2 fiber rings, 5 SWs, 3 optical splitters, 11 FBGs, and 6 SBOADMs. SW1, SW2, SW4, and SW5 were set to the parallel status, whereas SW3 was set to the cross status. The employed SWs and optical splitters allow the proposed transport system to establish additional optical pathways and to reconfigure the optical pathway in case of a fiber link failure. The 11 FBGs (FBG1–FBG11) deployed along with each span of single mode fibers (SMFs) were employed to remotely sense the health of each fiber link. The fiber failure detection system in CO usually sends out 11 dedicated optical detection signals to the transport system, and each FBG sensor was supported to reflect a dedicated optical detection signal back to the CO along the reverse downstream transmission routing pathway. When all detection signals were reflected back to the CO, they were routed by a four-port OC and reflected again by another set of 11 FBGs before they were received by the fiber failure detection system. Apart from the detection system, SBOADMs were also employed to drop and add dedicated optical signals among the fiber ring and connected RoF groups. To avoid interference between the detection system and RoF transmission groups, the reflection wavelengths of FBGs should be far from the optical carriers of RoF groups. In the network architecture, the CO was located 20 km away from remote nodes 1 (RN1) and RN2, and the distance between each pair of SBOADM was maintained at 1 km. In this case, downstream optical signals were sent from the CO to SBOADM1 via SW1, a 20-km trunk fiber 1, SW2 in RN1, and a 1-km SMF1. The optical signal belonging to RoF group 1 was dropped by SBOADM1. The other downstream optical signals passed through SBOADM1 and directed to SBOADM5, SBOADM4, SBOADM3, SBOADM2, and SBOADM6 sequentially.

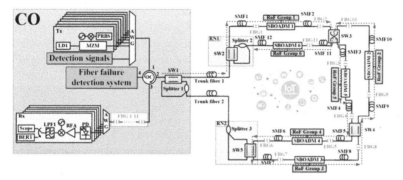

Figure 1. The proposed two-ring based radio over fiber (RoF) transport system with a self-healing functionality.

To drop dedicated RoF signals into the connected RoF groups in both CW and CCW directions, the SBOADM published in [13] was modified as shown in Figure 2. A 2*2 SW and optical switch trigger (OST) were inserted into each SBOADM to bridge the two add/drop ports with the connected RoF group. The OST is a simple optical receiver model that can output high voltage when receiving optical signals and output low voltage when no optical signal is fed. For example, when the downstream light waves were fed into the SBOADM from the right-hand side, the optical pathway of the dropped, passed, and added optical signals are indicated by the red, blue, and green auxiliary lines, respectively, in Figure 2. In case the optical pathways were reconfigured due to fiber link failure, the downstream light waves were fed into the SBOADM from the left-hand side as shown in Figure 3. The dropped light waves were initially directed toward the OST because the SW in SBOADM is in the cross status. These light waves stimulate the OST to output a high-voltage pulse that changes the SW status to parallel status. Afterward, the downstream light waves can be dropped to relative RoF groups, and the added light wave can be sent back to the CO along the reverse transmission direction of the downstream light waves. When the optical pathways are reconfigured again, the light waves were fed into the SBOADM from the right-hand side. The dropped light wave was also fed into the OST along the parallel status SW, after which the status of SW changes back to cross status. Consequently, the dropped light wave can be redirected to the connected ROF group as shown in Figure 2.

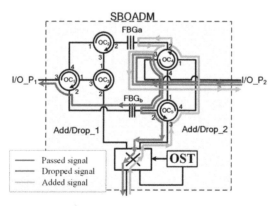

Figure 2. Structure of a single-line bidirectional optical add/drop multiplexer (SBOADM; the optical signals are fed from the right-hand side).

Sensors **2019**, *19*, 4201

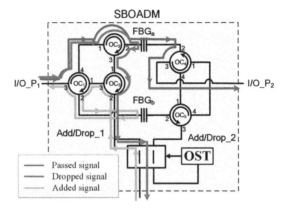

Figure 3. Structure of SBOADM (the optical signals are fed from the left-hand side).

3. Two-Step Fiber Link Failure Detection and Self-Healing Algorithm

3.1. First Detection Step of the Proposed Algorithm

To ensure quality of service (QoS), a two-step fiber link failure detection and self-healing algorithm was developed to monitor the fiber link health, locate the failure fiber links, and bypass the impact of fiber link failures on the proposed transport system. To evaluate the algorithm, the commercial software, VPI transmission maker, was employed to simulate the signal transmissions in the two-ring-based RoF transport system. In the simulation setup, the insertion losses of OC and SW are 0.8 dB and 0.7 dB, respectively. The reflection ratio and reflection bandwidth of the employed 11 FBG sensors (FBG1–FBG11) are 99% and 20 GHz, respectively. The reflection center wavelengths of these FBGS are 1550.017 nm, 1550.417 nm, 1550.817 nm, 1551.217 nm, 1551.617 nm, 1552.017 nm, 1552.417 nm, 1552.817 nm, 1553.217 nm, 1553.617 nm, and 1554.017 nm, respectively. In this case, when optical signals insert from the input/output port1 (I/O_P1) of the SBOADM to I/O_P2 or from I/O_P2 to I/O_P1, they will suffer roughly 2.4 dB insertion loss which is caused by passing through FBG one time and OC three times.

The first detection step of the proposed algorithm was designed to detect, locate, and bypass any fiber link failure except for trunk fiber 1. As shown in Figure 4, the CO periodically sends out 11 detection signals ($\lambda 1$–$\lambda 11$) every T-minutes to monitor the health of the two-ring-based RoF transport system in the detection phase (as indicated by the gray background in Figure 4). The preinstalled 11 FBG sensors (FBG1–FBG11) should be able to reflect each detection signal; otherwise, one or more fiber link failures may occur in the transport system. Generally, if detection signals send out the CO with 0 dBm, the reflected $\lambda 1$ to $\lambda 11$ in Figure 5 will suffer an attenuation of roughly 12.6 dB to 45.5 dB. Nevertheless, if none of the detection signals is received, then the trunk fiber 1 or SMF1 may reach failure. The detection system can execute the second detection step to find, locate, and bypass unknown fiber link failures as will be discussed later.

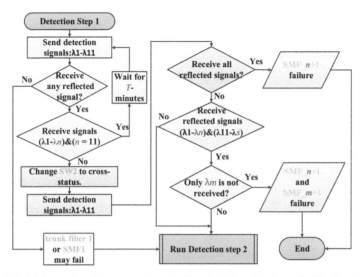

Figure 4. First detection step of the proposed two-step fiber link failure detection and self-healing algorithm.

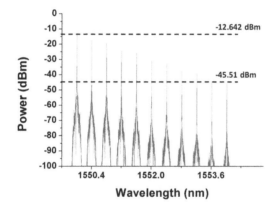

Figure 5. Detection signals received by the central office (CO) when no breakpoint is present in the transport system.

If parts of the detection signals are not received, that is, only $\lambda1$ to λn are received and $n < 11$, then one or more fiber links may not function properly. The detection system can execute the second phase of the first detection step (indicated by the blue background in Figure 4) to locate and bypass any fiber link failure. In this step, the SW2 in RN1 switches to cross status and resends the 11 detection signals to the transport system. The CO can utilize low-power wide-area network (LPWAN) technique, such as long range (LoRa) or narrowband-internet of things (NB-IoT), to remote switch the SW2 status. The processes should be able to be finish in a few seconds since the response time of commercial SW is roughly 10 ms. For example, if the SMF5 does not work properly, as shown in Figure 6, only $\lambda1$ to $\lambda4$ will be received by the detection system in the first phase of the first detection step. In this case, n was set to 4. Subsequently, when SW2 switches from the parallel status to the cross status in the second phase of the first detection step, the downstream detection signals are split and transmitted by two pathways, namely, pathway1-1 and pathway1-2, in the RN1 as indicated by the blue and red auxiliary lines, respectively, in Figure 6. The optical signals in pathway1-2 would go through both SBOADM1 and SBOADM5, whereas those in pathway1-1 would go through SBOADM6, SBOADM2,

SBOADM3, and SBOADM4 sequentially. All 11 optical detection signals would be reflected back to the CO as presented in Figure 7. In other words, the impact of the failure fiber link would be bypassed, and the breakpoint could be found at SMFn+1 (n = 4), which is consistent with our assumptions.

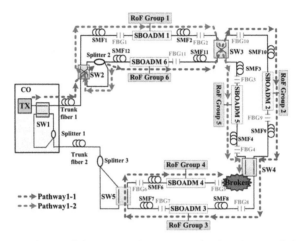

Figure 6. Routing pathways of downstream detection signals when the second phase of the first detection step is executed.

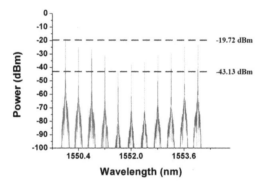

Figure 7. Detection signals received by the CO when the second phase of the first detection step is executed.

The first step of the proposed algorithm can deal with any fiber link failure except for that of trunk fiber 1. When trunk fiber 1 or any other two fiber links are blocked, switching only SW2 cannot heal the transport system and parts of the detection signals cannot be reflected back to the CO. To address these problems, the third phase of the first detection step (indicated by the purple background in Figure 4) was executed. In this phase, the detection system checks whether the two blocked fiber links are located on both sides of the SBOADM or not. The CO may only receive $\lambda 1$ to λn and $\lambda 11$ to λs ($s \geq 1$). $\lambda 1$ to λn were reflected by pathway1-2, whereas $\lambda 11$ to λs were reflected by pathway1-1. If two failure fiber links are present in both sides of the SBOADM, then only λm would not be received by the CO and the breakpoints could be found at SMFn+1 and SMFm+1. For instance, the SMF6 in Figure 6 is also broken simultaneously, and only $\lambda 1$ to $\lambda 4$ and $\lambda 11$ to $\lambda 6$ would be reflected back to the CO. Given that only $\lambda 5$ was not received, the two failure fibers could be found at SMFn+1 (n = 4) and SMFm+1 (m = 5). Nevertheless, if more than one detection signal is not received by the CO, then the detection system should execute the second detection step of the proposed algorithm to determine the locations of breakpoints.

3.2. Second Detection Step of the Proposed Algorithm

The flowchart of the second detection step of the proposed algorithm is shown in Figure 8. This step was mainly designed to deal with trunk fiber 1 failure and two fiber link failure conditions and would be executed after the first detection step. The first phase of the second detection step (indicated by the gray background in Figure 8) switched the SW1 in the CO and the SW5 in the RN2 from the parallel status to the cross status and then resent the 11 detection signals. In an extreme case, if both trunk fibers 1 and 2 are not functional, then no detection signal would be reflected by the FBG sensors. In this case, the two-step fiber link failure detection and self-healing algorithm could not recover all connections. Apart from the extreme case, if all detection signals were reflected back to the CO and the noise level of each received signal does not exceed that of the detection signals received in the first detection step, then the failure fiber link could be found at trunk fiber 1 or at both SMFn+1 and SMFs. Depending on the processing results obtained in the first detection step, if no detection signal was received, then the failure fiber could be found at trunk fiber 1. Otherwise, the failure fiber links could be found at SMFn+1 and SMFs, where n and s are defined in the first detection step. For instance, if SMF2 and SMF7 do not work, then n and s are set to 1 and 7, respectively, after executing the first detection step (only the SW2 changes to cross status). After executing the first phase of the second detection step, SW1 and SW5 switched to the cross status and the downstream detection signals were transmitted to pathway2-1, pathway2-2, pathway2-3, and pathway2-4 as shown in Figure 9. In this case, both the failure fiber links were bypassed, and the CO received all the detection signals as shown in Figure 10. The failure fiber links could be found at SMFn+1 (n = 1) and SMFs (s = 7).

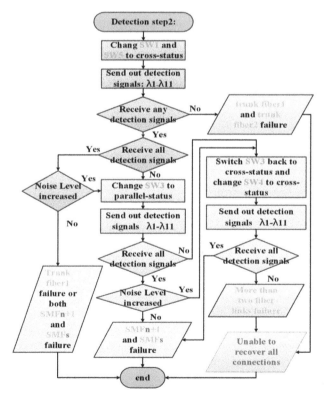

Figure 8. Second detection step of the proposed two-step fiber link failure detection and self-healing algorithm.

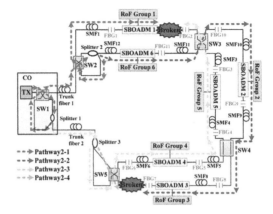

Figure 9. Routing pathways of the downstream detection signals during the execution of the first phase of the second step of the detection step.

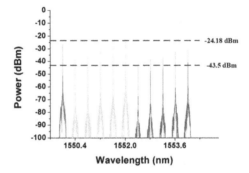

Figure 10. Detection signals received by the CO during the first phase of the second step of the detection step.

In some cases, the CO still could not locate the failure fiber links even if all the detection signals were reflected back to the CO. Given that downstream detection signals go through the transport system along the four pathways when executing the first phase of the second detection step, parts of the downstream detection signals in some pathways may go through the two-ring-based RoF transport system and be injected into the CO without being blocked by any failure fiber link. In this case, the CO would receive all the detection signals, but the noise levels of these detection signals would be higher than those received in the first detection step. In this case, the second phase of the second detection step (indicated by the green and blue backgrounds in Figure 8) should be executed to locate the breakpoints. In parallel, if only parts of the detection signals were received by the CO in the first phase of the second detection step, then the second phase of the second detection step should also be executed to locate the breakpoints.

In the second phase of the second detection step, either SW3 or SW4 could be switched to determine the failure fiber link locations. For example, SW3 can be initially switched from the cross status to the parallel status before resending the detection signals to the transport system. If all detection signals were received, then the location of the failure fiber links could be found at SMFn+1 and SMFs; otherwise, SW3 could be switched back to the cross status and SW4 could be switched from the parallel status to the cross status. If all the detection signals were received by the CO, then the failure fiber links could also be found at SMFn+1 and SMFs. However, if the CO still could not receive all the detection signals, then more than two breakpoints might be present in the transport system and the two-steps fiber link failure detection and self-healing algorithm would be unable to recover all connections.

To evaluate the second phase of the second detection step, SMF10 and SMF7 were assumed to be blocked as shown in Figure 11. According to the first detection step of the proposed algorithm, the parameters n and s were set to 6 and 10, respectively. The CO received all the detection signals after executing the first phase of the second detection step. The optical spectra of the received detection signals are shown in Figure 12. The noise levels of the obtained detection signals were obviously larger than those shown in Figure 5 because the detection signals in pathway3-1 and pathway3-4 were transmitted through the RoF transport system and were directly routed back to the CO. The second phase of the second detection step should be executed to find the actual breakpoints. If SW3 in Figure 11 switched to the parallel status, then the detection signals in pathway3-1 and pathway3-2 would be routed back to the CO without being blocked by any failure fiber link. The noise level of the obtained detection signals would also be enlarged; therefore, SW3 should be switched back to the cross status and SW4 should be switched from the parallel status to the cross status as shown in Figure 13. The locations of all FBG sensors were reached by pathway3-1, pathway3-2, and pathway3-4. The optical spectra of the reflected detection signals in the CO are shown in Figure 14. Given that all the detection signals were received and their noise levels do not show any obvious increase, the blocked fiber links were assumed to be found at SMFn+1 (n = 6) and SMFs (s = 10).

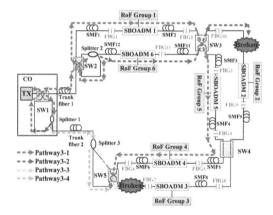

Figure 11. Routing pathways of the downstream detection signals when the first phase of the second detection step is executed.

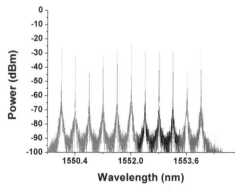

Figure 12. Detection signals received by the CO during the execution of the first phase of the second detection step.

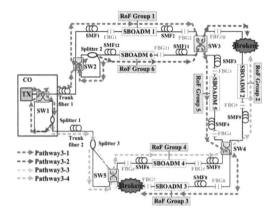

Figure 13. Routing pathways of downstream detection signals when the second phase of the second detection step is executed.

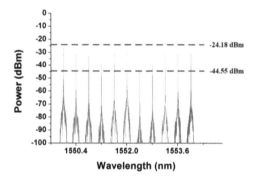

Figure 14. Detection signals received by the CO when the second phase of the second detection step is executed.

The proposed two-ring-based RoF transport system with the assistance of the FBG remote sensing technique and the two-step fiber link failure detection and self-healing algorithm can automatically recover network connections in the presence of fiber link failures. Except in some extreme conditions, the proposed detection algorithm can locate up to two fiber link failures and restore all network connections before the blocked fiber links are repaired.

4. Conclusions

RoF transport systems have been utilized to develop a 5G mobile network. The network traffic loading and QoS requirements in a 5G system are much higher than those in a 4G environment. Providing RoF transport systems with self-healing functionality to prevent interruptions in network services is crucial. To contribute to the extant knowledge in this field, we develop a two-ring-based RoF transport system based on a self-developed SBOADM and a two-step fiber link failure detection and self-healing algorithm. With the assistance of the FBG remote sensing technique, a fiber failure detection system in CO can utilize optical detection signals to monitor the health of each SMF span in the RoF transport system. When one or more fiber link failures take place in the RoF transport system, the detection system follows the processes of the proposed two-step fiber link failure detection and self-healing algorithm to detect the failure fiber links and to bypass their impact. Simulation results show that apart from some extreme situations, the proposed algorithm can achieve its function properly when the blocked fiber links are less than three. Furthermore, the wavelength reflection

Sensors **2019**, *19*, 4201

ranges of the FBGs used in the SBOADMs are employed to drop or add RoF communication signals, but the FBG sensors were employed to detect the health of the fiber links only. The reflection center wavelengths of FBG sensors can be set apart from the range of communication wavelengths (optical carriers of RoF groups) to avoid interference with each other and the power levels of the detection signals can be adjusted to avoid non-linear effects such as four-wave mixing or stimulated Brillouin scattering (SBS). Although the grating pitch of the FBG sensors may vary with temperature or strain, increasing the FBG reflection window or adding a proper thermal control/protection will increase the stability of the proposed scheme. The proposed RoF transport system can then guarantee QoS by automatically detecting the location of blocked fiber links and recovering all network connections before repairing these links.

Author Contributions: Conceptualization, C.-H.C., W.-S.T.; formal analysis, W.-S.T., C.-H.C., Z.-G.L., and D.-Y.L.; investigation, W.-S.T., C.-H.C., Z.-G.L., D.-Y.L., and T.-Y.Y.; project administration, C.-H.C. and W.-S.T.; writing—original draft, W.-S.T.; writing—review and editing, C.-H.C. and W.-S.T.

Funding: This research was funded by the Ministry of Science and Technology (MOST) of Taiwan under Grant Nos. 107-2221-E-415-009-, 108-2221-E-415-023-, and 106-2221-E-131-018-MY2.

Conflicts of Interest: The authors declare no conflict of interest.

References

1. Singh, S. Performance investigation on DWDM optical ring network to increase the capacity with acceptable performance. *Optik* **2014**, *125*, 5750–5752. [CrossRef]
2. Olmedo, M.I.; Suhr, L.; Prince, K.; Rodes, R.; Mikkelsen, C.; Hviid, E.; Monroy, I.T. Gigabit access passive optical network using wavelength division multiplexing—GigaWaM. *J. Lightwave Technol.* **2014**, *32*, 4285–4293. [CrossRef]
3. Yang, B.; Chen, X.; Shen, C.; Huang, X.; Ma, Z.; Bei, J.; Li, M.; Zhu, Q.; Na, T. Smile OAN: A long reach hybrid WDM/TDM passive optical network for next generation optical access. In Proceedings of the Asia Communications and Photonics Conference, Wuhan, China, 2–5 November 2016.
4. Feng, C.; Gan, C.; Gao, Z.; Wu, C. Novel WDM access network featuring self-healing capability and flexible extensibility. In Proceedings of the International Conference on Electrical, Computer Engineering and Electronics, Jinan, China, 29–31 May 2015.
5. Feng, K.M.; Wu, C.Y.; Yan, J.H.; Lin, C.Y.; Peng, P.C. Fiber Bragg grating-based three-dimensional multipoint ring-mesh sensing system with robust self-healing function. *IEEE J. Sel. Top. Quantum Electron.* **2012**, *18*, 1613–1621. [CrossRef]
6. Imtiaz, W.A.; Waqas, M.; Mehar, P.; Khan, Y. Self-healing hybrid protection architecture for passive optical networks. *Int. J. Adv. Comput. Sci. Appl.* **2015**, *6*, 144–148.
7. Pires, J.J.O. Constraints on the design of 2-fiber bi-directional WDM rings with optical multiplexer section protection. In Proceedings of the Digest of LEOS Summer Topical Meetings: Advanced Semiconductor Lasers and Applications, Copper Mountain, CO, USA, 30 July–1 August 2001.
8. Akhtar, A.; Turukhin, A.V.; Parsons, E.; Bakhshi, B.; Cardoso, J.; Vejas, M.; Savidis, H.; Derr, N.; Kovsh, D.; Golovchenko, E.A. First field demonstration of fault resilience in a regional undersea OADM network. In Proceedings of the OFC/NFOEC, Los Angeles, CA, USA, 4–8 March 2012.
9. Matsuura, M.; Oki, E. Optical carrier regeneration for wavelength reusable multicarrier distributed OADM network. In Proceedings of the Conference on Lasers and Electro-Optics, San Jose, CA, USA, 16–21 May 2010.
10. Sun, X.; Chan, C.K.; Wang, Z.; Lin, C.; Chen, L.K. A single-fiber bi-directional WDM self-healing ring network with bi-directional OADM for metro-access applications. *IEEE J. Sel. Areas Commun.* **2007**, *25*, 18–24. [CrossRef]
11. Park, S.B.; Lee, C.H.; Kang, S.G.; Lee, S.B. Bidirectional WDM self-healing ring network for hub/remote nodes. *IEEE Photonics Technol. Lett.* **2003**, *15*, 1657–1659. [CrossRef]
12. Rashed, A.N.Z. Transmission performance evaluation of optical add drop multiplexers (OADMs) in optical telecommunication ring networks. *Am. J. Eng. Technol. Res.* **2011**, *11*, 23–37.
13. Chang, C.H.; Lu, D.Y.; Lin, W.H. All-passive optical fiber sensor network with self-healing functionality. *IEEE Photonics J.* **2018**, *10*, 1–10. [CrossRef]

14. Zhao, L.; Huang, X.; Jia, J.; Zhu, Y.; Cao, W. Detection of broken strands of transmission line conductors using fiber Bragg grating Sensors. *Sensors* **2018**, *18*, 2397. [CrossRef]

15. Pham, T.; Bui, H.; Le, H.; Pham, V. Characteristics of the fiber laser sensor system based on etched-Bragg grating sensing probe for determination of the low nitrate concentration in water. *Sensors* **2017**, *17*, 7. [CrossRef]

16. Chen, D.; Huo, L.; Li, H.; Song, G. A fiber bragg grating (FBG)-enabled smart washer for bolt pre-load measurement: Design, analysis, calibration, and experimental validation. *Sensors* **2018**, *18*, 2586. [CrossRef] [PubMed]

17. Posada-Roman, J.; Garcia-Souto, J.; Poiana, D.; Acedo, P. Fast interrogation of fiber Bragg gratings with electro-optical dual optical frequency combs. *Sensors* **2016**, *16*, 2007. [CrossRef] [PubMed]

18. Li, W.; Xu, C.; Ho, S.; Wang, B.; Song, G. Monitoring concrete deterioration due to reinforcement corrosion by integrating acoustic emission and FBG strain measurements. *Sensors* **2017**, *17*, 657. [CrossRef] [PubMed]

19. Sun, L.; Li, C.; Li, J.; Zhang, C.; Ding, X. Strain transfer analysis of a clamped fiber Bragg grating sensor. *Appl. Sci.* **2017**, *7*, 188. [CrossRef]

20. Peng, P.C.; Wang, J.B.; Huang, K.Y. Reliable fiber sensor system with star-ring-bus architecture. *Sensors* **2010**, *10*, 4194–4205. [CrossRef] [PubMed]

21. Zhaoxin, W.; Lin, C.; Chun-Kit, C. Demonstration of a single-fiber self-healing CWDM metro access ring network with unidirectional OADM. *IEEE Photonics Technol. Lett.* **2006**, *18*, 163–165. [CrossRef]

22. Zhu, M.; Zhang, J.; Sun, X. Centrally Controlled self-healing wavelength division multiplexing passive optical network based on optical carrier suppression technique. *Opt. Eng.* **2015**, *54*, 126105. [CrossRef]

23. Imtiaz, W.A.; Khan, Y.; Mahmood, K. Design and analysis of self-healing dual-ring spectral amplitude coding optical code division multiple access system. *Arab. J. Sci. Eng.* **2015**, *41*, 3441–3449. [CrossRef]

24. Peng, P.C.; Huang, K.Y. Fiber Bragg grating sensor system with two-level ring architecture. *IEEE Sens. J.* **2009**, *9*, 309–313. [CrossRef]

25. Gu, H.W.; Chang, C.H.; Chen, Y.C.; Peng, P.C.; Kuo, S.T.; Lu, H.H.; Li, C.Y.; Yang, S.S.; Jhang, J.J. Hexagonal mesh architecture for large-area multipoint fiber sensor system. *IEEE Photonics Technol. Lett.* **2014**, *26*, 1878–1881. [CrossRef]

26. Ogushi, I.; Araki, N.; Azuma, Y. A new optical fiber line testing function for service construction work that checks for optical filters below an optical splitter in a PON. *J. Lightwave Technol.* **2009**, *27*, 5385–5393. [CrossRef]

Article

Design and Implementation of a Novel Measuring Scheme for Fiber Interferometer Based Sensors

Chao-Tsung Ma [1,*], Cheng-Ling Lee [2] and Yan-Wun You [2]

[1] Department of Electrical Engineering, CEECS, National United University, Miaoli 36063, Taiwan
[2] Department of Electro-Optical Engineering, National United University, Miaoli 36063, Taiwan; cherry@nuu.edu.tw (C.-L.L.); lovesky791013@gmail.com (Y.-W.Y.)
* Correspondence: ctma@nuu.edu.tw; Tel.: +886-37-382482

Received: 28 August 2019; Accepted: 18 September 2019; Published: 21 September 2019

Abstract: This paper presents a novel measuring scheme for fiber interferometer (FI) based sensors. With the advantages of being small sizes, having high sensitivity, a simple structure, good durability, being easy to integrate fiber optic communication and having immunity to electromagnetic interference (EMI), FI based sensing devices are suitable for monitoring remote system states or variations in physical parameters. However, the sensing mechanism for the interference spectrum shift of FI based sensors requires expensive equipment, such as a broadband light source (BLS) and an optical spectrum analyzer (OSA). This has strongly handicapped their wide application in practice. To solve this problem, we have, for the first time, proposed a smart measuring scheme, in which a commercial laser diode (LD) and a photodetector (PD) are used to detect the equivalent changes of optical power corresponding to the variation in measuring parameters, and a signal processing system is used to analyze the optical power changes and to determine the spectrum shifts. To demonstrate the proposed scheme, a sensing device on polymer microcavity fiber Fizeau interferometer (PMCFFI) is taken as an example for constructing a measuring system capable of long-distance monitoring of the temperature and relative humidity. In this paper, theoretical analysis and fundamental tests have been carried out. Typical results are presented to verify the feasibility and effectiveness of the proposed measuring scheme, smartly converting the interference spectrum shifts of an FI sensing device into the corresponding variations of voltage signals. With many attractive features, e.g., simplicity, low cost, and reliable remote-monitoring, the proposed scheme is very suitable for practical applications.

Keywords: optical fiber sensor; fiber interferometer; fast measurement scheme

---

1. Introduction

In recent years, optical fiber based communication systems featuring low transmission attenuation, high frequency bandwidth, high stability, and immunity to electromagnetic interference, have been widely used in the Internet of Things (IOT) infrastructure, in the big data transmission applications and also the cloud storage systems, in order to meet the increasing bandwidth requirements and security concerns. In addition to its excellent application potentials in the field of data communication, optical fiber is also valued in the field of sensing technology. With proper design, a fiber can be made into a variety of sensing devices because it is highly sensitive to physical parameters such as pressure, temperature, bending, and even refractive index changes. This is because the variation of physical quantities or states will affect the characteristics of light waves transmitted inside the fiber, and thus the physical quantity changes the variations in environmental parameters that can be detected by analyzing the changes of the light wave through the fiber. It is important to note that fiber based sensors integrating optical fiber communication networks can transmit optical signals over long distances and achieve a reliable remote monitoring mechanism.

In the aspects of designing fiber based sensors, there is a huge number of sensing devices that are usually designed based on the theory of interferometers, and the measured phase shifts of the interference waveforms are commonly used to obtain minute changes in the physical parameters of the environment. In fact, the development of feasible fiber optic interferometers with all-fiber configuration has become a very popular research topic recently. These all-fiber interferometers not only have the advantages in measuring precision but also have many superior intrinsic features of optical fibers. For example, optical interference signals do not travel in free space and are not affected by the weather and electromagnetic waves. In the literature, the development of all-fiber interferometers with different configurations has yielded considerable results, such as fiber-optic Michelson interferometers (FMI) [1–4], fiber-optic Mach–Zehnder interferometers (FMZI) [5–8], fiber-optic Fabry–Perot interferometers (FFPI) [9–17], and fiber-optic Fizeau interferometers (FFI) [18,19]. Typical FI based sensing systems and application examples have been categorized into 13 types, according to the measured parameters, i.e., temperature [20,21], mechanical vibration [22,23], acoustic wave [22,24], ultrasound [25,26], voltage [27,28], magnetic field [29,30], pressure [20,31], strain [21,32], flow velocity [33,34], humidity [35,36], gas [37,38], liquid level [39,40], and the refractive index (RI) [41,42]. Li et al. [20] proposed a cascaded-cavity FFPI to simultaneously sense air pressure and temperature. Huang et al. [21] presented a new sensing mechanism based on a strategically designed micro-cavity FMZI and FFPI for achieving simultaneous and cross-sensitivity free measuring of temperature and strain. The authors of [22] demonstrated that simultaneous measurement of mechanical and acoustic vibration can be realized, using a novel flexible FFPI. Up to 20 kHz acoustic vibration can be measured. In [23], up to 1.5 kHz mechanical vibration was measured using a commercial interrogator and fast Fourier transform algorithms. Low-frequency (around 13 Hz) acoustic pressure was successfully measured by Z. Gong et al. using a simple FFPI [24]. In [25], a high-intensity focused ultrasound (HIFU) was measured by utilizing an ultra-compact, low-temperature crosstalk, two-wave interferometer sensor. W. Zhang et al. [26] designed a miniature FFPI based sensor with spectral sideband filtering for ultrasound and image sensing. Chen et al. [27] used an FMI based sensor to measure voltage and the IEC standard was used to verify the performance of the proposed sensor. In [28], a FFPI driven by electric field forces was able to sense voltage with only 0.1 ms delay. In [29], an AC magnetic field was measured with high sensitivity and correction of temperature crosstalk using an elastic FFPI. A compact magnetic field sensor was designed based on a S-taper and an up-fusion-taper multimodal interference [30]. An absolute pressure sensor was designed based on an external FFPI enclosed in a vacuum cell [31]. Tang et al. [32] designed a dual-tapered photonic crystal fiber (PCF) based FMZI for strain sensing. Zhang et al. [33] proposed an FFPI for low liquid velocity, measured with a high sensitivity of 0.0016 nm/(μL/min). In [34], a hot cavity FFPI based flow sensor with SMF-CDF-SMF structure and with temperature self-calibration was proposed. In [35], a humidity sensor based on a three-fiber, core-offset FMZI was presented. Sensitivities of 0.104 dB/%RH and 0.0272 nm/%RH and 99.61% with correlation coefficients of 99.21% have been demonstrated at 30% and 60% relative humidity (RH), respectively. A humidity sensor designed with a multicore fiber, helical structure, and a gold film reflector was demonstrated by the authors of [36]. Two long period gratings (LPGs) were used to design an ultra-sensitive sensor for measuring underground mine toxic gases such as carbon monoxide (CO) and methane (CH4) [37]. In [38], the CO gas can be effectively detected with high sensitivity using a thin-core FMZI. Liu et al. [39] proposed a temperature-insensitive liquid level sensor based on FFPI with an error less than 0.4% of full scale. Dong et al. [40] used a D-shape fiber modal interferometer to design a liquid level sensor for achieving low temperature cross sensitivity. In [41], a RI sensor with 6.02×10^{-6} detection limit and 1.3320 to 1.3465 RIU range was designed with FMZI and Sagnac interferometer. In [42], RI was measured by using an open-microhole FFPI based sensor and Fourier band-pass filtering techniques.

In the above reviewed FI based sensing systems, the design of hardware systems and measuring mechanisms for achieving the desired sensing purposes can be summarized into three combinations: (1) the integration of broadband light source (BLS), optical spectral analyzer (OSA), circulators or power splitters and personal computer (PC) [20,21,30,31,33,35–37,40,41]; (2) the integration of an amplified spontaneous emission (ASE) or ASE with tunable optical filter (TOF), photodiode module (PM), data acquisition module (DAQM), circulators or optical couplers (OC) and PC [22,27,38]; (3) the integration of distributed feedback laser (DFBL) or DFBL with the erbium doped fiber amplifier (EDFA), PM or OSA, OC with wavelength-division multiplexer (WDM) module and PC with DAQM [24–26,28,29,32,34,39]. The common disadvantage of the existing systems and sensing methods mentioned above include high-cost, as well as being bulky and relatively slow in measuring speed if OSA is used. The above drawbacks have strongly handicapped FI based sensors being widely used in practical applications. To solve this problem, this paper proposes a cost-effective measuring scheme, in which a commercial laser diode (LD) and a photo detector module (PDM) are strategically used to detect the equivalent changes of optical power corresponding to the interference spectrum shifts caused by the variation in measuring parameters. To demonstrate the feasibility of the proposed measuring scheme, a temperature (T) and relative humidity (RH) measurement system based on a polymer microcavity fiber Fizeau interferometer (FMCFFI) [43] is presented, and the related design details are described. The merits of the proposed measuring scheme are low-cost and very easy to be integrated into fiber optic communication systems. In the proposed measurement system, instead of using OSA, a pair of commercial laser diodes (LD) and a photodetector (PD) are used to detect optical power corresponding to the interference spectrum shifts, and an optical/electrical signal processing system with a set of derived algorithms is utilized to convert the spectrum shifts into the equivalent output voltages. In this arrangement, the spectrum scanning time of the OSA can be saved and a low-cost, simple measurement system based on high-sensitivity fiber optic interferometer sensing device can be achieved. To demonstrate the proposed design idea, following this introduction section, the content of the second section reviews the principle and development of a PMCFFI acting as a T and RH sensor. The third section describes the details of the proposed new measuring mechanism for the PMCFFI based T and RH sensing device. A conclusion is then given in the last section.

2. Polymer Micro Cavity Fiber Fizeau Interferometer

2.1. PMCFFI Manufacturing Steps

The PMCFFI used in this paper adopts a common single mode fiber (SMF-28), the core diameter of which is 8.2 μm. First, we take a piece of SMF, use a fiber stripper to strip the jacket, wipe it off with an alcohol wipe, and then flatten the end face with a Fujikura fiber cutter CT-30. Next, we drop Norland Products optical glue NOA 61 onto a slide and evenly smooth it to control the thickness of the glue on the SMF end face. After that, we slowly dip the surface of the SMF into NOA 61 and observe it with an optical microscope to make sure that NOA 61 forms a curved surface that serves as the interference cavity. Based on the experiments, we are able to successfully produce many Fizeau interferometer components with different cavity lengths using special dipping techniques. The NOA 61 dipped fiber end face is then exposed by 9.85 W/cm², 320–480 nm UV light. After some time, the liquid monomer molecules are converted into a stable solid polymer resonator. The component is then placed into a 50 °C environment for 12 h to form the required chemical bond between the NOA 61 and the SMF for achieving an optimal bonding. Figure 1a shows the schematic diagram of the PMCFFI, and Figure 1b shows a photo of a PMCFFI component under an optical microscope.

NOA61 is highly hygroscopic and temperature-sensitive; it is a transparent, colorless, liquid monomer that cures under UV light. Through using NOA 61, the pre-mixing, drying or thermal curing typically required in other optical bonding techniques can be avoided, and the curing speed can be extremely fast. It has excellent light transmission, low shrinkage and slight elasticity relative to other optical bonding materials. These characteristics are important to ensure that users can obtain high-quality optical components. In particular, long-term characteristics can be maintained when environmental conditions change. It also has excellent adhesion and solvent resistance after being fully cured by UV light, but it has not yet reached the best adhesion to glass. One week of ageing is required for the chemical bond to be formed between the glass and NOA 61 for optimum adhesion. Alternatively, the best adhesion can be obtained by aging for 12 h at a temperature of 50 °C. When used for glass bonding, NOA 61 can withstand temperature changes from −15 °C to 60 °C before aging. After full aging, it can withstand temperature changes from −150 °C to 125 °C, making it ideal for harsh environment applications.

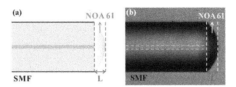

Figure 1. Polymer microcavity fiber Fizeau interferometer (PMCFFI): (**a**) schematic diagram; (**b**) actual photo.

2.2. PMCFFI Interference Principle

The PMCFFI adopted in this paper has the advantages of a simple and micro structure, as well as being easy to manufacture, and having a low cost. The main mechanism of its interferometer is described as follows: the light propagates into the SMF, and its mode field distribution is mainly concentrated in the core; when the light reaches the first interface between the SMF and the NOA61 polymer, a portion of the light will reflect first (r_1), and the other part will be transmitted to the end face of the NOA61 and then reflect (r_2). Since this configuration adopts two beam interference principles, the reflective surfaces are at the two ends of the polymer. Figure 2a,b shows the conceptual interference diagram and the simulation result of the transmission of the designed sensing device with the PMCFFI (27 μm).

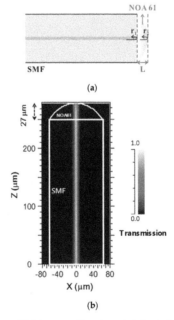

Figure 2. (**a**) Schematic diagram of light propagation characteristics in the resonant cavity of a PMCFFI; (**b**) The transmission of the designed sensing device with the PMCFFI (27 μm).

Based on the following simple cavity interference equation, we can derive the sensing characteristics of this PMCFFI. The light travels 2 nL in the NOA 61 polymer resonator, so the cavity phase δ is expressed as

$$\delta = (2\pi/\lambda) \times 2nL\delta = (2\pi/\lambda) \times 2nL, \tag{1}$$

where n represents refractive index of cavity, and L represents resonant cavity length. When $\delta = 2\,m\pi$, the interference spectrum shift is at its extremum. As a result, we get Equations (2) and (3):

$$(2\pi/\lambda) \times 2nL = 2m\pi; \tag{2}$$

$$2nL = m\lambda. \tag{3}$$

In (2) and (3), m is an integer. In the interference spectrum, the change of surrounding temperature and relative humidity (RH) will cause the cavity length, refractive index, and then spectrum wavelength change, denoting as $L' = L + \Delta L$, $n' = n + \Delta n$, and $\lambda'_m = \lambda_m + \Delta\lambda_m$, respectively. We can, therefore, get Equations (4) and (5):

$$\delta = (2\pi/\lambda'_m) \times 2n'L' = (2\pi/\lambda_m) \times 2nL; \tag{4}$$

$$\Delta(nL)/nL = \Delta\lambda_m/\lambda_m \tag{5}$$

Therefore, when the temperature and/or RH rise, the NOA 61 resonant cavity length will increase accordingly. Based on this, the interference spectrum shifts to a longer wavelength (redshift). On the other hand, when the temperature and/or RH drop, the interference spectrum will shift to a shorter wavelength (blueshift). This process proves that the optical path difference changes according to temperature and RH changes, which can be measured by analyzing the interference spectrum shifts.

2.3. Experimental Measurement Configuration and Results

For deriving a general calculating algorithm for the proposed measuring system, some experimental tests are performed and the related parameters are obtained. The arrangement of experimental measurement is shown in Figure 3. The designed PMCFFI is placed in a temperature/humidity controlled chamber (THCC). A BLS is used as the signal source and connected to the PMCFFI through a set of 2 to 2 optical couplers. When the light passes through the end face of the PMCFFI, the reflected light will return to the optical coupler and then enter an OSA for interference spectrum analysis. Here, the cavity lengths of PMCFFI are developed at 10, 15, 27 and 42 μm, respectively. Figure 4a shows the interference spectra of L = 10 μm PMCFFI with only RH variations. A complete set of RH sensing results with these components is obtained and found to have a linear response in phase shift, as shown in Figure 4b. As can be seen, different sensitivities are observed in different cavity lengths. To observe the free spectral range (FSR) of the measured interference, a set of FFT results of the PMCFFI (L = 10 μm) with only RH variations (from 20% to 90%) are shown in Figure 4c.

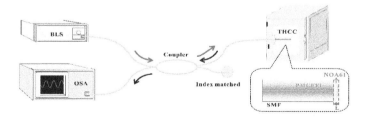

Figure 3. Schematic diagram of the PMCFFI experimental measurement configuration.

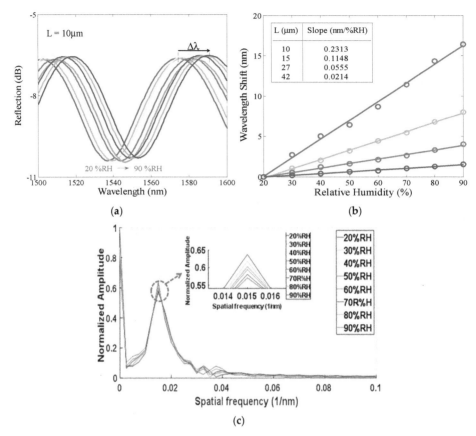

Figure 4. (**a**) Interference spectra of the PMCFFI (L = 10 μm) with only relative humidity (RH) variations; (**b**) Sensitivity comparison of different L based on wavelength shift due to RH changes; (**c**) the FFT result of the PMCFFI (L = 10 μm) with only RH variations (from 20% to 90%).

3. Development of the Proposed Measurement System Based on PMCFFI Sensing Device

3.1. Measurement Principle and Implementation

As addressed in the previous section, the output signals of the proposed PMCFFI sensing device are interference spectrum shifts, or equivalently, the variations in optical power at a specified wavelength. With this concept in mind, one can use a laser diode (LD) with the desired wavelength and a photodetector (PD) acting as the signal detector together with a general fiber network to construct a simple and cost effective measuring system. In practice, through a signal processing circuit, the measured optical power signals (dBm) can be converted into electrical signals (mV) and input to a computer for spectrum phase shift calculation and displaying the measuring RH. The structure and operating mechanism of the proposed measurement system are shown in Figure 5.

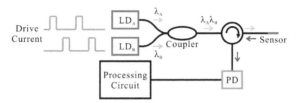

Figure 5. System configuration of the proposed fast measurement system with a PMCFFI sensor.

To have a clear picture of the proposed measuring concepts, Figure 6 shows the schematic diagrams of spectrum shift monitoring at two wavelengths. First, the interference spectrum (usually a sine-like or cosine-like periodic function) of the PMCFFI sensing device at a given temperature (20 °C) is first measured using OSA. Next, a certain period in the interference spectrum near the optical communication band is selected for sensing. In this case, a pair of LDs with wavelengths of λ_A and λ_B are used to monitor reflective optical powers of the two wavelengths on the spectrum, as shown in Figure 6a. The change of the sensed parameters will cause interference spectrum shifts. By monitoring the optical powers (dBm or voltage values in mV) at λ_A and λ_B, the interference spectrum shifts can be precisely calculated, and thus the changes in the monitored parameters can be obtained. Since the reflective power spectrum is approximate to a sine wave, we can obtain better linearity within the range of around $\pi/2$; this should be taken into consideration when selecting the wavelength of LDs, so that the selected LDs can monitor, with better linearity, during the different states of spectrum phase shifts. To determine if the spectrum shift exceeds the feasible detecting range in a certain design (theoretically less than π), we can observe the reflective power change of the two wavelengths: if the two power variation polarities remain opposite to each other, the spectrum shift has not reached its extremum, as shown in Figure 6b. If the two power variation polarities become the same, the spectrum shift has reached its extremum, as shown in Figure 6c. After the test is done, a microprocessor with the derived calculating algorithms, along with some signal processing circuits, are integrated to complete the hardware prototype of the proposed measuring system.

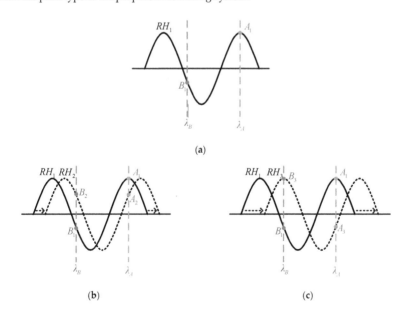

Figure 6. Schematic diagrams of spectrum shift monitoring on two wavelengths: (a) initial state; (b) right shift within feasible detecting range; (c) when the spectrum shift has reached its extremum.

3.2. System Prototype

Based on the theoretical analysis and the required sensing of the displaying functions, we have designed and practically constructed the hardware system as shown in Figure 7, where we can see (1) is the power supply port and the main power switch, (2) is a dc power control module, (3) is a voltage converting circuit for the PD and the thermistor, (4) is a liquid crystal display (LCD), (5) is a USB port, (6) is a 110V_{AC}/12V_{DC} adapter, (7) is the Arduino Mega 2560 control board, (8) is a FC/FC adaptor, (9) is the thermistor sensing port, and (10) is the power switch. In Figure 7d, we can see that the upper layer contains (11) fiber pigtail, (12) optical circulator, (13) photodetector (Thorlabs DET01CFC), and (14) the optic coupler. The two ports on one end of the 2 to 2 optical couplers are connected to the FC/APC, and the other two ports are connected to optical circulator inputs. The two optic circulator outputs are connected to the proposed PMCFFI device and the PD input, respectively.

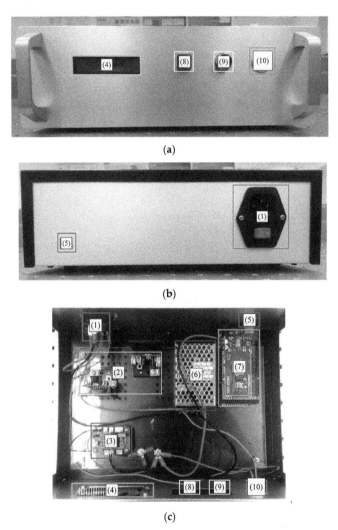

(a)

(b)

(c)

Figure 7. *Cont.*

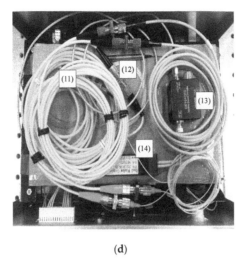

(d)

Figure 7. The developed prototype: (a) front view; (b) back view; (c) lower layer; (d) upper layer.

3.3. Measurement Results

After the measuring system is completed, the initial spectrum of the PMCFFI with L = 27 μm is firstly measured by the OSA. The two LD monitoring wavelengths are then selected at λ_B = 1571 nm and λ_A = 1591 nm, as shown in Figure 8.

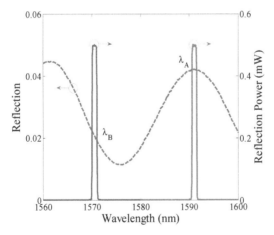

Figure 8. Using 2 laser diodes (LDs) to monitor the reflective power spectrum at 20 °C.

As shown in s we can have the following equations:

$$V_1 = a(RH_0 - b_1) \tag{6}$$

$$V_2 = a(RH_0 - b_2) \tag{7}$$

where V_1/V_2 represent the measured voltage values at initial RH_0 = 20% and for T_1/T_2 respectively, and *a* represents the voltage-RH slope, b_1/b_2 represent the two x-intercepts for T_1/T_2 respectively. To derive a general equation describing the relationships among the voltage, V, T and RH, the following equation can be obtained from Figure 9:

$$V_T = a(RH - b_T) \tag{8}$$

Based on the geometric relationship shown in Figure 9, b_T can be derived as follows:

$$b_T = b_1 + (b_2 - b_1)(T - T_1)/(T_2 - T_1) \tag{9}$$

Using Equations (6), (7) and (9), b_T can be rewritten as:

$$b_T = (RH_0 - V_1/a) + (-V_2/a + V_1/a)(T - T_1)/(T_2 - T_1) \tag{10}$$

Substituting Equations (10) into (8) gives the following relationship:

$$V_T = a(RH - RH_0) + V_1 + (V_2 - V_1)(T - T_1)/(T_2 - T_1) \tag{11}$$

For a given operating temperature range ($T_1 \sim T_2$), the related parameters, a, RH_0, V_2, V_1, T_1 and T2 can be decided by performing some initial experiment tests. It follows that the monitored RH value can be achieved and shown in a calibrated voltage (V_T) for any temperature between T_2 and T_1.

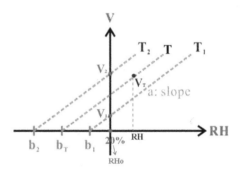

Figure 9. Relationship between voltage and RH at different temperatures.

To test the performance of the proposed measuring system, the PMCFFI with 27 μm cavity is chosen for some practical measurements. First, the temperature is fixed at 20 °C, and the RH varies from 20% to 90%. It should be noted that the NOA61's temperature operating range is from −150 °C to 125 °C; however, for demonstration purposes the temperature range of 20 °C to 45 °C is tested in this study. In order to maintain good monitoring linearity, we observe optic power of λ_B for temperatures ranging from 20 °C to 30 °C; for the temperature range of 35 °C to 45 °C, we observe optic power of λ_A. Figure 10a shows the calculated and experimental values at 20, 25, and 30 °C, and Figure 10b shows the calculated and experimental values at 35, 40 and 45 °C. As can be observed, the experimental results are very close to calculation results. Finally, the experimental values of V_1, V_2, and the calculated a are used in the derived voltage calculating Equation (11) and written into the Arduino Mega 2560 control board. This completes parameter setting of the measuring system for the chosen PMCFFI sensing device. In this study, a thermistor is used to measure the operating temperature (T) in order to correctly determine the voltages corresponding to the monitored RHs. Figure 11 shows a set of detailed measurement results.

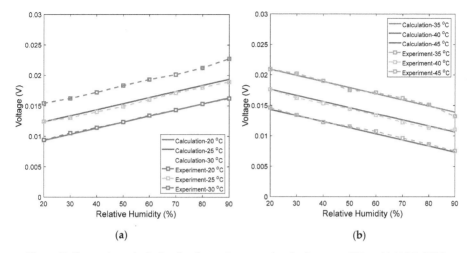

Figure 10. Comparison of calculated and experiment results of voltages vs. RHs at (**a**) 20 °C, 25 °C, and 30 °C, with λ_B; (**b**) voltages vs. RHs at 35 °C, 40 °C, and 45 °C, with λ_A.

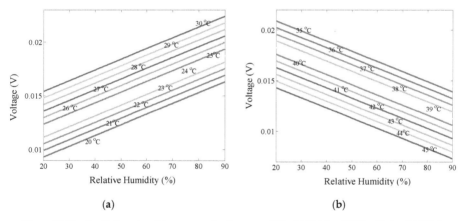

Figure 11. The detailed results of the measured voltages vs. RHs: (**a**) 20 °C to 30 °C, with λ_B; (**b**) 35 °C to 45 °C, with λ_A.

4. Conclusions

This paper has presented a novel measuring scheme for a PMCFFI based fast and sensitive temperature and RH measurement system, suitable for long-distance monitoring system states or environmental parameters. The measurement system eliminates expensive devices, such as BLS and OSA, that are normally required to measure the interference spectrum shifts of fiber interferometers. In the proposed measuring mechanism, with the derived Ts, RHs and voltages converting algorithms, commercially available LD and PD can be used to measure optical power, and an optical/electrical signal processing unit has been adopted to convert the optical powers, corresponding to a certain spectrum shifts, into the equivalent voltages. It is worthwhile noting that, although the aging effect with regard to long term laser exposure is not obvious in the designed PMCFFI, the laser power should be optimally regulated to further minimize the possible aging effect in long-term operations. The authors believed that using the proposed measuring scheme, most of the optical fiber sensors on FI developed in the literature can be easily implemented, and that a low-cost, reliable and easy-to-use measurement system can be developed for a variety of practical applications. It is also reasonable to note that the proposed sensing mechanism has significant commercial potential.

Author Contributions: This work was carried out in collaboration between all authors. The corresponding author, C.-T.M. verified the design concept and mathematical methods, analyzed the results, wrote the original draft, revised and polished the final manuscript. C.-L.L. verified the testing methods and results. Y.-W.Y., a student in the Department of EOE, performed the experiments and managed data and figures.

Funding: MOST Taiwan: MOST 108-2221-E-239-007.

Acknowledgments: The authors would like to thank the Ministry of Science and Technology of Taiwan for financially support the research regarding new sensing methods for advanced energy storage systems.

Conflicts of Interest: The authors declare no conflict of interest.

References

1. Xu, L.; Jiang, L.; Wang, S.; Li, B.; Lu, Y. High-temperature sensor based on an abrupt-taper Michelson interferometer in single-mode fiber. *Appl. Opt.* **2013**, *52*, 2038–2041. [CrossRef] [PubMed]
2. Deng, M.; Sun, X.; Han, M.; Li, D. Compact magnetic-field sensor based on optical microfiber Michelson interferometer and Fe_3O_4 nanofluid. *Appl. Opt.* **2013**, *52*, 734–741. [CrossRef] [PubMed]
3. Lee, C.-L.; Lee, C.-F.; Li, C.-M.; Chiang, T.-C.; Hsiao, Y.-L. Directional anemometer based on an anisotropic flat-clad tapered fiber Michelson interferometer. *Appl. Phys. Lett.* **2012**, *101*, 23502. [CrossRef]
4. Zhang, X.; Bai, H.; Pan, H.; Wang, J.; Yan, M.; Xiao, H.; Wang, T. In-Line Fiber Michelson Interferometer for Enhancing the Q Factor of Cone-Shaped In wall Capillary Coupled Resonators. *IEEE Photon. J.* **2018**, *10*, 6801808.
5. Tan, Y.; Sun, L.-P.; Jin, L.; Li, J.; Guan, B.-O. Microfiber Mach-Zehnder interferometer based on long period grating for sensing applications. *Opt. Express* **2013**, *21*, 154–164. [CrossRef] [PubMed]
6. Xu, Y.; Qin, Z.; Harris, J.; Baset, F.; Lu, P.; Bhardwaj, V.R.; Bao, X. Vibration sensing using a tapered bend-insensitive fiber based Mach-Zehnder interferometer. *Opt. Express* **2013**, *21*, 3031–3042. [CrossRef] [PubMed]
7. Hsu, J.-M.; Lee, C.-L.; Chang, H.-P.; Shih, W.C.; Li, C.-M. Highly Sensitive Tapered Fiber Mach–Zehnder Interferometer for Liquid Level Sensing. *IEEE Photon. Technol. Lett.* **2013**, *25*, 1354–1357. [CrossRef]
8. Zhang, R.; Pu, S.; Li, Y.; Zhao, Y.; Jia, Z.; Yao, J.; Li, Y. Mach-Zehnder Interferometer Cascaded with FBG for Simultaneous Measurement of Magnetic Field and Temperature. *IEEE Photon. J.* **2019**, *19*, 4079–4083. [CrossRef]
9. Tian, J.; Lu, Y.; Zhang, Q.; Han, M. Microfluidic refractive index sensor based on an all-silica in-line Fabry-Perot interferometer fabricated with microstructured fibers. *Opt. Express* **2013**, *21*, 6633–6639. [CrossRef]
10. Jiang, M.; Li, Q.-S.; Wang, J.-N.; Jin, Z.; Sui, Q.; Ma, Y.; Shi, J.; Zhang, F.; Jia, L.; Yao, W.-G.; et al. TiO_2 nanoparticle thin film-coated optical fiber Fabry-Perot sensor. *Opt. Express* **2013**, *21*, 3083–3090. [CrossRef]
11. Wu, X.; Solgaard, O. Short-cavity multimode fiber-tip Fabry-Pérot sensors. *Opt. Express* **2013**, *21*, 14487–14499. [CrossRef]
12. Wang, Y.; Wang, D.N.; Wang, C.; Hu, T. Compressible fiber optic micro-Fabry-Pérot cavity with ultra-high pressure sensitivity. *Opt. Express* **2013**, *21*, 14084–14089. [CrossRef]
13. Lee, C.L.; Lee, L.H.; Hwang, H.E.; Hsu, J.M. Highly sensitive air-gap fiber Fabry–Pérot interferometers based on polymer-filled hollow core fibers. *IEEE Photon. Technol. Lett.* **2012**, *24*, 149–151. [CrossRef]
14. Lee, C.L.; Zheng, Y.C.; Ma, C.L.; Chang, H.J.; Lee, C.F. Dynamic micro-air-bubble drifted in a liquid core fiber Fabry-Pérot interferometer for directional fiber-optic level meter. *Appl. Phys. Lett* **2013**, *102*, 193504. [CrossRef]
15. Lee, C.-L.; Hsu, J.-M.; Horng, J.-S.; Sung, W.-Y.; Li, C.-M. Microcavity Fiber Fabry–Pérot Interferometer with an Embedded Golden Thin Film. *IEEE Photon. Technol. Lett.* **2013**, *25*, 833–836. [CrossRef]
16. Lee, C.-L.; Hung, C.-H.; Li, C.-M.; You, Y.-W. Simple air-gap fiber Fabry–Perot interferometers based on a fiber endface with Sn-microsphere overlay. *Opt. Commun.* **2012**, *285*, 4395–4399. [CrossRef]
17. Wang, B.; Tian, J.; Hu, L.; Yao, Y. High Sensitivity Humidity Fiber-Optic Sensor Based on All-Agar Fabry–Perot Interferometer. *IEEE Photon. J.* **2018**, *18*, 4879–4885. [CrossRef]

18. Li, E.; Peng, G.-D.; Ding, X. High spatial resolution fiber-optic Fizeau interferometric strain sensor based on an in-fiber spherical microcavity. *Appl. Phys. Lett.* **2008**, *92*, 101117. [CrossRef]
19. Chan, J.Y.; Le, C.L.; Ha, P. Dynamic mitigation of EV charging stations impact on active Distribution Networks with Distributed BESSs. In Proceedings of the 2018 23rd Opto-Electronics and Communications Conference (OECC), Jeju Island, Korea, 2–6 July 2018.
20. Li, Z.; Tian, J.; Jiao, Y.; Sun, Y.; Yao, Y. Simultaneous Measurement of Air Pressure and Temperature Using Fiber-Optic Cascaded Fabry–Perot Interferometer. *IEEE Photon. J.* **2019**, *11*, 7100410. [CrossRef]
21. Huang, B.; Xiong, S.; Chen, Z.; Zhu, S.; Zhang, H.; Huang, X.; Feng, Y.; Gao, S.; Chen, S.; Liu, W.; et al. In-Fiber Mach-Zehnder Interferometer Exploiting a Micro-Cavity for Strain and Temperature Simultaneous Measurement. *IEEE Photon. J.* **2019**, *19*, 5632–5638. [CrossRef]
22. Wu, S.; Wang, L.; Chen, X.; Zhou, B. Flexible Optical Fiber Fabry–Perot Interferometer Based Acoustic and Mechanical Vibration Sensor. *J. Light. Technol.* **2018**, *36*, 2216–2221. [CrossRef]
23. Leandro, D.; Lopez-Amo, M. All-PM Fiber Loop Mirror Interferometer Analysis and Simultaneous Measurement of Temperature and Mechanical Vibration. *J. Light. Technol.* **2018**, *36*, 1105–1111. [CrossRef]
24. Gong, Z.; Chen, K.; Zhou, X.; Yang, Y.; Zhao, Z.; Zou, H.; Yu, Q. High-Sensitivity Fabry-Perot Interferometric Acoustic Sensor for Low-Frequency Acoustic Pressure Detections. *J. Light. Technol.* **2017**, *35*, 5276–5279. [CrossRef]
25. Gao, R.; Lu, D.; Cheng, J.; Qi, Z. Ultrasonic Detection of High-Intensity Focused Ultrasound Field using Quadrature Point Phase Step in a Fiber Optic Interferometric Sensor. *J. Light. Technol.* **2019**, *37*, 2694–2699. [CrossRef]
26. Zhang, W.; Qiao, X.; Shao, Z.; Wang, R.; Rong, Q.; Guo, T.; Li, J.; Ma, W. An optical fiber Fabry-Perot interferometric sensor based on functionalized diaphragm for ultrasound detection and imaging. *IEEE Photon. J.* **2017**, *9*, 1–8. [CrossRef]
27. Chen, X.; He, S.; Li, D.; Wang, K.; Fan, Y.; Wu, S. Optical fiber voltage sensor based on Michelson interferometer using phase generated carrier demodulation algorithm. *IEEE Photon. J.* **2016**, *16*, 1. [CrossRef]
28. Zhou, L.; Huang, W.; Zhu, T.; Liu, M. High Voltage Sensing Based on Fiber Fabry–Perot Interferometer Driven by Electric Field Forces. *J. Light. Technol.* **2014**, *32*, 3337–3343. [CrossRef]
29. Chen, X.; Wu, S.; Zeng, Y.; Zhou, B.; Wang, L.; Liu, L.; He, S. Elastic Optical Fiber Fabry–Perot Interferometer for Highly Sensitive AC Magnetic Field Measurement. *IEEE Photon. J.* **2018**, *18*, 5799–5804. [CrossRef]
30. Wu, W.; Cao, Y.; Zhang, H.; Liu, B.; Zhang, X.; Duan, S.; Liu, Y. Compact Magnetic Field Sensor Based on a Magnetic-Fluid-Integrated Fiber Interferometer. *IEEE Magn. Lett.* **2019**, *10*, 1–5. [CrossRef]
31. Ghildiyal, S.; Ranjan, P.; Mishra, S.; Balasubramaniam, R.; John, J. Fabry–Perot Interferometer-Based Absolute Pressure Sensor with Stainless Steel Diaphragm. *IEEE Sens. J.* **2019**, *19*, 6093–6101. [CrossRef]
32. Tang, Z.; Lou, S.; Wang, X.; Zhang, W.; Yan, S.; Xing, Z. High-Performance Bending Vector and Strain Sensor Using a Dual-Tapered Photonic Crystal Fiber Mach–Zehnder Interferometer. *IEEE Photon. J.* **2019**, *19*, 4062–4068. [CrossRef]
33. Zhang, Q.; Hou, D.; Wang, L.; Zhao, Y.; Zhao, C.; Kang, J. F-P Interferometer for Low Liquid Velocity Measurement with Non-contact and High Sensitivity. In Proceedings of the 2018 Asia Communications and Photonics Conference (ACP), Hangzhou, China, 26–29 October 2018.
34. Zhou, B.; Jiang, H.; Lu, C.; He, S. Hot Cavity Optical Fiber Fabry–Perot Interferometer as a Flow Sensor with Temperature Self-Calibrated. *J. Light. Technol.* **2016**, *34*, 5044–5048. [CrossRef]
35. Liu, S.; Meng, H.; Deng, S.; Wei, Z.; Wang, F.; Tan, C. Fiber Humidity Sensor Based on a Graphene-Coated Core-Offset Mach–Zehnder Interferometer. *IEEE Sens. J.* **2018**, *2*, 1–4. [CrossRef]
36. Liu, Y.; Zhou, A.; Yuan, L. Gelatin-Coated Michelson Interferometric Humidity Sensor Based on a Multicore Fiber with Helical Structure. *J. Light. Technol.* **2019**, *37*, 2452–2457. [CrossRef]
37. Singh, M.; Raghuwanshi, S.K.; Prakash, O. Ultra-Sensitive Fiber Optic Gas Sensor Using Graphene Oxide Coated Long Period Gratings. *IEEE Photon. Technol. Lett.* **2019**, *31*, 1473–1476. [CrossRef]
38. Feng, W.; Deng, D.; Yang, X.; Liu, W.; Wang, M.; Yuan, M.; Peng, J.; Chen, R. Trace Carbon Monoxide Gas Sensor Based on PANI/Co3O4/CuO Composite Membrane-Coated Thin-Core Fiber Modal Interferometer. *IEEE Sens. J.* **2018**, *18*, 8762–8766. [CrossRef]
39. Liu, T.; Zhang, W.; Wang, S.; Jiang, J.; Liu, K.; Wang, X.; Zhang, J. Temperature Insensitive and Integrated Differential Pressure Sensor for Liquid Level Sensing Based on an Optical Fiber Fabry–Perot Interferometer. *IEEE Photon- J.* **2018**, *10*, 1–8. [CrossRef]

40. Dong, Y.; Xiao, S.; Xiao, H.; Liu, J.; Sun, C.; Jian, S. An Optical Liquid-Level Sensor Based on D-Shape Fiber Modal Interferometer. *IEEE Photon. Technol. Lett.* **2017**, *29*, 1067–1070. [CrossRef]

41. Li, X.; Warren-Smith, S.C.; Ebendorff-Heidepriem, H.; Zhang, Y.-N.; Nguyen, L.V. Optical fiber refractive index sensor with low detection limit and large dynamic range using a hybrid fiber interferometer. *J. Light. Technol.* **2019**, *37*, 1. [CrossRef]

42. Zhang, W.; Liu, Y.; Zhang, T.; Yang, D.; Wang, Y.; Yu, D. Integrated Fiber-Optic Fabry-Pérot Interferometer Sensor for Simultaneous Measurement of Liquid Refractive Index and Temperature. *IEEE Sens. J.* **2019**, *19*, 5007–5013. [CrossRef]

43. Lee, C.-L.; You, Y.-W.; Dai, J.-H.; Hsu, J.-M.; Horng, J.-S. Hygroscopic polymer microcavity fiber Fizeau interferometer incorporating a fiber Bragg grating for simultaneously sensing humidity and temperature. *Sens. Actuators B Chem.* **2016**, *222*, 339–346. [CrossRef]

Article

Nonlinearity Correction in OFDR System Using a Zero-Crossing Detection-Based Clock and Self-Reference

Shiyuan Zhao [1,2], Jiwen Cui [1,2,*] and Jiubin Tan [1,2]

1   Center of Ultra-Precision Optoelectronic Instrument, Harbin Institute of Technology, Harbin 150080, China
2   Key Lab of Ultra-Precision Intelligent Instrumentation, Harbin Institute of Technology,
    Ministry of Industry and Information Technology, Harbin 150080, China
*   Correspondence: cuijiwen@hit.edu.cn; Tel.: +86-451-8641-2041

Received: 30 July 2019; Accepted: 20 August 2019; Published: 22 August 2019

Abstract: Tuning nonlinearity of the laser is the main source of deterioration of the spatial resolution in optical frequency-domain reflectometry (OFDR) system. In this paper, we develop methods for tuning nonlinearity correction in an OFDR system from the aspect of data acquisition and post-processing. An external clock based on a zero-crossing detection is researched and implemented using a customized circuit. Equal-spacing frequency sampling is, therefore, achieved in real-time. The zero-crossing detection for the beating frequency of 20 MHz is achieved. The maximum sensing distance can reach the same length of the auxiliary interferometer. Moreover, a nonlinearity correction method based on the self-reference method is proposed. The auxiliary interferometer is no longer necessary in this scheme. The tuning information of the laser is extracted by a strong reflectivity point at the end of the measured fiber. The tuning information is then used to resample the raw signal, and the nonlinearity correction can be achieved. The spatial resolution test and the distributed strain measurement test were both performed based on this nonlinearity correction method. The results validated the feasibility of the proposed method. This method reduces the hardware and data burden for the system and has potential value for system integration and miniaturization.

Keywords: optical fibers; Rayleigh scattering; optical frequency-domain reflectometry; strain measurement

1. Introduction

An optical frequency-domain reflectometry (OFDR)-based distributed sensing system was initially proposed by Froggatt et al. in 1998 and utilized in distributed disturbance measurements such as strain or temperature measurements because of its high spatial resolution and sensitivity [1–3]. In an OFDR system, the interference signals are collected as a function of the optical frequency of a tunable laser source (TLS). A fast Fourier transformation (FFT) is then used to convert this optical frequency-domain information to a desired spatial information, where it is required that the interference signals are sampled at an equal interval of optical frequencies [4]. However, any frequency tuning nonlinearity of a TLS gives rise to a non-uniform sampling interval of the optical frequency when the signal is sampled by an equal time interval, which in turn results in spreading of the reflection peak energy, deteriorating the spatial resolution [4].

Two kinds of methods were developed in recent years to solve the problem of the laser tuning nonlinearity. The first involves utilizing an auxiliary interferometer to produce an external sampling clock as a data acquisition trigger [5]. Although this method occurs real time and does not need post-processing, the maximum measurement length is limited by the time delay of the auxiliary interferometer in order to satisfy the Nyquist law [6]. This limits the measurement range of the system. Some researches proposed in-phase quadrature detection (IQ), such as the 3 × 3 coupler [7] or optical hybrid receiver [8], which can double the detection length. However, in these schemes, a complex operation in the demodulation is introduced, which would decrease the real-time demodulation of the system.

The second method to correct the tuning nonlinearity is to perform post-signal processing after acquiring OFDR data. Normally, this technique involves acquiring an auxiliary interferometer signal along with the OFDR signal and extracting the TLS phase information from the auxiliary interferometer signal, compensating nonlinearity in the OFDR signal using a correction algorithm [4]. One of the algorithms involves the resampling technique that resamples the main interference signals with an accurate equidistant optical frequency grid based on the optical frequency information of the TLS using interpolation algorithms. Badar et al. proposed a self-correction scheme in which only one detector is contained in the measurement system. In their proposed scheme, an intentional beating signal is introduced at the beginning of the OFDR spectrum, which is treated as an auxiliary interferometer to acquire tunable laser phase information for post-signal processing [9]. However, two drawbacks exist in their scheme. One lies in the fact that an extra delay for the intentional beating signal is introduced in the system, which increases the instability. Another lies in the fact that the intentional beating signal is at the beginning of the spatial domain. This design makes the optical path difference (OPD) of the main interferometer much longer than that of the auxiliary interferometer. Therefore, much interpolation is implemented, which increases the probability of false information, and also significantly increases the data volume. Moreover, the intentional beating signal occupies one segment at the low-frequency position, which sacrifices a part of the effective measurement range.

In this paper, tuning nonlinearity correction methods in an OFDR system were developed from the aspect of data acquisition and post-processing. On one hand, a new hardware-based method for real-time sampling is presented, which was implemented by designing an external clock to provide triggers at zero-crossing positions with uniform frequency spacing. The limited measurement range, which is equal to one-half of the OPD of the auxiliary interferometer in the conventional OFDR acquisition mode, can double and extend to the same length with the OPD of the auxiliary interferometer. On the other hand, the nonlinearity correction based on a single interferometer and self-reference is demonstrated. The tuning information of the laser is obtained from a PC connector at the end of the measured fiber. In this case, no extra delay fiber is needed because the fiber to be measured also plays the role of the optical path of the PC-constituted interferometer. Additionally, since the PC connector is at the end of the measured fiber, the OPD of the PC-constituted interferometer is longer than that of the main interferometer. That means that the beating frequency of the PC connector is larger than that of the reflectivity points (served as the sensing part) before the PC connector. Therefore, in the algorithm demonstrated later, substantial phase subdivision is not needed to resample the interferometer fringe signal. This makes the effect of the nonlinearity correction more stable.

The rest of this paper is organized as follows: Sections 2 and 3 describe two approaches for correcting nonlinearity of the laser. The first is the external clock based on zero-crossing detection. The circuit design and scheme are demonstrated. This method is validated by the spatial resolution test and maximum measurement distance. The second is nonlinearity correction using self-reference. The principle of the method is described. The OFDR trace is analyzed, and the distributed strain is measured based on the self-reference method. Section 4 concludes this paper and gives a brief comparison of the two methods proposed in this paper.

2. Method 1: External Clock Based on a Zero-Crossing Detection

2.1. Method Description

Ignoring the phase noise of the laser, the interference pattern of the auxiliary interferometer with a Mach–Zehnder scheme can be written as

$$I_a = \cos[2\pi\gamma(t)\tau_a t + \phi_0]. \tag{1}$$

In Equation (1), $\gamma(t)$ is the tuning speed of the laser, τ_a is the time delay of the Mach–Zehnder interferometer, and ϕ_0 is the initial phase. Equation (1) is equal to 0 when the phase is

$$2\pi\gamma(t)\tau_a t + \phi_0 = \frac{\pi}{2} + k\pi. \tag{2}$$

In Equation (2), k is an integer.

According to $t = \frac{v(t)}{\gamma(t)}$, where $v(t)$ is the optical frequency, Equation (2) can be written as

$$2\pi v(t)\tau_a + \phi_0 = \frac{\pi}{2} + k\pi \ (k = 1, 2, \ldots). \tag{3}$$

Then, the optical frequency at the zero-crossing points can be expressed as

$$v(t_k) = \frac{k}{2\tau_a} + v_c. \tag{4}$$

In Equation (4), v_c refers to a constant. It can be seen from Equation (4) that the optical frequency increment is equal and related to the OPD of the auxiliary interferometer at each zero-crossing point.

Next, we analyze the signal of the main interferometer. The interference intensity, which is interfered by the reflectivity point D on the measured fiber and the local light from the reference arm in the main interferometer, can be written as

$$I_D = \cos[2\pi v(t)\tau_D + \phi_d]. \tag{5}$$

In Equation (5), τ_D is the time delay between point D and the local light from the reference arm, and ϕ_d is the initial phase. The zero-crossing positions of the interferometer pattern of the auxiliary interferometer can be used to resample the beat signal expressed as Equation (5). When resampling the interference signal I_D with an interval of $\Delta v = \frac{1}{2\tau_a}$ as indicated in Equation (4), the resampled signal can be written as

$$I_{D,M} = \cos[\pi\frac{M}{\tau_a}\tau_D + \phi_d] \ (M = 1, 2, \ldots). \tag{6}$$

The trigger frequency of the auxiliary interferometer is $f_a = 2\gamma\tau_a$. The beating frequency of the main interferometer is $f_D = \gamma\tau_D$. To satisfy the Nyquist criterion, it is necessary that $f_a > 2f_D$ and $\tau_a > \tau_D$. Therefore, the measurable distance could reach the distance of the auxiliary interferometer. However, most commercial data acquisition cars (DAQs) can only sample at the rising or falling edge if using the external clock mode. This results in the measurable distance being limited to half the OPD of the auxiliary interferometer.

2.2. Hardware Implementation

As demonstrated by the principle above, the core task lies in designing a circuit which can detect all the zero-crossing points of the interferometer signal of the auxiliary interferometer (AI). Considering the tuning nonlinearity of the laser, the sinusoidal signal has a frequency fluctuation centered at its nominal beating frequency. The requirement for the circuit is high speed and low time delay. Furthermore, since the nominal beating frequency depends on the tuning speed of the laser, the maximum frequency of the AI signal is determined by the required measurement speed of the system. In our system, the maximum frequency of the AI signal is designed to be 20 MHz. That means that the maximum input frequency of the circuit for zero-crossing detection is 20 MHz. The total time delay of the circuit should be less than half a period of the AI signal. Under this rule, the trigger signal of the circuit can be considered to track and reflect the changing-frequency AI signal with success. The half-period is 25 ns when the frequency is 20 MHz. Therefore, it should result in the final trigger signal having a total time delay less than 25 ns on the consideration of the device selection for the circuit.

Figure 1 shows the circuit scheme of zero-crossing detection for the AI signal. Figure 2a gives the signals at each node. The AI signal is inversed firstly. Then, the original AI signal and its inversion signal are sent into the two comparators. The output of the two comparators goes through a high level every half-period of the AI signal. Then, the output of the comparators is sent to an XOR gate, after which the square wave appears at each half-period of the AI signal. The differentiation unit is used to convert the square wave to a narrow pulse. The narrow pulse serves as the trigger signal. The rising edge of the trigger signal appears after the zero-crossing of the AI signal. The time delay ΔT depends on the time delay sum of each component delay in the circuit.

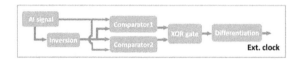

Figure 1. Circuit scheme of the zero-crossing detection for the auxiliary interferometer (AI) signal.

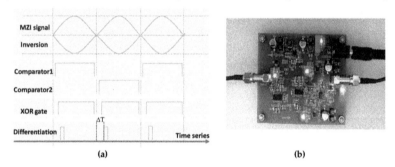

(a)   (b)

Figure 2. (**a**) Time-series analysis for each node on the circuit; (**b**) photograph of the circuit board based on the proposed zero-crossing detection scheme.

2.3. Experiment

A conventional OFDR system is shown in Figure 3. The light from the laser (LUNA, Phoenix 1400 with 3 MHz linewidth) is split into two paths by a 10:90 optical coupler, with 10% light sent to an auxiliary interferometer with a delay fiber of 250 m. The measured fiber has a length of 122 m; thus, its round trip is about 250 m. The end of the fiber is an APC connector immersed in a refractive index matching liquid for the sake of reducing reflectivity. The laser sweeps from 1540 nm to 1560 nm; thus, the two-point spatial resolution Δz of the system is 40 µm calculated by $\Delta z = c/2n\Delta v$, where n is the refractive index of the fiber under test (FUT), and Δv is the optical frequency tuning range of the TLS. The customized clock circuit is added after the PD. The sampling mode of the data acquisition card (DAQ) is set to use the rising edge of the external clock as the trigger source.

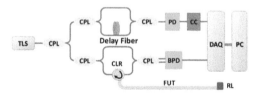

Figure 3. Optical frequency-domain reflectometry (OFDR) system. The auxiliary interferometer is an unbalanced Mach–Zehnder interferometer with a 250-m reference delay fiber. TLS: tunable laser source; CPL: fiber couple; CLR: fiber circular; PD: photo detector; BPD: balanced photo detector; CC: clock circuit; DAQ: data acquisition card; PC: personal computer; FUT: fiber under test; RL: refractive index matching liquid.

Firstly, the input–output timing sequence of the customized clock circuit was tested. Different tuning speeds were set so that the nominal beating frequencies of the AI signal were different. The beating frequency satisfied $f_{beat} = \gamma\tau$, where γ is the tuning speed with a unit of Hz/s, and τ is the time delay of the auxiliary interferometer. The tuning speeds of 32 nm/s, 80 nm/s, and 128 nm/s corresponded to beating frequencies of 5 MHz, 12.5 MHz, and 20 MHz, respectively. The input–output time sequences are shown in Figure 4. It can be seen that the time delay between the zero-crossing point and its following trigger signal were all within one half-period of the AI signal. Furthermore, this time delay was almost the same and can be considered as constant if the time delay difference of the two comparators used in the circuit is negligent.

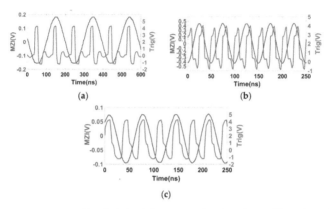

Figure 4. Input–output curve of the clock circuit for zero-crossing detection at different input frequencies. The blue curve is the AI signal and the pink curve is the output trigger signal of the clock circuit. (a–c) Nominal input frequencies of 5 MHz, 20 MHz, and 12.5 MHz, respectively.

Then, the OFDR trace was sampled by the DAQ with our customized clock circuit at the condition of 80 nm/s tuning speed. The nominal beating frequency of the AI signal was then 12.5 MHz. The original signal from the main interferometer was fast Fourier transformed to the spatial domain. The result is shown in Figure 5. It can be seen that the measurement length can reach a length equal to the OPD of the auxiliary interferometer. To evaluate the linearity of the system quantitatively, the full width at half maximum (FWHM) of the reflectivity peak of a fiber connector is generally used. The FWHM of the APC connector was 40 μm, which was the Fourier transform-limited spatial resolution [10]. At the end of the fiber, the FWHM of the APC connector decreased to about 3.48 mm. The resolution deterioration mainly came from the increasing phase noise of the laser, which increased with the length, and also from the immersion in the refractive index matching liquid.

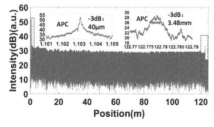

Figure 5. Measured OFDR trace with an APC connector end immersed in a refractive index matching liquid. The first APC connector and the APC at the end of the fiber are shown in the insets. The trace shows a good nonlinearity correction result. The broadened peak of the final APC connector mainly results from the phase noise of the laser and the effect of the refractive index matching liquid.

3. Method 2: Nonlinearity Correction Based on Self-Reference

3.1. Method Description

Figure 6 describes the nonlinearity correction process using the self-reference method. The raw signal in the optical frequency domain was sampled using a fixed sampling rate, then converted to the spatial domain by fast Fourier transformation. A rectangle band-pass filter was applied on the data in the spatial domain. The band should cover the central band of the reference point which may be a PC connector. The reminder of the spectrum was set to zero. After that, the filtered data were inverse fast Fourier-transformed back to the optical frequency domain. After that, the Hilbert transformation (HT) and arc tangent operation were used to extract the phase; then, the phase was unwrapped. The nonlinearity information was obtained. Next, the unwrapped phase was divided into equal-spaced segments with an equal phase internal. These positions were then used to resample the raw signal. Finally, the nonlinearity-corrected interferometer signal was obtained. It should be noted that, in the process mentioned, no interpolation was introduced.

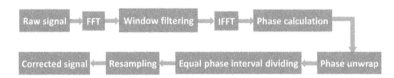

Figure 6. Procedure for the self-reference method for the laser tuning nonlinearity correction.

The mathematical operation for the method is demonstrated below.

After filtering, the temporal component of the PC-constituted interference is separated from the original signal. The detected interference pattern interfered by the strong reflectivity point (e.g., a PC connector) with the local light can be written as

$$I_{pc}(t) = \cos[2\pi\gamma(t)\tau t + \xi_0] = \cos[2\pi v(t)t + \xi_0]. \tag{7}$$

In Equation (7), $\gamma(t)$ is the tuning speed of the laser, τ is the time delay between the PC connector and the local light from the reference arm, and ξ_0 is the initial phase. The Hilbert transformation (HT) of $I_{pc}(t)$ can be expressed as

$$HT\{I_{pc}(t)\} = \sin(2\pi\tau v(t) + \xi_0). \tag{8}$$

The phase of the interference pattern is, therefore, represented as

$$\phi(t) = 2\pi\tau v(t) + \xi_0 = \tan^{-1}[HT\{I_{pc}(t)\}/I_{pc}(t)]. \tag{9}$$

Since the arctangent function maps the phase angle to the range $[-\pi, \pi]$, the phase needs to be unwrapped. Derived from Equation (9), the relationship between the change in optical frequency of the laser and the change in phase of the interference pattern can be represented as

$$\Delta v(t) = \frac{1}{2\pi n} \frac{c}{\Delta L} \Delta\phi(t), \tag{10}$$

where ΔL is the OPD between the PC connector and the local light from the reference arm. It can be seen from Equation (10) that the resampled points at each equal optical phase interval represent equal optical internal frequencies. The relationship between the maximum correction range L_m and the optical phase interval $\Delta\phi_\pi = \pi/K$ (K is the subdividing number and is a positive integer) can be represented as

$$L_m = K \times L_{pc}, \tag{11}$$

where L_{pc} is the distance between the PC connector and the starting position. Generally, a phase increment π or $\pi/2$ can meet the nonlinearity correction requirement, and the maximum sensing distance will not be lower than the OPD of the PC connector.

3.2. Simulation

To simplify the simulation, we assumed that the optical frequency of the laser satisfied the quadratic function modeled by

$$v(t) = \gamma t + a t^2,\tag{12}$$

where γ is the tuning speed and a is the nonlinearity coefficient [11].

The simulation parameters for the tuning process were as follows: tuning range = 2500 GHz; tuning speed = 5000 GHz/s; sampling rate = 10 MS/s; sampling number = 5 MS; optical frequency deviation at the middle optical frequency = −1 GHz.

Two reflection points on the measured fiber were simulated and their OPDs (roundtrip) were 20 m and 24 m. The second reflection point was the reference point used to correct nonlinearity. Figure 7a shows the OFDR trace in the spatial domain. It can be seen that the peaks were broadened centered at 10 and 12 m, resulting from the tuning nonlinearity of the laser. The spatial resolution was, therefore, deteriorated. In Figure 7b, the blue curve represents the ideal interference pattern in the temporal domain, which is the interference between the ideal reflection at the position of 24 m and the local light. The red curve represents the result after the inverse fast Fourier transformation (IFFT) operation (before the phase calculation) shown in Figure 6. It can be seen that the recovered interference pattern had the same phase as the ideal interference pattern. Therefore, the self-reference method had the same effect as the conventional auxiliary interferometer shown in Figure 3. Figure 7c compares the results with and without nonlinearity correction using the self-reference method. It can be seen that, after the nonlinearity correction, the Fourier transform-limited spatial resolution can be achieved.

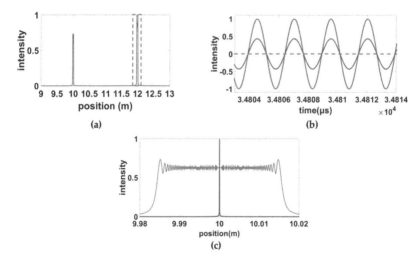

Figure 7. (a) Two reflection peaks in the spatial domain broadened by the nonlinearity tuning of the laser. (b) The interference patterns over a period of time. The blue curve is the ideal interference pattern in the temporal domain. The red curve is the signal recovered by the band-pass filtered signal shown in (a). (c) The pink and blue curves are the first reflection peak without and with the nonlinearity correction using the self-reference method. All intensities are normalized.

3.3. Experiment

The OFDR system using self-reference for the nonlinearity correction was similar to the system shown in Figure 3. However, as shown in Figure 8a, only the main interferometer was kept. The TLS swept from 1540 nm to 1560 nm, and its nominal tuning speed was 40 nm/s. The sampling rate of the data acquisition card (DAQ) was10 MS/s. Two configurations were investigated. The first is shown in Figure 8b. All segments were composed of a single-mode fiber. The second is shown in Figure 8c. The fiber part between the −3-dB attenuator and the APC connector was a Ge-doped fiber with 1-m-long dense weak FBG arrays inscribed in the middle. Each single FBG had a length of 8 mm and a gap of 2 mm. The weak FBG arrays were made using a phase mask and UV light. The nominal central wavelength was 1550 nm. A 3 dBattenuator was inserted before the sensing segment to reduce the incident light power. The PC connectors in the two configurations were used as the reference reflectivity point in the process of nonlinearity correction.

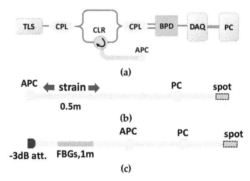

(a)

(b)

(c)

Figure 8. OFDR system using the self-reference method for nonlinearity correction. (**a**) Basic OFDR interrogation system. (**b**) Configuration 1, composed of a single-mode fiber. (**c**) Configuration 2, where the fiber part between the attenuator and the APC connector is a Ge-doped fiber with 1-m-long dense weak FBG arrays inscribed in the middle. The end of the fiber is knotted for reducing of the reflectivity. A PC connector is used for the self-reference reflectivity point.

Firstly, the spatial resolution tests for these two configurations were performed. The equal phase intervals in the self-reference for the nonlinearity correction were all set to $\pi/2$. Figure 9 shows the measured OFDR trace of configuration 1 without nonlinearity correction. The FWHM of the PC connector was about 1.5 m. Then, the band-pass filter shown in the red dashed box in Figure 9 was used to extract the optical phase information of the laser. The correction process followed the process shown in Figure 6. The corrected OFDR trace is shown in Figure 10. The inset is the reflection of the PC connector. The FWHM of the PC connector was about 40 μm, which was equal to the Fourier transform-limited spatial resolution. Figure 11 shows the measured OFDR trace of configuration 2 after nonlinearity correction. The left inset in Figure 11 is the attenuator. The distance between two APC reflectivity planes was about 2 cm, which is in agreement with the practical value. The right inset in Figure 11 is the beginning of the FBG arrays. It can be seen that the FBG was about 10 dB higher than the Rayleigh scatter level. The interval and reflectivity strengths were even. From the results of the spatial resolution tests, the nonlinearity correction method using self-reference was effective. The structures on the measured fiber were all distinguished and measured with high accuracy and resolution.

Then, the distributed sensing tests were implemented. The distributed strain was measured using the conventional demodulation method demonstrated in References [1,12]. The nonlinearity correction was achieved using the self-reference method. The gauge length was set to 1 cm. The stretching part was a single-mode fiber with acrylate coating as shown in Figure 8b. The measured fiber was stretched by a nanometer stage as shown in Figure 12a. The device could generate a standard strain calculated by $\varepsilon = \Delta L/L$, where ΔL is the elongation length of the fiber and L is the original length. The distributed strains are shown in Figure 12b. Thus, it is certain that the self-reference method does not influence the classical distributed strain demodulation in OFDR.

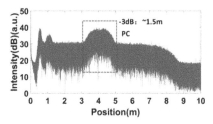

Figure 9. Measured OFDR trace without nonlinearity correction. The FWHMof the PC connector is about 1.5 m.

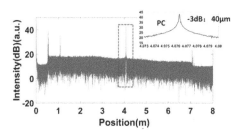

Figure 10. OFDR trace of configuration 1 with nonlinearity correction. The inset is the reflection of the PC connector.

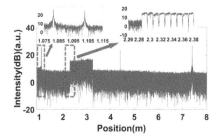

Figure 11. OFDR trace of configuration 2 with nonlinearity correction.

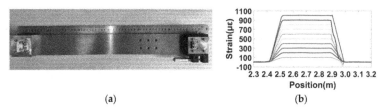

| (a) | (b) |

Figure 12. (a) Linear displacement stage to apply a certain strain to the fiber. (b) The distributed strain upon increasing strain from 100 με to 1000 με with 100-με intervals.

4. Conclusions

In summary, we developed methods for tuning nonlinearity correction in an OFDR system from the aspect of data acquisition and post-processing. Based on their principles, these two methods both took advantage of the auxiliary interferometer information (in the self-reference method, a PC-constituted interferometer served as the auxiliary interferometer) to find the equal-spacing frequency position.

Sensors **2019**, *19*, 3660

The difference was that the former triggered the acquisition only at the position of zero-crossing, while the latter extracted the phase information to obtain the continuous phase changing of the laser. Therefore, in the second method, a smaller frequency interval can be set, and this makes it possible to achieve nonlinearity correction for a longer measurable range. Another difference between these two nonlinearity correction methods is that the correction method implemented by the hardware is high-speed and in real time. The correction using post-processing is not in real time, although it can approach real time with the usage of high-performance computing equipment. The advantage of the self-reference method lies in that, compared to the conventional post-processing method, the self-reference method can reduce the hardware and data burden for the system, and it is expected to have potential value in system integration and miniaturization.

Author Contributions: S.Z. conceptualized the research, developed the algorithm, performed the experiment, and wrote the manuscript. J.C. supervised the research, and J.T. reviewed the manuscript.

Funding: This research was funded by the National Natural Science Foundation of China under grant number 51575140, and the Heilongjiang Province Outstanding Youth Science Fund Project under grant number HSF20190040.

Conflicts of Interest: The authors declare no conflicts of interest.

References

1. Froggatt, M.; Moore, J. High-spatial-resolution distributed strain measurement in optical fiber with Rayleigh scatter. *Appl. Opt.* **1998**, *37*, 1735–1740. [CrossRef] [PubMed]
2. Gifford, D.K.; Kreger, S.T.; Sang, A.K.; Froggatt, M.E.; Soller, B.J. Swept-wavelength interferometric interrogation of fiber Rayleigh scatter for distributed sensing applications. In Proceedings of the Fiber Optic Sensors and Applications V. International Society for Optics and Photonics, Boston, MA, USA, 10–12 September 2007; Volume 6770.
3. Ding, Z.; Wang, C.; Liu, K.; Jiang, J.; Yang, D.; Liu, T. Distributed optical fiber sensors based on optical frequency domain reflectometry: A review. *Sensors* **2018**, *18*, 1072. [CrossRef] [PubMed]
4. Moore, E.D.; McLeod, R.R. Correction of sampling errors due to laser tuning rate fluctuations in swept-wavelength interferometry. *Opt. Express* **2008**, *16*, 13139–13149. [CrossRef] [PubMed]
5. Feng, B.; Liu, K.; Liu, T.; Jiang, J.; Du, Y. Improving OFDR spatial resolution by reducing external clock sampling error. *Opt. Commun.* **2016**, *363*, 74–79. [CrossRef]
6. Song, J.; Li, W.; Lu, P.; Xu, Y.; Chen, L.; Bao, X. Long-range high spatial resolution distributed temperature and strain sensing based on optical frequency-domain reflectometry. *IEEE Photonics J.* **2014**, *6*, 6801408. [CrossRef]
7. Kim, Y.; Kim, M.J.; Rho, B.S.; Kim, Y.H. Measurement Range Enhancement of Rayleigh-Based Optical Frequency Domain Reflectometry with Bidirectional Determination. *IEEE Photonics J.* **2017**, *9*, 7106308. [CrossRef]
8. Gabai, H.; Botsev, Y.; Hahami, M.; Eyal, A. Optical frequency domain reflectometry at maximum update rate using I/Q detection. *Opt. Lett.* **2015**, *40*, 1725–1728. [CrossRef] [PubMed]
9. Badar, M.; Lu, P.; Buric, M.; Ohodnicki, R. Self-correction of nonlinear sweep of tunable laser source in OFDR. In Proceedings of the Fiber Optic Sensors and Applications XVI, Baltimore, MD, USA, 16–17 April 2019; Volume 11000.
10. Xie, W.; Zhou, Q.; Bretenaker, F.; Xia, Z.; Shi, H. Fourier transform-limited optical frequency-modulated continuous-wave interferometry over several tens of laser coherence lengths. *Opt. Lett.* **2016**, *41*, 2962–2965. [CrossRef] [PubMed]
11. Deng, Z.; Liu, Z.; Li, B.; Liu, Z. Precision improvement in frequency-scanning interferometry based on suppressing nonlinear optical frequency sweeping. *Opt. Rev.* **2015**, *22*, 724–730. [CrossRef]
12. Cui, J.; Zhao, S.; Yang, D.; Ding, Z. Investigation of the interpolation method to improve the distributed strain measurement accuracy in optical frequency domain reflectometry systems. *Appl. Opt.* **2018**, *57*, 1424–1431. [CrossRef] [PubMed]

Article

Dynamic Deformation Reconstruction of Variable Section WING with Fiber Bragg Grating Sensors

Zhen Fu [1], Yong Zhao [2], Hong Bao [1,*] and Feifei Zhao [1]

[1] Key Laboratory of Electronic Equipment Structure Design of Ministry of Education, Xidian University, Xi'an 710071, China
[2] School of New Energy Vehicles, Henan Mechanical and Electrical Vocational College, Zhengzhou 451191, China
* Correspondence: hbao@xidian.edu.cn; Tel.: +86-029-8820-3040

Received: 27 June 2019; Accepted: 26 July 2019; Published: 30 July 2019

Abstract: In order to monitor the variable-section wing deformation in real-time, this paper proposes a dynamic reconstruction algorithm based on the inverse finite element method and fuzzy network to sense the deformation of the variable-section beam structure. Firstly, based on Timoshenko beam theory and inverse finite element framework, a deformation reconstruction model of variable-section beam element was established. Then, considering the installation error of the fiber Bragg grating (FBG) sensor and the dynamic un-modeled error caused by the difference between the static model and dynamic model, the real-time measured strain was corrected using a solidified fuzzy network. The parameters of the fuzzy network were learned using support vector machines to enhance the generalization ability of the fuzzy network. The loading deformation experiment shows that the deformation of the variable section wing can be reconstructed with the proposed algorithm in high precision.

Keywords: deformation reconstruction; fuzzy system; inverse finite element method; strain modification; variable section wing

---

1. Introduction

Through the real-time perception of the deformation of the wing structure, the flight attitude is adjusted in real time by the actuation and control systems to ensure the safety of the aircraft. This is one of the key technologies for the next generation of intelligent space vehicles. Using the strain sensors arranged on the wing structure to obtain the strain data of the structure in real time, and reconstructing the deformation information of the structure using the specific strain displacement relation [1], that is, shape sensing, it has become a research hotspot of the wing structure deformation. Due to the advantages of small size, anti-electromagnetic interference and the ability to form a large-scale quasi-distributed sensing network [2], the FBG strain sensors are widely used in the field of shape sensing.

The key to shape sensing is constructing a relationship between the structural deformation and the strain measurement. There are many research methods of strain-based deformation reconstruction proposed by domestic and foreign scholars, such as the modal transformation method, piecewise linearization method, and finite element method. The modal transformation method can accurately reconstruct the deformation of plate and beam structure [3], but it needs accurate finite element model. Another problem is that the algorithm is difficult to apply to the large deformation of geometric nonlinear structures because it is based on the principle of linear superposition [4]. The Ko method is based on the classical Euler Bernoulli beam theory [5]. By integrating the discrete surface strain measurements with piecewise continuous polynomials, high-precision reconstruction of the beam element in a one-dimensional direction can be achieved. The Ko method can be used for deformation reconstruction of cantilever beam structure, and ground experiment for deformation of wing structure has been realized [5,6]. The scheme is only suitable for one-dimensional deformation of the beam structure, but it is difficult to estimate the element deformation under multidimensional complex loads because the scheme requires a large number of strain sensors.

In order to meet the requirements of deformation monitoring, the deformation reconstruction algorithm should be able to adapt to complex topology and boundary conditions, and maintain accuracy and stability under a wide range of loads or elastic inertia changes of materials. Tessler and Spangler have proposed an inverse finite element method (iFEM) [7], which is based on the least square variational principle and uses the measured strain data to estimate the deformation of plate structure [8–10]. Furthermore, they have applied it to shape sensing of plate and shell structure undergoing large displacements [11]. Based on inverse finite element theory, Gherlone realized static deformation reconstruction of wing plate structure [12]. Furthermore, the inverse finite beam element, which is suitable for beam/frame structure deformation, is constructed by combining iFEM with Timoshenko beam theory, and the three-dimensional deformation reconstruction of frame structure under static condition is realized [4]. Based on the theory of ZigZag and the idea of inverse finite element, Cerracchio constructed the deformation reconstruction model of plate element for composite sandwich structure [13]. Yong Zhao et al. have studied the effect of sensor distribution on the reconstruction accuracy in the process of inverse finite beam element reconstruction, and realized the 3D deformational reconstruction of the wing like frame structure [14–16]. Xinglin Pan et al. have deeply studied the effect of measurement strain error on reconstruction accuracy in the process of static deformation reconstruction of beam structure, and put forward the error correction using fuzzy network [17]. When the wing is regarded as a constant-section beam structure, the above research studies the deformation reconstruction of the wing structure with high precision [18,19]. For some variable-section wings, when the element division is dense enough, the partial element can be regarded as a constant-section beam, but the number of sensors used is increased, and furthermore, the error caused by element superposition is so big that it cannot be neglected [20]. Guanghong Chuan et.al analyzed the stiffness matrix error of the variable-section Timoshenko Beam for the different rates of the section areas [21]. With the above conclusion, the reconstruction model of the cross-section wing is constructed in this paper. Meanwhile, the above studies only focus on the structural deformation reconstruction under static load, but not on the structural deformation reconstruction under dynamic excitation. Furthermore, it is found that the reconstruction accuracy and stability of inverse finite element method will be affected by the accuracy of the deformation reconstruction model and the sensor placement.

For the sake of sensing the deformation of the cross-section wing accurately, a variable cross-section inverse finite beam element model is proposed in this paper. Meanwhile, this paper derives the relationship between strain and deformation of the cross-section wing based on inverse finite element method. The deformation reconstruction of a cross-section wing is accomplished with the strain measurement of the structure surface. Furthermore, in order to remove the influence of the strain sensors installation and dynamic un-modeled error for the dynamic deformation reconstruction, this paper proposes a self-structuring linear support vector regression algorithm fuzzy network (SSILSVRFN) to modify the strain measurements. The inverse finite element method is employed to reconstruct the deformation displacement with the modified strain, and the accuracy of the deformation reconstruction is improved. The content of this paper is divided into the following parts: firstly, based on the inverse finite element framework, the inverse finite element reconstruction equation for the variable cross-section beam is constructed for the deformation reconstruction of glass fiber wing model with variable cross section. Then, a fuzzy network based on iterative linear support vector regression is constructed to correct the real-time error of dynamic strain measurement. Finally, the vibration experiments on the fiberglass wing model show that the proposed algorithm can effectively improve the accuracy and stability of wing deformation reconstruction.

2. A Deformation–Reconstruction Model for Wing with Variable Cross-Section

2.1. Inverse Finite Element Model for Variable Cross-Section Beam

For a typical variable-section beam (Figure 1), the displacement of point B on the beam surface can be expressed with six kinematic variables $\left[u, v, w, \theta_x, \theta_y, \theta_z\right]$ on point A [4]

$$
\begin{aligned}
u_x(x, y, z) &= u(x) + z\theta_y(x) - y\theta_z(x) \\
u_y(x, y, z) &= v(x) - z\theta_x(x) \\
u_z(x, y, z) &= w(x) + y\theta_x(x)
\end{aligned}
\tag{1}
$$

where $u_x, u_y,$ and u_z are the displacements along the $x, y,$ and z axes, respectively, with $u(x), v(x),$ and $w(x)$ denoting the displacements at $y = z = 0$; $\theta_x(x), \theta_y(x),$ and $\theta_z(x)$ are the rotations about the three coordinate axes; positive orientations for the displacements and rotations are depicted in Figure 1. These kinematic assumptions neglect the effect of axial warping due to torsion, i.e., each cross-section remains flat and rigid with respect to thickness-stretch deformations along the y and z-axes. The six kinematic variables can be grouped in vector form as

$$
u(x) = \left[u, v, w, \theta_x, \theta_y, \theta_z\right]^T
\tag{2}
$$

In the finite element framework, the deformation of any point along the centroid-axis of the beam element can be obtained by interpolation of the determined shape function $N(x)$ and the node degrees of freedom u^ε.

$$
u(x) = N(x)u^\varepsilon
\tag{3}
$$

Based on the small strain hypothesis, the relationship between the section strain at any point along the centroid-axis and the node degree of freedom is

$$
e(u) = B(x)u^\varepsilon
\tag{4}
$$

where the matrix $B(x)$ (see Appendix A) contains the derivatives of the shape functions $N(x)$, and the section strain vector $e(u)$ is expressed as

$$
e(u) \equiv [e_1, e_2, e_3, e_4, e_5, e_6]^T
\tag{5}
$$

$$
e_1(x) \equiv u_{,x}(x) \quad e_4(x) \equiv w_{,x}(x) + \theta_y(x)
$$

$$
e_2(x) \equiv \theta_{y,x}(x) \quad e_5(x) \equiv v_{,x}(x) - \theta_z(x)
$$

$$
e_3(x) \equiv -\theta_{z,x}(x) \quad e_6(x) \equiv \theta_{x,x}(x)
$$

And the surface strain vector $[\ \varepsilon_x(x, y, z), \gamma_{xz}(x, y), \gamma_{xy}(x, y)\ ]^T$ of the beam element can be expressed with the above section strain vector $e(u)$ in theory:

$$
\begin{aligned}
\varepsilon_x(x, y, z) &= e_1(x) + ze_2(x) + ye_3(x) \\
\gamma_{xz}(x, y) &= e_4(x) + ye_6(x) \\
\gamma_{xy}(x, y) &= e_5(x) - ze_6(x)
\end{aligned}
\tag{6}
$$

While, the section strain vector $e(u)$ is calculated from the kinematic variables and cannot be directly measured. In the inverse finite element method, the least square error between the section strain $e(u)$ and the section strain e^ε obtained from the surface measured strain is constructed as following

$$
\varnothing(u) = \|e(u) - e^\varepsilon\|^2
\tag{7}
$$

when $\varnothing(u)$ obtains the minimum value, the section strain $e(u)$ is replaced with the section strain e^ε to form a relationship model between the beam element node degree of freedom and the section strain

$$k^e u^e = f^\varepsilon \tag{8}$$

where $k^e = \frac{L}{n} * \sum_{i=1}^{n} \left[B^T [x_i] B[x_i] \right], f^\varepsilon = \frac{L}{n} * \sum_{i=1}^{n} \left[B^T [x_i] e^{\varepsilon i} \right]$. Note that k^e resembles an element stiffness matrix of the direct finite element method and f^ε resembles the load vector; L is the element length (Figure 1); n and x_i ($0 \le x_i \le L$) are, respectively, the number and the axial coordinate of the locations where the section strains are evaluated, and the superscript εi is used to denote the section strains computed from the surface measured strain.

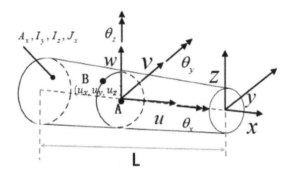

Figure 1. Schematic diagram of a variable section beam.

2.2. Calculation of Section Strain of Variable Section Beam Element

Correctly solving the section strain is the key to the element deformation reconstruction. For the constitutive cross-section beam (Figure 2), the relationship between the section strain e^ε and the external load force and moment can be expressed as [4]

$$
\begin{aligned}
N &= A_x e_1{}^\varepsilon & M_x &= J_x e_6{}^\varepsilon \\
Q_y &= G_y e_5{}^\varepsilon & M_y &= D_y e_2{}^\varepsilon \\
Q_z &= G_z e_4{}^\varepsilon & M_z &= D_z e_3{}^\varepsilon
\end{aligned}
\tag{9}
$$

where $A_x \equiv EA$ is the axial rigidity; $G_y \equiv k_y^2 GA$ and $G_z \equiv k_z^2 GA$ are the shear rigidities, with k_y^2 and k_z^2 denoting the shear correction factors; $J_x \equiv GI_p$ is the torsional rigidity; $D_y \equiv EI_y$ and $D_z \equiv EI_z$ are the bending rigidities; A is the section area; E (Young's modulus) and G (shear modulus) are the elastic constants. M is a moment acting in a certain direction on the unit, and its unit is $N*m$; the unit of Q is N.

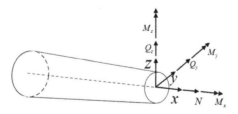

Figure 2. The force form of the beam.

Once the type of the external load is known, the format of the section strain e^ε can be determined with using Equation (9). For example, the concentrated load, bending moment and torsion performed on the beam element, the in-plane force N, Q_y, Q_z and torsion M_x of the element are constant along the axis direction, and the bending moment M_y and M_z change along the x-axis. The conclusion is valid for both constant-section beam and variable-section beam [21]. The section strain e_1^ε can be expressed as

$$e_1^\varepsilon(x) = \frac{C_1}{A_x(x)} \tag{10}$$

where C_1 is an unknown constant and $A_x(x)$ is the tensile stiffness of the element section. Similarly, the section strains e_4^ε, e_5^ε, and e_6^ε can be written as follows

$$
\begin{aligned}
e_4^\varepsilon(x) &= \frac{C_4}{G_z(x)} \\
e_5^\varepsilon(x) &= \frac{C_5}{G_y(x)} \\
e_6^\varepsilon(x) &= \frac{C_6}{I_p(x)}
\end{aligned}
\tag{11}
$$

where $G_y(x), G_z(x)$, and $I_p(x)$ are the bending stiffness and torsional stiffness of the variable section beam element [21]; C_4, C_5, and C_6 are unknown constants; C_4 and C_5 can be computed with the following equation in [4]:

$$
\begin{aligned}
C_4 &= Q_y = \frac{\partial D_z e_3^\varepsilon}{\partial x} \\
C_5 &= Q_z = \frac{\partial D_y e_2^\varepsilon}{\partial x}
\end{aligned}
\tag{12}
$$

Since the changes of bending moment M_y and M_z along the centroid-axis of the beam element are linear, M_y and M_z can be assumed as

$$
\begin{aligned}
M_y(x) &= C_5 x + C_2 \\
M_z(x) &= C_4 x + C_3
\end{aligned}
\tag{13}
$$

where C_2 and C_3 are unknown constants. Submitting the Equation (13) into Equation (9), e_2^ε and e_3^ε can be expressed as

$$e_2^\varepsilon(x) = \frac{C_5 x + C_2}{D_y(x)} \ , \ e_3^\varepsilon(x) = \frac{C_4 x + C_3}{D_z(x)} \tag{14}$$

For the variable-section wing (Figure 3), the relationship between the surface strain measurement ε^* and the section strain vector is ascertained with using the strain-tensor transformation from the (θ, X, r) to (X, Y, Z) [4,22,23]:

$$\varepsilon^* = \varepsilon_x \left(\cos^2 \beta - v \sin^2 \beta \right) + \gamma_{x\theta} \cos \beta \sin \beta \tag{15}$$

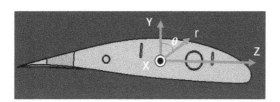

Figure 3. Orthogonal and cylindrical coordinate systems of the wing section.

Substituting $r = R_i$ in Equation (6), yields

$$\varepsilon_x = e_1 + e_2 R_i \sin\theta + e_3 R_i \cos\theta$$

$$\gamma_{x\theta} = e_4 \cos\theta - e_5 \sin\theta + e_6 R_i \tag{16}$$

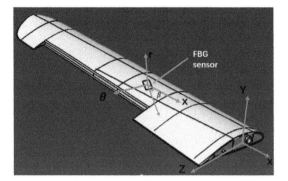

Figure 4. Location and coordinate system of a FBG sensor placed on the wing external surface.

Substituting (10), (11), (14), and (16) into (15) gives the relationship between the strain measurements and the section strain for the variable section beam element as (Figure 4)

$$
\begin{aligned}
\varepsilon^*(x_i, \theta_i, \beta_i) &= \frac{C_1}{A_x(x_i)}(\cos^2 \beta_i - v\sin^2 \beta_i) + \frac{C_5 x_i + C_2}{D_y(x_i)}(\cos^2 \beta_i - v\sin^2 \beta_i)R_i \sin \theta_i \\
&+ \frac{C_4 x_i + C_3}{D_z(x_i)}(\cos^2 \beta_i - v\sin^2 \beta_i)R_i \cos \theta_i + \frac{C_4}{G_z(x_i)}\cos \beta_i \sin \beta_i \cos \theta_i \\
&- \frac{C_5}{G_y(x_i)}\cos \beta_i \sin \beta_i \sin \theta_i + \frac{C_6}{I_p(x_i)}R_i \cos \beta_i \sin \beta_i
\end{aligned}
\tag{17}
$$

where $R_i = \sqrt[2]{y_i^2 + z_i^2}$; v is Poisson's ratio; R_i is the polar radius of the beam section; x_i, θ and β are the positions of the strain measurements on the surface of the beam element.

Therefore, the unknown parameters $C_1, C_2, C_3, C_4, C_5, C_6$ can be solved from six different surface measurement strain values $x_i, \theta_i, \beta_i, (i = 1, \ldots, 6)$ with using the Equation (17), and the section strains e^ε will be determined with Equations (10), (11), and (14).

3. Strain Error Correction

Because of the strain measurement system error and the model error, there is a big deviation between the actual displacement and the displacement computed from the strain measurement data with iFEM. The strain measurement system error consist of the location error resulted from the FBG sensors attachment process, measurement error of strain measurement instrument et al. The model error, including the dynamic un-modeled error caused by the difference between the static model and dynamic model, the connection and segmentation of the structure elements. For removing the above errors, the basic modification strategy is: (1) the actual strain value $\tilde{\varepsilon}$ is computed from actual deformation captured from the third-party measurement instrument with reconstruction model, Equation (8). (2) the fuzzy network of the self-structuring iterative linear support vector regression fuzzy network (SSILSVRFN) algorithm [24] is trained and fixed with the above actual strain value $\tilde{\varepsilon}$ and the corresponding measured strain values ε. (3) The actual deformation of the structure for any loading cases can be computed from the actual strain value $\tilde{\varepsilon}_A$ computed from the corresponding strain measurement ε_A with the above fixed fuzzy network.

For the deformation reconstruction based on strain measurement, there are many error factors in the measurement of structural surface strain. Thus, the difference exists between the measured strain and the actual strain; and the corresponding relationship between the measured strain and the actual strain is difficult to be accurately described with mathematical expressions. The self-structuring fuzzy network (SSFN) is attempted to approach this unknown relationship because the fuzzy network has the characteristic of infinite approximation to any function mapping relation. SSFN algorithm constructs the network from zero rule, and the adjustment of the rules number and structure is based on the training data as current network input. The training network is difficult to achieve the desired application effect in the testing phase because network structure is optimal for the current data. This paper proposes a SSILSVRFN using the support vector regression theory and cluster idea. The SSILSVRFN system is divided into the structure training phase and the parameter learning phase.

The structure training phase adopts the SSRG (self-structuring rule generation) algorithm for automatic fuzzy rule generation and initialization. According to the spatial distribution of the input data and the analysis of the overall data, a reasonable distribution of fuzzy sets center and width is achieved. Not only the size of the whole network is effectively reduced, but also the impact on the network structure of the data order is avoided.

The SSILSVRFN proposed in this paper is divided into five layers, as shown in Figure 5.

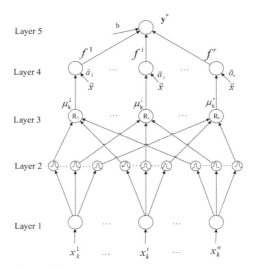

Figure 5. Structure of the self-structuring linear support vector regression algorithm fuzzy network (SSILSVRFN).

In parameter learning phase, the support vector regression (SVR) is based on the structural risk minimization principle; and SVR learning algorithm has a better performance in generalization and prevention of over-fitting. Therefore, SSILSVRFN adopts iterative linear SVR (ILSVR) to iteratively adjust parameter of fuzzy rules.

The SSILSVRFN training and learning system diagram is shown in Figure 6.

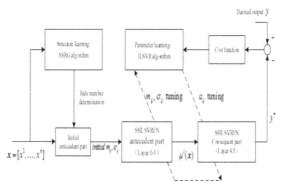

Figure 6. The SSILSVRFN training.

3.1. SSILSVRFN Structure Learning

The structure learning determines the fuzzy rule number and initial fuzzy set center m_{ij} and width σ_{ij}. Once the number of fuzzy rules is determined, the numbers of nodes in the layer 2, 3 and 4 will be accordingly determined, which are nr, r and r, respectively. In the structure learning phase, a fuzzy rule is regarded as a cluster that corresponds to a rule node of layer 3 in the input space. A SSRG algorithm is proposed to determine the suitable number of rules.

The flow chart (Figure 7) is as follows

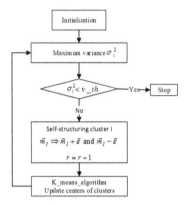

Figure 7. The flow chart of self-structuring rule generation SSRG.

The algorithm steps are presented as follows

Step 1: classify the training samples, and set the number of iteration $N_{ite} = 0$. Assign each input training data x to cluster P, which is calculated as follows

$$P = \arg\min_{1 \leq p \leq r} d(x, m_p) \tag{18}$$

$$d(x, m_p) = \sum_{j=1}^{n} |x^j - m_{pj}| \tag{19}$$

Step 2: calculate the maximum variance. For the first cluster, the variance is defined as follows

$$I = \arg\max_{1 \leq i \leq r} \overline{\sigma}_i^2 \tag{20a}$$

$$\overline{\sigma}_i^2 = \sum_{j=1}^{n} \overline{\sigma}_{ij}^2 \tag{20b}$$

$$\overline{\sigma}_{ij}^2 = \frac{1}{N_i} \sum_{x \in C_i} (x^j - m_{ij})^2 \tag{20c}$$

Step 3: if the maximum value of cluster variance $\overline{\sigma}_I^2$ is less than the threshold value v_{th} and the current iteration number is less than 20, stop; otherwise, the cluster with the largest difference will be split to generate a new cluster. The center point vectors of the new clusters are, respectively, set as $m_I + \varepsilon$ and $m_I - \varepsilon$, where ε is a constant vector with a small value. This paper sets the ith component of ε to be 1% of the domain of input variable x^i.

Step 4: update the centers of clusters. Recalculate the center of cluster **P** as

$$m_{\mathbf{P}} \leftarrow m_{\mathbf{P}} + \frac{1}{N_P} \sum_{k=1}^{N_P} (x_{\mathbf{k}} - m_{\mathbf{P}}) \tag{21}$$

where N_p is the number of samples in cluster **P**. The iteration number is updated as $N_{ite} = N_{ite} + 1$.

3.2. SSILSVRFN Parameter Learning

The iterative linear SVR (ILSVR) learning algorithm is used to adjust the antecedent and consequent parameters of fuzzy rules in the SSILSVRFN algorithm. In ILSVR algorithm, the purpose of learning parameters is to optimize antecedent and consequent parameters of fuzzy rules based on the cost function.

3.2.1. Consequent Parameter Learning

In this section, the fourth layer represented by the structure diagram in Figure 5 is used to calculate the output value of each rule. The calculation expression is as follows

$$f^i = \mu^i(x) \cdot \left(a_{i0} + \sum_{j=1}^{n} a_{ij}x^j \right) = \mu^i(x) \cdot \sum_{j=0}^{n} a_{ij}x^j, x^0 \Delta 1 \tag{22}$$

where $a_{i0} + \sum_{j=0}^{n} a_{ij}x^j$ is the corresponding consequent value of the node; and $\mu^i(x)$ is the corresponding firing strength.

The fifth layer represents the output variable of the output layer, which calculates as follows

$$y^* = \sum_{i=1}^{r} f^i + b \tag{23}$$

where b is the compensation constant and f^i is the output value of the node.

By substituting (22) into (23), the output of SSILSVRFN can be transformed as follows

$$y^*(x) = \sum_{i=1}^{r} \sum_{j=0}^{n} a_{ij}\mu^i x^j + b \tag{24}$$

Equation (24) shows that the output y^* is a linear function of $\mu^i x^j$ with weights a_{ij}. Therefore, linear SVR can be employed to learn parameters a_{ij}. After structure learning, the number of rules r and the initial rule antecedent part parameters are determined. The input data x_k is transformed to the following vector

$$\begin{aligned} \varphi(x_k) &= [\varphi^1(x_k), \dots, \varphi^{r(n+1)}(x_k)] \\ &= \left[\mu^1 x_k^0, \dots, \mu^1 x_k^n, \dots, \mu^r x_k^0, \dots, \mu^r x_k^n \right] \in \mathfrak{R}^{r(n+1)} \end{aligned} \tag{25}$$

The vector φ as input into a linear SVR, and the training data pairs are represented as follows

$$\begin{aligned} S &= \left\{ (x_1, y_1), \ (x_2, y_2), \ \cdots, \ (x_N, y_N) \right\} \\ &= \left\{ \{\varphi(x_1), y_1\}, \ \{\varphi(x_2), y_2\}, \ \cdots, \ \{\varphi(x_N), y_N\} \right\} \end{aligned} \tag{26}$$

The linear regression function $y^*(x)$ is given as

$$y^*(x) = \sum_{k=1}^{N} (\alpha_k - \hat{\alpha}_k)x_k x + b \tag{27}$$

According to Equations (26) and (27), the optimal linear regression function $y^*(x)$ is given as

$$y^*(x) = \sum_{k=1}^{N} (\alpha_k - \hat{\alpha}_k) \langle \varphi(x), \ \varphi(x_k) \rangle + b \tag{28}$$

where α_k c and $\hat{\alpha}_k$ are calculated by SVR software, i.e., Library for Support Vector Machines (LIBSVM) in this section. Based on Equation (25), Equation (28) can be represented as follows

$$
\begin{aligned}
y^*(\mathbf{x}) &= \sum_{k=1}^{N} (\alpha_k - \hat{\alpha}_k) \sum_{m=1}^{r(n+1)} \varphi^m(\mathbf{x})\varphi^m(\mathbf{x_k}) + b \\
&= \sum_{m=1}^{r(n+1)} \left[\sum_{k=1}^{N} (\alpha_k - \hat{\alpha}_k)\varphi^m(\mathbf{x_k}) \right]\varphi^m(\mathbf{x}) + b \\
&= \sum_{i=1}^{r} \sum_{j=0}^{n} \left[\sum_{k=1}^{N} (\alpha_k - \hat{\alpha}_k)\mu^i x_k^j \right]\mu^i x^j + b
\end{aligned}
\tag{29}
$$

Since Equation (29) is equivalent to Equation (24), by comparing these two expressions, the mathematical expression of the parameters a_{ij} can be obtained.

$$
a_{ij} = \sum_{k=1}^{N} (\alpha_k - \hat{\alpha}_k)\mu^i x_k^j, i = 1,\ldots,r, j = 0,\ldots,n
\tag{30}
$$

3.2.2. Antecedent Parameter Learning

The initial parameters m_{ij} and σ_{ij} in SSILSVRFN are determined by the SSRG algorithm. Further, these parameters are tuned based on the minimization of the cost function. The output function of SSILSVRFN in Equation (24) can be written as follows

$$
y^*(\mathbf{x}) = \sum_{i=1}^{r} \left\{ \exp\left[-\sum_{j=1}^{n} \left[\frac{x^j - m_{ij}}{\sigma_{ij}}\right]^2 \right] \right\} \cdot \left(\sum_{j=0}^{n} a_{ij}x^j \right) + b
\tag{31}
$$

The above equation shows that the antecedent parameters m_{ij} and σ_{ij} are not the linear combination coefficients of the fuzzy network output y^*. Therefore, the linear SVR cannot be directly applied. In response to this problem, this section uses Taylor series expansion to linearize it. Since the parameters m_{ij} and σ_{ij} are independent of each other, the fuzzy network output y^* is expanded as shown below.

$$
\begin{aligned}
y^*(\mathbf{p}) &= y^*(\mathbf{p}(0)) + \sum_{i=1}^{r} \sum_{j=1}^{n} \Delta m_{ij}\left(m_{ij} - m_{ij}(0)\right) \\
&+ \sum_{i=1}^{r} \sum_{j=1}^{n} \Delta \sigma_{ij}\left(\sigma_{ij} - \sigma_{ij}(0)\right) + y_h^*
\end{aligned}
\tag{32}
$$

where y_h^* is the remainder of the Taylor expansion, $P = [m_{11},\ldots,m_{rn}, \sigma_{11},\ldots\sigma_{rn}]$ denotes the antecedent parameters vector. The first partial derivative Δm_{ij} and $\Delta \sigma_{ij}$ are, respectively, written as follows

$$
\Delta m_{ij} = \frac{\partial y^*}{\partial m_{ij}} = f^i \cdot 2 \cdot \frac{x_j - m_{ij}}{\left(\sigma_{ij}\right)^2}
\tag{33}
$$

$$
\Delta \sigma_{ij} = \frac{\partial y^*}{\partial \sigma_{ij}} = f^i \cdot 2 \cdot \frac{\left(x_j - m_{ij}\right)^2}{\left(\sigma_{ij}\right)^3}
\tag{34}
$$

Let $\hat{y}^* = y^*(\mathbf{p}) - y^*(\mathbf{p}(0))$. Equation (32) can be further transformed as follows

$$
\hat{y}^* \approx \sum_{i=1}^{r} \sum_{j=1}^{n} \left[\Delta m_{ij}\left[m_{ij} - m_{ij}(0)\right] + \Delta \sigma_{ij}\left[\sigma_{ij} - \sigma_{ij}(0)\right] \right]
\tag{35}
$$

Equation (35) shows that $\hat{y}^*$ is a linear function with linear combination coefficients $m_{ij} - m_{ij}(0)$ and $\sigma_{ij} - \sigma_{ij}(0)$. Therefore, linear SVR can be used to optimize tuning parameters $m_{ij} - m_{ij}(0)$ and

$\sigma_{ij} - \sigma_{ij}(0)$. Being similar to the derivation of consequent parameter learning, the parameters m_{ij} and σ_{ij} can be solved.

$$m_{ij} = m_{ij}(0) + \sum_{k=1}^{N}\left[[\alpha_k - \hat{\alpha}_k]\Delta m_{ij}(x_{\mathbf{k}})\right]$$
$$\sigma_{ij} = \sigma_{ij}(0) + \sum_{k=1}^{N}\left[[\alpha_k - \hat{\alpha}_k]\Delta\sigma_{ij}(x_{\mathbf{k}})\right], i = 1,\ldots,r, j = 0,\ldots,n \tag{36}$$

The linearized function $\hat{y}^*$ in Equation (35) approximates the output of SSILSVRFN y^* with the assumption that the updated values of m_{ij} and σ_{ij} are very close to the original $m_{ij}(0)$ and $\sigma_{ij}(0)$. After the optimization of the linear SVR, the updated parameters may be too large to meet this assumption. Thus, ILSVR is used to solve this problem. After several linear SVR learning iterations, the parameter learning tends to converge. The updated values of m_{ij} and σ_{ij} can meet the assumption in a Taylor expansion because the change in the two parameters tends to zero.

Throughout the learning process of parameters, the steps of the entire ILSVR algorithm can be summarized as follows

Step 1: set the maximal iteration number T_{ite} and set the iteration index l to zero.

Step 2: calculate consequent parameters $a_{ij}(l)$, $i = 1,\ldots,r, j = 0,\ldots,n$ by using the linear SVR and Equation (30).

Step 3: based on the learned parameters $a_{ij}(l)$ in Step 2, calculate the antecedent parameters $m_{ij}(l)$ and $\sigma_{ij}(l)$ using the linear SVR and Equation (36).

Step 4: if $l \geq T_{ite}$, this algorithm stops; otherwise, update the iteration index $l = l + 1$ and go back to step 2.

The overall structure block diagram of the variable section wing dynamic deformation reconstruction method is shown in the Figure 8.

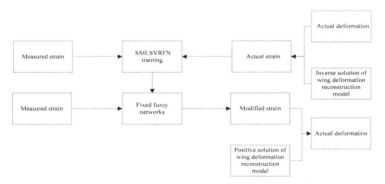

Figure 8. Structural block diagram of dynamic deformation reconstruction method for variable section wing.

4. Verifications through Simulations and Experimentation

In order to assess the accuracy and effectiveness of the measurement scheme proposed in this paper, simulations and model experimentation are performed. In Section 4.1, a finite element model of a variable-section wing is constructed through direct finite element (FE) analyses software (ABAQUS), used to assess the accuracy of the reconstructing model of the variable-section wing. In Section 4.2, a physical wing model is tested under different dynamic loads, in order to assess the feasibility of measurement scheme proposed in this paper.

4.1. Simulation Test

The FE model of the whole variable section wing (Figure 9A) is directly obtained from the CAD model of the wing (Figure 9B) with using ABAQUS. The whole wing is combined the skin with the framework, and the framework mainly consists of glass fiber rib plate and glass fiber sheet. The single wingspan is 1500 mm long. The whole wing structure is divided into two segments for deformation reconstruction: the first section is the wing root part which is 600 mm; the second section is the aileron which is 900 mm. For the largest cross-sectional area of the wing root, the length and the width are 500 mm and 70 mm, respectively. For the smallest cross-sectional area of the wing tip, the length is 200 mm and the width is 28 mm (Figure 10). The cross section area decreases along the span direction, and the shape of each section is the similar (Figure 10B). For the wing skin, the Young's modulus is $E = 8.3$ GPa, the Poisson ratio is $v = 0.22$, and the density is $\rho = 4500$ kg/m^3. For the glass fiber rib plate and the glass fiber sheet, the Young's modulus is $E = 7.3$ GPa, the Poisson ratio is $v = 0.22$, and the density is $\rho = 5000$ kg/m^3. The wing is connected to the airframe with two thin-walled carbon fiber beams. For the carbon fiber beam, the Young's modulus is $E = 210$ GPa, the Poisson ratio is $v = 0.307$, and the density is $\rho = 4000$ kg/m^3. The longer beam length is 757 mm, the external radius is 18 mm and the thickness is 2 mm, and the shorter beam length is 345 mm, the external radius is 7.5 mm and the thickness is 1 mm.

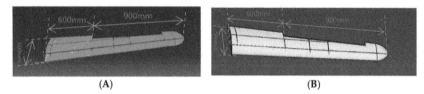

(A) (B)

Figure 9. The finite element (FE) model and the CAD model of the variable section wing. (A) The FE model; (B) The CAD model.

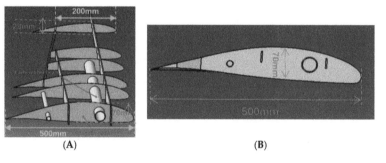

(A) (B)

Figure 10. Wing structure. (A) Framework of the wing; (B) wing cross section.

Two different loads are performed on the FE model. Herein, the point C (Figure 11) is selected as the target point to assess the accuracy of the reconstructing model. In Figure 11, the hexahedron element is used to divide the wing structure, and the total number of the element is 4546. The comparison between the deformation computed from the strain values with using the reconstruction model (Equation (7)) and the deformation analyzed with using the FE model for the point C is shown in Table 1. The strain values are obtained from the FE analysis.

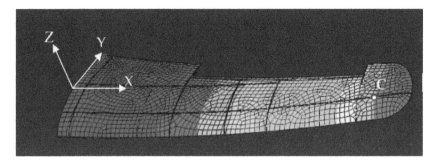

Figure 11. The contour plot of the wing deformation.

The comparisons of Table 1 demonstrate that the reconstructing model presents higher accuracy, for the main deformation w_z under the two loading cases, the percent errors remain is below 6.0%. Although the percentage errors are relatively high for the displacements in the other two directions and the rotations for three directions, the absolute errors are very small. During the flight, the main deformation of the wing is the displacement along the vertical direction, Z (Figure 11).

Table 1. The comparison between the FE analysis and the reconstructing using inverse finite element method (iFEM). The displacements are expressed in millimeter and rotations are expressed in radian.

Loading		u_x	v_y	w_z	θ_x	θ_y	θ_z
135 N	FE analysis	0.81	2.43	124.06	−0.1292	0.0039	0.0011
	IFEM	0.63	2.08	116.93	−0.0846	0.0035	0.0032
	Absolute error	0.18	0.35	7.13	0.0446	0.0004	0.0021
	Percent error	22.2%	14.4%	5.7%	34.5%	10.3%	190.9%
200 N	FE analysis	1.52	4.27	168.18	−0.2154	0.0066	0.0018
	IFEM	1.17	3.72	158.09	−0.1492	0.0058	0.0047
	Absolute error	0.35	0.55	10.09	0.0662	0.0008	0.0029
	Percent error	23.0%	12.8%	6.0%	30.7%	12.1%	161.1%

4.2. Physical Model Test

For the aim of assessing the effectiveness of the measurement scheme proposed in this paper, a dynamic loading test is performed on the variable-section wing physical model. The size of the physical wing (Figure 12) is same to the CAD model of the wing (Figure 9B).

Figure 12. Variable-section wing physical model.

In the experiment, the strain data are obtained from the strain measurement system composed of FBG strain sensors (Fiber Bragg Grating| os1100, Micron Optics, Atlanta, GA, USA) and the FBG interrogator (Optical Sensing Instrument| Si 155, Micron Optics, Atlanta, GA, USA). Twelve FBG strain sensors (the range of initial wavelength is (1527 nm, 1564 nm)) are placed at different locations along the wing and used to capture the surface strain. The placement of the strain sensors is shown in Figure 13.

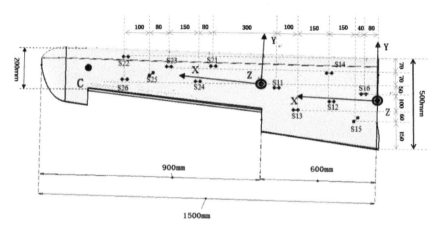

Figure 13. Finite element model of wing with variable cross section.

In order to evaluate the accuracy of the deformation reconstructed from the algorithm proposed in this paper, the actual deformation of the wing structure is captured from the 3D optical measurement instrument (see Figure 14A, NDI Optrotrak Certus) which is abbreviated as NDI. Several position sensors which send the infrared lights to the CCD cameras of the NDI to reflect the deformation of the wing structure. The accuracy of NDI is 0.1 mm in its measurement range. The whole experiment system and the coordinate can be seen in Figure 14B.

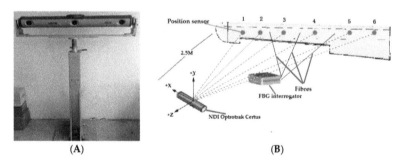

| (A) | (B) |

Figure 14. Wing tests. (**A**) NDI Optrotrak Certus; (**B**) experiment system.

In the experiment, twenty different static loadings are performed on the end of the wing, which used for the fuzzy network training. The total weight of the loading is 20 kg. The dynamic load is caused by the sudden removal of the load applied on the wing tip (Figure 15).

Sensors **2019**, *19*, 3350

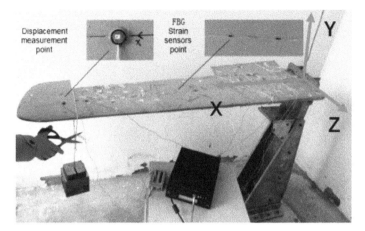

Figure 15. Dynamic load performed on the variable section wing model.

Under the working condition, the aircraft is mainly subjected to the vertical lifting force. Therefore, in this experiment, the load along the Y-axis is mainly performed on the wing model. When the kinematic variable u^e of the wing is solved by the measured strain through the Equation (8), any deformation of the wing surface can be calculated by using the Equation (1). To verify the accuracy of the reconstruction, the root mean square error RMS (Root Mean Square) is applied.

$$RMS = \sqrt[2]{\sum_{i=1}^{j} (disp^{NDI}(x_i) - disp^{iFEM}(x_i))^2 / j} \qquad (37)$$

where, $disp(x_i)$ is the displacement of one node along the wing in one direction; the superscript 'NDI' refers to the deformation values captured from NDI; 'iFEM' refers to the predicted values computed from strain data with iFEM; and j is the number of the nodes used to describe the wing deformation. In the test, the number of nodes used to describe the beam deformation is 6, and these nodes are placed on the surface of the wing (Figure 14B). Meanwhile, the node of maximum deformation, point C (Figure 13) is taken account. The tracking of point C for dynamic loading cases can be seen in Figure 16, and the comparison between the deformations of the point C computed from the strain data are shown in Tables 2 and 3. The maximum percentage error among the individual nodal displacements is shown in Table 4. The comparison between the deformations of the whole wing computed from the strain data are shown in Table 5.

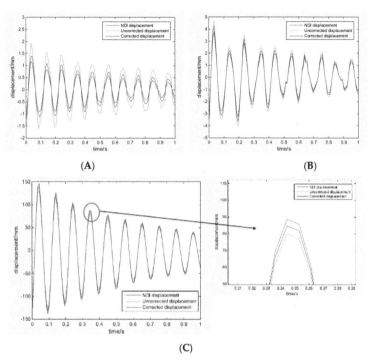

Figure 16. Comparison between deformation displacements of point C for the different time. (**A**) X-axis; (**B**) Z-axis; and (**C**) Y-axis.

Table 2. Comparison of deformation along the y-axis for different time by removing the load of 20 kg loaded at the wing tip, the measured strain is obtained from the fiber Bragg grating (FBG) sensor measurement, and the actual strain is calculated from the NDI measurement with iFEM. C_Y^U and C_Y^M are, respectively, the deformation computed from the unmodified and modified strain measurements with iFEM at point C.

Time/s	Measured Strain	Actual Strain	Modified Strain	C_Y^U/mm	C_Y^{NDI}/mm	C_Y^M/mm	Percentage Error
0.04	0.001671	0.002137	0.002183	132.01	149.16	138.89	6.7%
0.14	0.001428	0.001945	0.00187	113.07	127.52	124.97	5.2%
0.24	0.001334	0.001659	0.001603	95.32	105.29	99.72	4.5%
0.34	0.001079	0.001271	0.001261	78.48	89.13	86.29	4.2%
0.44	0.001235	0.001357	0.001391	67.86	76.11	69.70	5.3%
0.54	0.000934	0.001099	0.001087	57.46	64.37	62.79	5.6%
0.64	0.00083	0.001075	0.001035	48.95	56.28	52.86	4.5%
0.75	0.000644	0.000723	0.000714	46.53	53.30	49.95	5.1%
0.85	0.000654	0.000779	0.000742	38.67	44.32	40.75	3.9%
0.95	0.000627	0.000696	0.0007	33.83	37.78	36.58	5.2%

Table 3. Comparison of deformation along the *x*-axis and *z*-axis for different time by removing the load of 20 kg loaded at the wing tip.

Time/s	C_X^U/mm	C_X^{NDI}/mm	C_X^M/mm	Percentage Error	C_Z^U/mm	C_Z^{NDI}/mm	C_Z^M/mm	Percentage Error
0.04	1.82	1.26	1.42	12.70%	4.75	3.87	4.11	6.20%
0.14	1.54	0.89	1.03	15.73%	2.37	1.92	2.06	7.29%
0.24	1.46	0.92	0.99	7.61%	3.29	2.87	3.15	9.76%
0.34	1.27	0.75	0.82	9.33%	2.46	2.14	2.33	8.88%
0.44	1.14	0.67	0.73	8.96%	2.34	2.03	2.18	7.39%
0.54	1.08	0.56	0.65	16.07%	2.26	1.98	2.09	5.56%
0.64	1.03	0.51	0.6	17.65%	2.12	1.85	1.97	6.49%
0.75	0.89	0.47	0.52	10.64%	2.06	1.74	1.88	8.05%
0.85	0.82	0.43	0.51	18.60%	1.84	1.58	1.69	6.96%
0.95	0.78	0.37	0.44	18.92%	1.79	1.62	1.72	6.17%

Table 4. The maximum percentage error among the individual nodal displacements, $disp_Y^U$ and $disp_Y^M$ are, respectively, the displacement computed form the unmodified and modified strain measurements with iFEM.

Node	$disp_Y^U$/mm	$disp_Y^{NDI}$/mm	$disp_Y^M$/mm	Maximum Percentage Error
1	132.01	149.06	138.97	6.7%
2	79.57	94.38	88.63	6.0%
3	41.6	55.42	52.83	4.6%
4	19.41	29.31	27.79	5.1%
5	9.78	11.5	10.96	4.7%
6	3.66	4.77	4.57	4.1%

It is found that, the deformation compute from the modified strain data with iFEM on point C is close to the result captured from the NDI. From the Table 2, it is shown that the minimum value of the percentage error of the C point displacement is 3.9%, and the maximum value is 6.7% along the *y*-axis. Although the percentage errors are relatively high for the displacements in the other two directions, deformations along the *x*-axis and *z*-axis are very small from the Table 3. From the Table 4, it is shown that the maximum percentage error among the individual nodal displacements is 6.7%, and the maximum value is 4.1%.

From the Table 5, it is shown that the accuracy of the deformation reconstruction with iFEM is increased when the strain measurements are modified with using SSILSVRFN algorithm proposed in this paper. The RMS of the reconstruction for the maximum deformation is 12.01 mm when the strain measurements are unmodified, while RMS is 3.97 mm when the strain measurements are modified with using SSILSVRFN algorithm. The reduction percentage of RMS is 66.8% at least.

Table 5. Comparison of deformation along the *y*-axis for different time, RMS_U and RMS_M are, respectively, the accuracy of the deformation computed form the unmodified and modified strain measurements with iFEM.

Time/s	RMS_U/mm	RMS_M/mm	Percentage Reduced
0.04	12.01	3.97	66.8%
0.14	11.13	3.18	71.4%
0.24	10.22	3.38	66.9%
0.34	9.42	3.71	60.5%
0.44	9.51	3.20	66.3%
0.54	9.31	3.29	64.6%
0.64	7.05	2.16	69.4%
0.75	6.21	2.03	67.2%
0.85	6.15	2.50	59.2%
0.95	5.83	2.18	62.5%

During the flight, the main deformation of the wing is the displacement along the vertical direction (Figures 11 and 12). From the Tables 1 and 2, numerical and experimental studies show that:

1. By comparing between the FE analysis and the reconstructing with using IFEM, numerical studies show that the percent error of the deformation reconstruction along the main direction remains below 6.0%.
2. Because of the strain measurement system error and the model error, experimental studies show that the percent error of the deformation reconstruction along the main direction computed from the unmodified strain measurements with iFEM remains below 13%.
3. Experimental application of the proposed method shows that: the percent error of the deformation reconstruction along the main direction computed from the modified strain measurements remains below 6.7%.

The dynamic deformation reconstruction algorithm proposed in this paper shows high precision for the wing structure with variable section.

5. Conclusions

In view of the effects of the dynamic un-modeled error and the sensor placement error on the cross-section wing deformation sensing, a dynamic deformation reconstruction algorithm for wing structure with variable section is proposed in this paper. For the sake of reducing the influence of the strain measurement error, the in situ strain data are modified with using SSILSVRFN in real time. The test performed on the wing shows that, the dynamic deformation of the variable cross-section wing can be accurately reconstructed from the strain measurement with the scheme proposed in this paper. The measurement scheme proposed in this paper can be applied to the dynamic deformation reconstruction of the structure from the variable section beam. If the measurement objects are different, the reconstruction model and the correction network need to be re-established. Nevertheless, the dynamic loading in this paper is caused by sudden removal of the load applied on the wing tip. The further work is performing a dynamic experiment on the shaking platform to simulate the dynamic deformation caused by different airflow excitations, to assess the effectiveness and accuracy of the reconstruction algorithm.

Author Contributions: Z.F. carried out the majority of work on the paper, proposing a deformation-reconstruction model for wing with variable cross-section, fuzzy correction method and writing the article; H.B. analyzed the method; Y.Z. and F.Z. gave advice regarding the writing of the article.

Funding: This research was funded by [Key Technology Research on Steerable 110-m Aperture Radio Telescope] grant number [2015CB857100], [Xinjiang Astronomical Observatory] grant number [2014KL012], and [Shanghai Aerospace Science and Technology Innovation Fund] grant number [SAST201413].

Conflicts of Interest: The authors declare no conflict of interest.

Appendix A

The interpolation shape function in [4] is only applicable to constant-section beam elements. This paper also uses interdependent interpolations (Tessler and Dong, 1981) to derive the interpolation shape function suitable for the variable section beam element.

$$u(\xi) = \sum_{i=1,r,2} L_i^{(2)}(\xi)u_i$$

$$v(\xi) = \sum_{j=1,q,s,2} L_j^{(3)}(\xi)v_i - \sum_{j=1,q,s,2} N_j^{(4)}\theta_{zj}$$

$$w(\xi) = \sum_{j=1,q,s,2} L_j^{(3)}(\xi)w_i + \sum_{j=1,q,s,2} N_j^{(4)}\theta_{yj}$$

$$\theta_x(\xi) = \sum_{i=1,r2} L_i^{(2)}(\xi)\theta_{xi} \tag{A1}$$

$$\theta_y(\xi) = \sum_{j=1,q,s,2} L_j^{(3)}\theta_{yj}$$

$$\theta_z(\xi) = \sum_{j=1,q,s,2} L_j^{(3)}\theta_{zj}$$

The shape function is expressed as follows

$$[L_1^{(2)}, L_r^{(2)}, L_2^{(2)}] = \tfrac{1}{2}[\xi(\xi - 1), 2(1 - \xi^2), \xi(\xi + 1)]$$

$$[L_1^{(3)}, L_2^{(3)}] = \tfrac{1}{16}((9\xi^2 - 1))[(1 - \xi), (1 + \xi)]$$

$$[L_q^{(3)}, L_s^{(3)}] = \tfrac{9}{16}(\xi^2 - 1)[(3\xi - 1), -(3\xi + 1)] \tag{A2}$$

$$[N_1^{(4)}, N_2^{(4)}] = \tfrac{L^e}{128}[(9\xi^2 - 1)(\xi^2 - 1), -(9\xi^2 - 1)(\xi^2 - 1)]$$

$$[N_q^{(4)}, N_s^{(4)}] = \tfrac{3L^e}{128}[-(9\xi^2 - 1)(\xi^2 - 1), (9\xi^2 - 1)(\xi^2 - 1)]$$

An element is referred to a local axial coordinate $x \in [0, L]$, where L denotes the element length. Furthermore, a non-dimensional coordinate $\xi \equiv \left(\frac{2x}{L} - 1\right) \in [-1, 1]$ is used to define the element shape function (Figure A1).

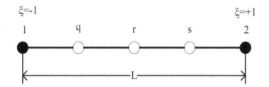

Figure A1. Inverse finite element geometry and nodal topology.

References

1. Xie, C.; Wu, Z.; Yang, C. Aeroelastic analysis of flexible wing with high aspect ratio. *J. Beijing Univ. Aeronaut. Astronaut.* **2003**, *29*, 1087–1090.
2. Shang, B.; Song, B.; Wan, F. Application of optical fiber sensor in structural health monitoring of aircraft. *Fiber Opt. Cable Appl. Technol.* **2008**, *3*, 7–10.
3. Foss, G.; Haugse, E. Using modal test results to develop strain to displacement transformations. In Proceedings of the 13th International Conference on Modal Analysis, Nashville, TN, USA, 13–19 February 1995; pp. 112–128.
4. Gherlone, M.; Cerracchio, P.; Mattone, M.; di Sciuva, M.; Tessler, A. Shape sensing of 3D frame structures using an inverse Finite Element Method. *Int. J. Solids Struct.* **2012**, *49*, 3100–3112. [CrossRef]
5. Van Tran Fleischer, W.L.K. *Extension of ko Straight-Beam Displacement Theory to Deformed Shape Predictions of Slender Curved Structures*; NASA: Edwards, CA, USA, 2011.
6. Jutte, C.V.; Ko, W.L.; Stephens, C.A.; Stephens, C.A.; Bakalyar, J.A.; Rechards, W.L.; Parker, A.R. *Deformed Shape Calculation of a Full-Scale Wing Using Fiber Optic Strain Data from a Ground Loads Test*; NASA: Hampton, VA, USA, 2011.
7. Jan, L.; Spangler, A.T. *A Variational Principle for Reconstruction of Elastic Deformations in Shear Deformable Plates and Shells*; NASA CASI: Hampton, VA, USA, 2003.

8. Gherlone, M.; Cerracchio, P.; Mattone, M.; Di Sciuva, M.; Tessler, A. Beam shape sensing using inverse finite element method: Theory and experimental validation. In Proceedings of the 8th International Workshop on Structural Health Monitoring, Stanford, CA, USA, 13–15 September 2011.

9. Kefal, A.; Oterkus, E. Displacement and stress monitoring of a Panamax containership using inverse finite element method. *Ocean Eng.* **2016**, *119*, 16–29. [CrossRef]

10. Kefal, A.; Oterkus, E. Displacement and stress monitoring of a chemical tanker based on inverse finite element method. *Ocean Eng.* **2016**, *112*, 33–46. [CrossRef]

11. Tessler, A.; Roy, R.; Esposito, M.; Surace, C.; Gherlone, M. Shape Sensing of Plate and Shell Structures Undergoing Large Displacements Using the Inverse Finite Element Method. *Shock Vib.* **2018**, *2018*, 8076085. [CrossRef]

12. Gherlone, M.; Cerracchio, P.; Mattone, M. Shape sensing methods: Review and experimental comparison on a wing-shaped plate. *Prog. Aerosp. Sci.* **2018**, *99*, 14–26. [CrossRef]

13. Cerracchio, P.; Gherlone, M.; di Sciuva, M.; Tessler, A. A novel approach for displacement and stress monitoring of sandwich structures based on the inverse Finite Element Method. *Compos. Struct.* **2015**, *127*, 69–76. [CrossRef]

14. Zhao, Y.; Du, J.; Bao, H.; Xu, Q. Optimal Sensor Placement for Inverse Finite Element Reconstruction of Three Dimensional Frame Deformation. *Int. J. Aerosp. Eng.* **2018**, *2018*, 1–10. [CrossRef]

15. Zhao, Y.; Du, J.; Bao, H.; Xu, Q. Optimal Sensor Placement based on Eigenvalues Analysis for Sensing Deformation of Wing Frame. *Sensors* **2018**, *18*, 2424. [CrossRef] [PubMed]

16. Zhao, F.; Bao, H.; Xue, S.; Xu, Q. Multi Objective Particle Swarm Optimization of Sensor Distribution Scheme with Consideration ofthe Accuracy and the Robustness for Deformation Reconstruction. *Sensors* **2019**, *19*, 1306. [CrossRef] [PubMed]

17. Pan, X.; Bao, H.; Zhang, X. The in situ strain measurements modification based on Fuzzy nets for frame deformation reconstruction. *J. Vib. Meas. Diagn.* **2018**, *38*, 360–364.

18. Gherlone, M.; Cerracchio, P.; Mattone, M. An inverse finite element method for beam shape sensing: theoretical framework and experimental validation. *Smart Mater. Struct.* **2014**, *23*, 045027. [CrossRef]

19. Wang, Y.; Zhu, Z.; Chen, Z. A test method suitable for flexible UAV wing deformation. *Comput. Meas. Control* **2012**, *20*, 2894–2896.

20. Forgit, C.; Lemoine, B.; le Marrec, L. A Timoshenko-like model for the study of three-dimensional vibrations of an elastic ring of general cross-section. *Acta Mech.* **2016**, *227*, 2543–2575. [CrossRef]

21. Chuan, G.; Chen, Y.; Tong, G. Element Stiffness Matrix of Timoshenko Beam with Variable Section. *Chin. J. Comput. Mech.* **2014**, *31*, 266–272.

22. Lurie, A.I. *Theory of Elasticity*; Springer: Berlin/Heidelberg, Germany; New York, NY, USA, 2005.

23. Mainçon, P. Inverse FEM I: load and response estimates from measurements. In Proceedings of the 2nd International Conference on Structural Engineering, Mechanics and Computation, Cape Town, South Africa, 5–7 July 2004.

24. Feng, S. Research on the Fuzzy Network Method for Measuring the Deformation of the Long Flexible Base Antenna of the Base Wing. Master's Thesis, Xi'an University, Xi'an, China, 2015.

Article

Metal Forming Tool Monitoring Based on a 3D Measuring Endoscope Using CAD Assisted Registration

Lennart Hinz *, Markus Kästner and Eduard Reithmeier

Insitute of Measurement and Automatic Control, Leibniz University Hannover, Nienburger Straße 17, 30167 Hannover, Germany; markus.kaestner@imr.uni-hannover.de (M.K.); eduard.reithmeier@imr.uni-hannover.de (E.R.)
* Correspondence: lennart.hinz@imr.uni-hannover.de; Tel.: +49-511-762-3235

Received: 28 February 2019; Accepted: 1 May 2019; Published: 5 May 2019

Abstract: In order to provide timely, reliable, and comprehensive data for the maintenance of highly stressed geometries in sheet-bulk metal forming tools, this article features a possible setup by combining a 3D measuring endoscope with a two-stage kinematic. The measurement principle is based on the projection of structured light, allowing time-effective measurements of larger areas. To obtain data of proper quality, several hundred measurements are performed which then have to be registered and finally merged into one single point cloud. Factors such as heavy, unwieldy specimens affecting precise alignment. The rotational axes are therefore possibly misaligned and the kinematics and the hand-eye transformation remain uncalibrated. By the use of computer-aided design (CAD) data, registration can be improved, allowing a detailed examination of local features like gear geometries while reducing the sensitivity to detect shape deviations.

Keywords: endoscopy; maintenance; fringe projection; registration; metrology

---

1. Introduction

In the context of increasing automation and digitalization of integrated industrial manufacturing processes, monitoring and automated quality assurance are essential factors, strengthening the importance of new measurement approaches. Optical measurement techniques are contactless and can meet current and future emerging requirements like significantly reduced measuring times while increasing the resolution of the data [1,2].

Sheet-bulk metal forming is a newly emerging forming process, producing complex geometries due to a combination of deep drawing and bulk metal forming [3]. Possible areas of application are given by gearing or carrier elements for transmissions in the automotive industry [4]. The tools of sheet-bulk metal forming plants are stressed during operation. A challenge for maintenance service is to evaluate the condition of the tool without having accurate information about possible deviations. A quantitative assessment of possible damage is therefore mostly unfeasible. By using endoscopic devices, geometries can be imaged which would not be assessable for classical measuring devices. Additionally, recent research strives to extend the capabilities of endoscopic inspection devices from 2D to true 3D imaging [5]. This article shall give an overview of a possible forming tool inspection setup, combining a two-stage kinematic with an endoscopic measuring device, based on the fringe projection approach. By using a structured light projection and a camera-based observation, object geometries can be measured within seconds or below with a higher resolution than e.g., sensors based on the time-of-flight approach [6].

2. In Situ and Offline Inspection of Forming Tools

The measuring system was developed within the last years and is intended to perform measurements in geometries which would be unreachable for most common systems. This is done by using a small and flexible measuring head. The system is based on the fringe projection approach in order to quantify the condition of sheet-bulk metal forming tools in an industrial environment [7]. An endoscopic sensor is needed to enable measurements while the tool remains inside the forming plant. Based on the maximum available space (~10 cm high), the sensor features a small measuring head being mounted onto a robot guided arm. An overview of a planned application is given in Figure 1, showing a possible measurement inside a forming plant. The general objective of this setup is to enable in situ measurements of certain highly stressed features between a certain number of forming cycles. This ensures continuous monitoring of the forming tool without stopping the process or partially disassemble the forming tool. Measurement times of less than 10 s are therefore the aim of current research. This includes the positioning of the measuring head. The system shall offer inspections for a wide range of different specimens, including flat gearing geometries as shown in Figure 1 or more complex geometries with multiple inner features (as shown in Figure 2) which are hard to access for most measurement approaches.

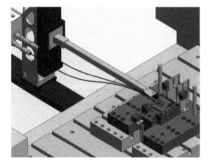
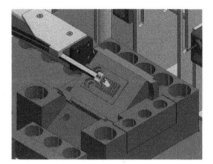

(a) Measuring with the endoscopic fringe projection system

(b) Detailed view

Figure 1. Overview of a possible in-situ measurement inside a metal forming plant (the top elements are hidden for a better perspective).

Figure 2. Specimen and measuring head while projecting structured light.

Since the in situ determination of deviations and the quantification of wear is limited to certain single features, representing the condition of the entire specimen, it is further necessary to precisely measure the entire geometry of each forming tool enabling comprehensive analysis. Due to the given geometrical constraints and the limited bending radii (~20 cm) of the fibers, the workspace of the kinematics robot is restricted, leading to an offline inspection setup which shall be presented here. To guarantee comparable results for both principle tasks, it is important to use the same measurement system with identical optics.

Based on this first industrial application, introducing a robust system for time-effective and automated full-field 3D measurements at high resolution, a multitude of possible applications and future research is conceivable. The given setup is expected to enable a much better characterization of abrasion and wear, leading to deeper process understanding and helps to optimize certain process parameters. Additionally, it provides the basis for in situ inspections by identifying highly stressed geometries.

3. Problem and Approach

The approach is developed for measuring a specimen with multiple inside-lying geometries. Such a specimen is illustrated in Figure 3 and serves as a prerequisite for further investigations. The specimen features a height of 66 mm, an average diameter of 82.9 mm, 84 involute teeth and has a mass of ∼10 kg. Since most of the specimen's features are hard to access for most other measurement approaches, the usage of the endoscopic fringe projection technique seems reasonable. Furthermore, the specimen could be utilized for possible future in situ inspections (mentioned in Section 2) of single features.

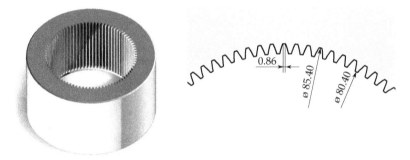

Figure 3. Overview and geometrical features of the specimen.

By combining several hundred measurements from different poses, the approach given here is trying to obtain detailed point cloud data and could therefore potentially measure all features. Since the whole inspection process shall be automated and less time consuming, a multi-axis kinematic is appropriate for traveling to all desired poses. A particular focus is therefore given to the alignment of all measurements. Given the fact of heavy, unwieldy specimens, various challenges affect precise alignment based on positioning data. This is especially relevant when the rotational axes are slightly misaligned. Moreover, the kinematics and the hand-eye transformation remain uncalibrated in order to meet the requirements of rapid inspection.

Figure 4a shows the deviations by aligning all measurements based on the positioning data only. It can be observed that some areas are affected by systematic misalignment, leading to meaningless results. The calculated deviation is based on an Euclidean metric, given in Section 7.4. It would, therefore, be appropriate to use a registration algorithm in order to optimize the overall alignment. A key challenge for stitching large datasets is to ensure that recursive serial registration is robust against drifting effects, caused by an accumulation of uncorrected errors. Despite featuring big overlapping areas in the to be registered point clouds, a conventional registration approach has not performed well. An example is given in Figure 4b. With an almost ideal set of starting values, the algorithm attempted to register all radial datasets in an anti-clockwise direction, leading to an accumulated error.

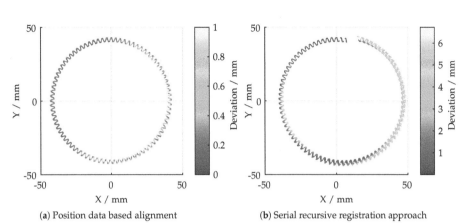

(a) Position data based alignment   (b) Serial recursive registration approach

Figure 4. Comparison of different alignment approaches.

The experimental setup is designed to analyze the condition of sheet metal forming tools after certain cycles. The specimens are machined with high precision and reference geometry data is available. Therefore, the assumption was made that form deviations are negligibly small and computer-aided design (CAD) data based registration can be used to align the measured data. While this should allow a detailed examination of local features like gear geometries, it may reduce sensitivity to detect global deviations. The following sections aim at giving an overview of all crucial data processing algorithms and show first results.

4. Experimental Setup

In Figure 5 an overview of the experimental setup is given. The specimen is mounted onto a URS150BCC rotation stage, supplied by Newport Corporation (Irvine, CA, USA) with a typical accuracy of ±15 mdeg. The measuring head is mounted onto a M-IMS300V motorized vertical stage with a typical accuracy of ±5 μm which is also supplied by Newport Corporation. To measure the entire specimen, 588 separate measurements were performed. This includes 84 steps in radial and 7 steps in the axial direction ($7 \times 84 = 588$). The entire testing procedure lasts a maximum of 60 min (not including data fusion).

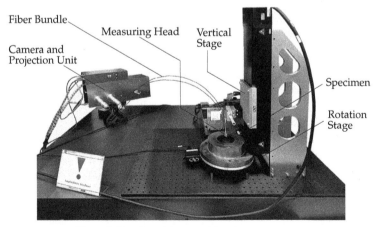

Figure 5. Experimental setup with major components.

5. Hardware Components

Figure 6 shows the major components of the camera and projector unit. Polychromatic, Gaussian distributed green light with a peak wavelength of 528 nm or 505 nm, depending on the diode in use, is emitted by a high power LED (OSRAM Licht AG, Munich, Germany). A Koehler Illumination setup is used to create a homogeneous spot which is imaged (using an additional mirror) onto a digital micro mirror device (DMD) by Texas Instruments (Dallas, TX, USA) with 1024 × 768 individual adjustable mirrors. The DMD is used to create the projected patterns. Previous beam-shaping setups utilized a laser beam which was homogenized by a rotating diffusor and two microlens arrays (fly eye). Unfortunately, this approach led to higher measurement uncertainties, mainly related to speckle interference [8].

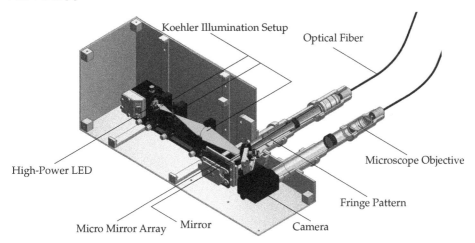

Figure 6. Overview of all major components of the light forming, fringe forming and camera unit.

The fringes, created by the DMD, are injected into an endoscopic fiber by the use of a microscope objective and a tube lens. The fiber itself is a combination of multiple flexible glass fiber cores, combined to a fiber bundle in a common cladding. The fiber in use features 100,000 individual fiber cores (pixels) at a diameter of 1.7 mm and a length of 1 m and is supplied by Fujikura Ltd. (Tokio, Japan). The fiber is connected to a measuring head where the pattern is projected onto the specimen by using a gradient-index (GRIN) rod lens (Grintech GmbH, Jena, Germany). Due to the small apertures of the fiber, the light source has to feature a comparatively small chip diameter, in order to reduce the loss of light. The maximum optical power output (white projector image), measured in focus, is approximately 5.8 mW [8]. The projection is observed by a Point Grey GS3-U3-23S6M-C industrial CMOS camera (FLIR Integrated Imaging Solutions GmbH, Ludwigsburg, Germany) through another identical fiber and GRIN rod lens. Since the projection is green, an additional optical bandpass filter is provided.

Depending on the reflectance characteristics of the specimen, typical exposure times range from 10 to 50 ms. Since technical surface geometries exhibit highly varying reflectivity, the dynamic range of the sensor can be enhanced by combining differently exposed images to a high dynamic range image. Usually, eight different projection patterns are required to perform a measurement.

Different measuring head configurations have emerged throughout the development, each optimized for a certain set of requirements. This section provides an overview of the different types and shall outline each individual scope of application. Figure 7a shows a measuring head with a 10 mm working distance (wd). Between the two pairs of fibers and optics, a 30° triangulation base is formed. The measuring volume is approximately 6 mm × 6 mm × 3 mm [9]. This measuring head features the lowest measurement uncertainty and the highest magnification alike. Figure 7c shows the design, using gradient-index rod lenses with 20 mm working distance, combined with additional mirror prisms. Therefore, a more compact design is realized, allowing a parallel arrangement of the fibers. The measuring volume is approximately 10 mm × 10 mm × 4 mm [9]. To enhance the depth of field of the measuring head with 10 mm working distance, additional liquid lenses can be added, which lead to the design of Figure 7b.

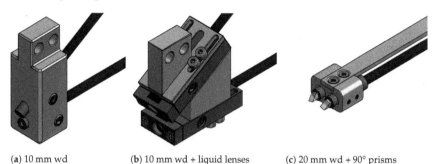

(a) 10 mm wd   (b) 10 mm wd + liquid lenses   (c) 20 mm wd + 90° prisms

Figure 7. Overview of the measuring heads currently in use (scale 1:1).

The measuring head, using two liquid lenses, offers a good compromise between maximum focus and depth of field. Since two liquid lenses add ten more optic interfaces to the optical system, a loss of approximately 50 to 55% optical power was measured in focus. Other drawbacks include the need for a time-effective autofocus, the electrical control setup, the much more extensive calibration and the larger physical size of the measuring head itself.

The specimen requires a depth of field of approximately 2.5 mm (compare Figure 7c) which is close to the limit of the 10 mm GRIN rod lens measuring head. Because of the parallel and rotated arrangement of the fibers, the only measuring head allowing movement into the specimen is the one using the design in Figure 7c. Because of the larger working distance and lower magnification, a bigger measuring field also reduces the required number of measurements to acquire the entire shape of the specimen but also leading to higher measurement uncertainties.

6. System Capabilities and Limitations

To quantify the accuracy of a single measurement, several external factors affecting the results should be taken into account:

- Optical properties of the technical surface
- Working distance, field curvature and depth of field (DOF) of the optics
- Size and form of the specimen
- Orientation between measuring head and the specimen
- Stray light and light scattering

The evident limitation of triangulation-based optical measurement approaches is given by the optical properties of the specimen. If the optical beam path is not unambiguously reconstructable, a measurement cannot be performed. This applies particularly to (partially) transparent surfaces. However, specular reflection can also be considered as a problem. By the use of HDR-imaging, a measurement can be performed in the highlight spot as long as the specimen does not feature structures causing multi reflections. Particularly due to the involute tooth flanks in combination with the fine surface properties of the specimen, it would be impossible to triangulate useful results. To overcome this, anti-glossy spray is applied. This leads to changes of the geometric form of the specimen and shall later be evaluated.

As mentioned in Section 5, all measuring heads feature different focus ranges. Those are also highly correlated to the radial position in the field of view due to field curvature. Since the 10 mm GRIN rod lens features the highest magnification, the best results can be produced. However, an off-focus orientation or different lens setups can also reduce the measurement uncertainty since noise or other artifacts are blurred. This is strongly dependent upon the spatial characteristics of the specimen. Other factors, regarding the form of the specimen, are connected to the geometrical orientation of each surface and the fringe frequency. If fringes are projected at an acute angle to a surface (like the lateral gear surfaces), aliasing effects occur due to the finite number of camera pixels and fiber bundles and can produce noise, reducing the accuracy. To overcome aliasing effects, the generation of adaptive projection patterns based on local frequency modulated fringes [10] is supposed to be implemented in the future.

7. Data Processing

The flowchart, given in Figure 8, provides an overview of all critical data processing algorithms. The routine can later be parallel processed, especially against the background of reducing data volume. Currently, 80 GB of data must be cached and then processed.

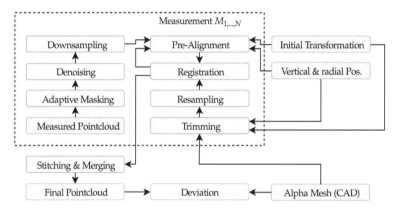

Figure 8. Flowchart of all significant data processing operations (green: input, blue: data processing, orange: output).

At first, a rough initial transformation $T_{initial}$ is performed to help the registration algorithm converging by aligning the first measurement with the CAD-originated shape data, from now on named alpha point data. A detailed explanation of $T_{initial}$ is given in Section 7.1.

The alpha coordinate system is placed at the center of rotation. By using the radial position data of the stage ϕ_{curr}, any additional measured point cloud can be roughly pre-aligned by the use of a rigid body transformation T_{curr} according to Equation (1). Vertical positioning is considered by shifting the data along the rotation axis.

$$T_{curr} = T_{initial} \cdot \begin{bmatrix} cos(\phi_{curr}) & -sin(\phi_{curr}) & 0 & 0 \\ sin(\phi_{curr}) & cos(\phi_{curr}) & 0 & 0 \\ 0 & 0 & 1 & z_{curr} \\ 0 & 0 & 0 & 1 \end{bmatrix} \tag{1}$$

Before starting the registration operation, a denoising algorithm is applied, removing separately scattered points. The algorithm is based on the work of [11] and computes the distribution of point neighbors distances. The points with mean distances outside a threshold are trimmed from the data. To further improve convergence and to speed up calculations, a random downsampling is useful and is set to 10% of the original samples (leading to ~100,000 samples in each dataset).

After pre-aligning, the denoised and downsampled measured data is registered to the alpha data. To enhance performance, only features which exist in the measured point cloud shall be taken into account. The current transformation T_{curr} is therefore used to trim the alpha data based on an region of interest (ROI) which originates from the center of mass (see Equation (3)) of the pre-aligned data. The size of the ROI was determined to 10 mm in axial and 0.2 rad in radial direction. Further, as discussed in Section 7.3, the alpha data is then randomly resampled in order to obtain a point cloud, to which the measured data is registered, according to Section 7.4. The amount of created alpha samples is comparable to the number of samples in the measured point cloud.

After having aligned the first measurement with the alpha point cloud, the progress is repeated, starting with trimming and resampling of the next alpha geometries, based on the current rigid body transformation T_{curr}. In this example, a single alpha template could be used for all registration operations. To ensure that this registration approach is robust against misalignment of the rotational axes or minor movements of the measuring head during the measurement, the initial transformation is continuously updated, based on the last registration operation. This could potentially lead to error drifting by an accumulation of minor errors during registration. A possible reason could be distortion which has not perfectly been removed due to calibration. To suppress this, big overlapping areas are featured in the to be registered point clouds. Another potential source of drifting effects is given by one remaining degree of freedom in each registration step. This is caused by the extruded shape of the specimen in the axial direction. As shown in Equation (2), the current registered point cloud is corrected by shifting the axial center of gravity to the intended position z_{ref} (based on the position data of the vertical stage).

$$pc_{shift} = \begin{bmatrix} x_1 & y_1 & z_1 \\ \vdots & \vdots & \vdots \\ x_n & y_n & z_n \end{bmatrix} + \underbrace{\begin{bmatrix} 0 & 0 & \left(z_{ref} - \frac{1}{n}\sum_{i=1}^{n} z_i\right) \end{bmatrix}}_{\text{Z-Shift}} \tag{2}$$
$$\underbrace{\phantom{\begin{bmatrix} x_1 & y_1 & z_1 \\ \vdots & \vdots & \vdots \\ x_n & y_n & z_n \end{bmatrix}}}_{\text{Point Cloud}}$$

The process of stitching and merging combines two registered point clouds with each other by removing points, mainly in the overlapping area. This procedure is repeated for all data sets. Currently, this is done by using the Computer Vision System Toolbox which is part of Matlab, supplied by The MathWorks Inc. (Natick, MA, USA).

The process of merging is illustrated in Figure 9. For each voxel $V_i \in \overline{V}$, where $\overline{V}$ is the volume in which all points are scattered, all points within V_i are merged to one single point $\mathbf{p}_{m,i} \in V_i$ according to the center of mass (see Equation (3)). The size of each voxel is determined by the grid step size s.

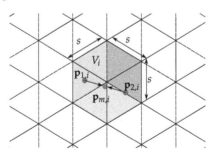

Figure 9. Process of merging points inside one voxel.

The final deviation map is then calculated, based on the distance metric presented in Section 7.4 which was already used in each registration operation.

7.1. Initial Transformation

The determination of the initial transformation is the only operation requiring user input. By selecting corresponding point sets P and X from both point clouds, a rigid transformation is computed. The goal behind this is, as shown in Figure 10, to transform each following measurement as good as possible from the camera coordinate system into the alpha coordinate system, which is given by CAD data.

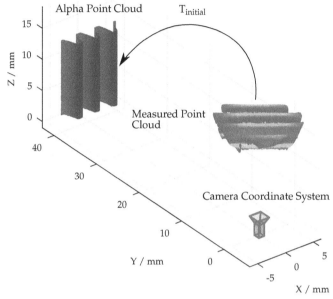

Figure 10. Initial transformation from camera coordinate system with respect to the alpha point cloud.

The algorithms are based on the work of [12] and [13], starting by calculating the center of mass of all point sets P and X according to Equation (3).

$$\mu_p = \frac{1}{n_p} \sum_{i=1}^{n_p} \mathbf{p}_i \quad \text{and} \quad \mu_x = \frac{1}{n_x} \sum_{i=1}^{n_x} \mathbf{x}_i \tag{3}$$

The square cross-covariance matrix Σ_{px} is given according to Equation (4) [12] and is used to find the optimum rotation through computing singular value decomposition: $\Sigma_{px} = U\Sigma V^T$. The resulting rigid body transformation is given in Equation (5). It should be mentioned that this approach can compute results in which the determinant of the rotation matrix R is less than zero. In this special reflection case, the third column of R should be multiplied by -1.

$$\Sigma_{px} = \frac{1}{n_p} \sum_{i=1}^{n_p} [(\mathbf{p}_i - \mu_p)(\mathbf{x}_i - \mu_x)^T] = \frac{1}{n_p} \sum_{i=1}^{n_p} [\mathbf{p}_i \mathbf{x}_i^T] - \mu_p \mu_x^T \tag{4}$$

$$T_{\text{initial}} = \begin{bmatrix} \overbrace{VU^T}^{R} & \overbrace{\mu_x - VU^T \mu_p}^{t} \\ \mathbf{0}^T & 1 \end{bmatrix} \tag{5}$$

7.2. Outlier Removal

To improve the quality of the measurement and to remove non-plausible data, each point cloud is masked. Since each point of the point cloud stems from one pixel in the camera coordinate system, the use of a two-dimensional mask can be applied to the measured point cloud. A manually defined region of interest is not useful for very large datasets. In addition, a fixed mask is not robust against misalignments since features are consequently not always at a fixed position in the camera coordinate system. Due to a misalignment of the rotation axes of the stage and the specimen, this could lead to inaccurately trimmed data. Therefore, the camera data is used to quantify the condition of each pixel and it's corresponding triangulated point based on two-dimensional signal characteristics. The following approach has proven to reduce the measurement uncertainty by eliminating outlier points where triangulation did not work as well.

The presented system is based on the fringe projection approach (active stereo vision). The correspondence between camera and projector is determined by sinusoidal patterns being shifted through the scene. Figure 11 shows a cross-section, (perpendicular to the wavefronts) of all projected and phase shifted patterns of one frequency which are shown partially in Figure 12. The patterns were projected onto a diffuse plane. It can be observed, that the projected patterns are in focus within the range of 0–150 pixels, resulting in high amplitude oscillations. The cross-section also shows areas of bad illumination. By assuming that a high pixel-wise deviation within all shifted patterns of one frequency is a suitable criteria of validating the quality of each triangulated point, this approach can be adopted by calculating the local standard deviation per pixel through all phase shifts of each frequency which is shown in Equation (6), where $i_{c,\mu}(u_p, v_p)$ is the mean value of each pixel. The result is shown in Figure 13.

$$i_{c,\text{std}}(u_p, v_p) = \sqrt{\frac{1}{n}(\sum_{i=1}^{n} i_{c,i}(u_p, v_p) - i_{c,\mu}(u_p, v_p))^2} \qquad (6)$$

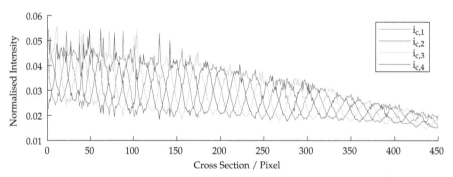

Figure 11. Cross-section of all captured images with the phase-shift of the highest frequency.

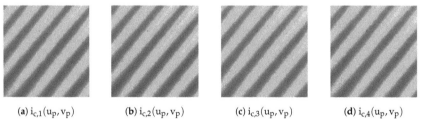

(a) $i_{c,1}(u_p, v_p)$ (b) $i_{c,2}(u_p, v_p)$ (c) $i_{c,3}(u_p, v_p)$ (d) $i_{c,4}(u_p, v_p)$

Figure 12. Cropped camera images showing shifted fringes of the highest frequency being projected onto a plane.

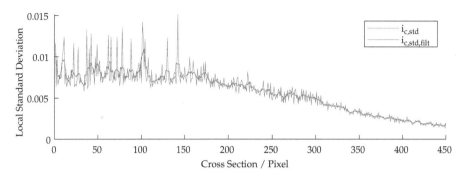

Figure 13. Calculated and filtered local standard deviation of the cross sections of Figure 11.

As can be seen in Figures 11–13, the oscillations carry additional high frequency noise when in focus. This effect is intensified by sampling the image by the use of fiber cores. Therefore, additional smoothing filters, based on a Gaussian approximation, are applied to $i_{c,std}(u_p, v_p)$. Finally, the data is thresholded and applied to each point cloud. Figure 14 shows the masking process. The threshold in use is rather less aggressive. However, as can be seen in Figure 14b, some lateral geometries have already been deleted. This is presumably due to the arrangement of the two fibers and the optics, because the projection comes from a certain lateral direction and so some structures remain badly illuminated due to shadowing.

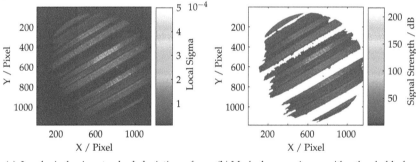

(a) Local pixel wise standard deviation of phase shift with the highest frequency

(b) Masked camera image with a threshold of $\sigma = 1.5 \cdot 10^{-5}$

Figure 14. Overview of signal based masking in camera coordinate space.

By merging all measured point clouds, captured from different perspectives, bigger holes in certain point clouds can be filled. The trimming threshold influences the results and conclusions of the inspection process. A more aggressive threshold leads to a decrease of the measurement uncertainty while reducing data density at areas of bad illumination. The determination of the threshold value should therefore always be done in accordance with the individual geometrical characteristics and possible wear of each specimen.

7.3. Resampling the Alpha Point Cloud

In order to calculate a deviation and to apply a registration algorithm, an alpha point cloud must be created from a common CAD file. The algorithms are optimized for the *.ply* file format, which stores the information in the face-vertex format. This is an object representation by listing all vertices, forming a face (face list) and listing all faces; each vertex is referring to (vertex list).

In order to enhance performance and to ensure that the registration algorithm guarantees comparable results for all given measurements with the same set of parameters (see Table 1), the reference alpha point cloud should match the shape of the measured point cloud as well as possible. Therefore, the alpha mesh is sliced into smaller geometries with the rough size of the measurement volume of the measuring head in use. Figure 15a shows an extracted part of the alpha geometry. Figure 15b,c show the stored polygonial and vertex data. In this case, the entire structure is periodical in the radial and axial direction, so the sliced template could be used for all registration operations. The trimming of a template out of the whole reference mesh is automated by using the initial transformation and the position data of the cinematics, each measurement is corresponding to.

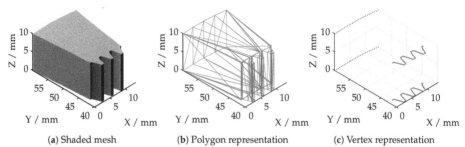

(a) Shaded mesh   (b) Polygon representation   (c) Vertex representation

Figure 15. The stored information in computer-aided design (CAD) reference geometry.

The first step is to trim all vertices and corresponding faces beyond the gear geometries. This is done by applying a radial threshold (50 mm) to the data. To create additional point cloud data, points on each polygon have to be sampled. The total number of sampling points for each polygon is calculated by comparing the area of each polygon with the total area of all polygons and the number of required samples. Since these polygons are triangles, the surface area in 3D space is given by the half cross product formula.

A method for generating unbiased random points with respect to the surface area was introduced by [14] and has been adopted. For a given triangle (with vertices A, B, C) a point on its surface can randomly be constructed by the following equation:

$$P = (1 - \sqrt{r_1})A + \sqrt{r_1}(1 - r_2)B + \sqrt{r_1}r_2C \tag{7}$$

By generating the two random numbers r_1 and r_2 and by taking the square root of r_1, a random point with respect to the surface area is constructed. r_1 sets the percentage from vertex A to the opposing edge $\overline{BC}$, leading to another edge on which r_2 defines the final point P. The sampling of 100,000 points lasted for about half a second. It has been experimentally proven to be important to generate uniformly distributed random numbers. Otherwise, this approach does not produce useful results and sampled points seem to accumulate in one edge of each triangle.

7.4. Distance Metrics and Registration

By the use of a proper distance metric, a point to point distance between two point clouds can be calculated. The Minkowksi distance of order p between two point clouds $X = (\mathbf{x}_1, ..., \mathbf{x}_n)$ and $Y = (\mathbf{y}_1, ..., \mathbf{y}_m)$ is defined as:

$$d_{\mathrm{mk}} = \sqrt[p]{\sum_{i=1}^{n} |\mathbf{x}_i - \mathbf{y}_i|^p} \tag{8}$$

where p is ranging from 1 (City block distance) to ∞ (Chebychev distance). For $p = 2$, the Minkowski distance gives the Euclidean distance, for each pair of points. To get information about the deviation of the measured point cloud to the alpha point cloud, the smallest distance in $d_{\mathrm{mk}} \in \mathbb{R}^{n \times m}$ is obtained by sorting the distances in each column in ascending order (set of closest points) or extracting the smallest distance. As the data sets can be large, this calculation is time intense (~30 min for all stitched point clouds), despite using GPU computation.

The metric is also used for the registration of each point cloud by combining the approach based on the Euclidean distance with a rigid affine transformation to define the mean-squares objective function:

$$\mathcal{F}(R, \mathbf{t}) = \frac{1}{N_p} \sum_{i=1}^{n} ||\mathbf{x}_i - R \cdot \mathbf{y}_i - \mathbf{t}||^2 \tag{9}$$

Let N_p be the number of pairs, representing correspondences and R the rotation matrix and $\mathbf{t}$ the translation vector of the unknown affine transformation. The goal is to find the least squares rotation and translation by using any optimization method, such as steepest descent, conjugate gradient or simplex [15]. More recent approaches aim to reduce computation time by using the much faster quaternion method, introduced by [16,17].

In this case, the iterative closest points (ICP) algorithm from the computer vision system toolbox (mentioned in Section 7) is used for each registration operation. The main parameters are listed in Table 1. Based on the given inlier ratio, specifying the percentage of matched points considered as inliers, the algorithm is estimating the optimal rigid body transformation between both sets of points.

Table 1. Parameterisation of the iterative closest points (ICP) registration algorithm.

Inlier Ratio		0.8
Maximum Number of Iterations		25
Tolerance	R_{diff}	0.005°
	t_{diff}	0.001 mm

While the total number of iterations in each registration operation is limited to 25, the tolerance of absolute difference between consecutive ICP iterations (in translation and rotation) provides an additional stop criterion. In combination with good initial values, the algorithm aborts the calculation based on these termination criteria in most cases. A typical registration operation takes <10 s by calculating 5 iterations on the average. Figure 16 shows a registration operation, where the initially transformed point cloud is fitted, based on the given termination criteria, into the alpha data to calculate a possible deviation later. The results for the first three measurements are shown in Figure 17. The process is repeated for all measured point clouds and the initial transformation matrix is updated after each registration operation.

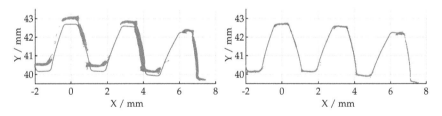

(**a**) Initially transformed point cloud   (**b**) Point cloud after registration

Figure 16. Result of registration operation (blue: measured point cloud, orange: alpha point cloud).

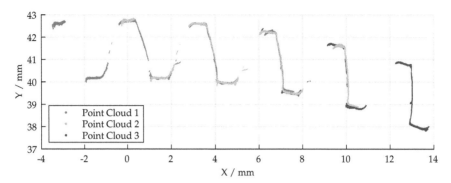

Figure 17. Result of registration approach for the first three point clouds.

8. Results

Figure 18 shows the final point cloud which was created according to the flowchart which is given in Figure 8. Each of the measured 588 point clouds was masked, based on the pixel-wise filtered standard deviation of the phase shift with the highest frequency, denoised and then downsampled to 10 % of the original size. By the use of an initial transformation and the positioning data of the stages, a registration algorithm was able to align each of the measurements by using an alpha point cloud which was resampled from trimmed CAD data. Finally, the data was merged, leading to ~42.1 mio. points. It is noticeable that the results are still fragmentary and gaps are mainly found at the side edges and back edges (originating from the rotation center). This is not involving the front surface on which the major influence of abrasion is expected. The point cloud can be loaded into common CAD software to supply actual geometry data. In order to secure wide compatibility, additional meshing might be required.

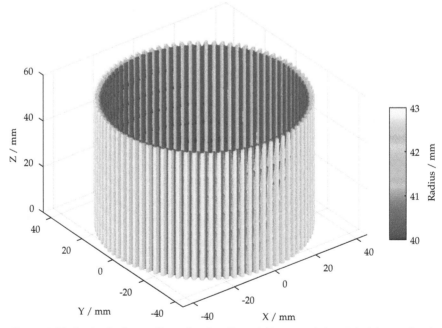

Figure 18. Final point cloud, according to flowchart Figure 8 (downsampled to 10 % of the actual size).

Figure 19 shows the remaining root mean square (RMS) error with respect to the Euclidean metric for all 588 registration operations after the ICP stopped due to its termination criterion (see Section 7.4). The layers correspond to the axial position of the stage. It can be assumed that the first layer is affected by minor drifting effects during registration. Shape deviations can presumably be excluded since all other layers seem to be unaffected. Uniform drifting effects concerning several layers could indicate significant shape deviations. The condition of certain features could still be quantified but the reconstruction of the whole shape would not be appropriate in this case with the given registration approach.

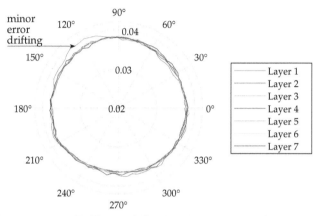

Figure 19. Root mean square (RMS) error of all registration operations with respect to each vertical layer and radial position

In order to make assessments with regard to indications of possible defects, the point cloud from Figure 18 is used to calculate the overall point-wise deviation based on the distance metric and alpha point cloud from Section 7. Despite using GPU acceleration, this last calculation is time-consuming (~30 min). This is due to large datasets. The calculated deviations are color-coded in Figure 20. The major deviations range between 0 and 0.2 mm. Noticeable deviations are mainly found at some back edges, presumably caused due to poor illumination and accumulations of anti-glossy spray. Except that, deviations are spread randomly throughout the entire point cloud. This result is supposed to provide comprehensive data to support maintenance. Furthermore, highly stressed features can be identified in order to enable possible future in situ inspection (introduced in Section 2).

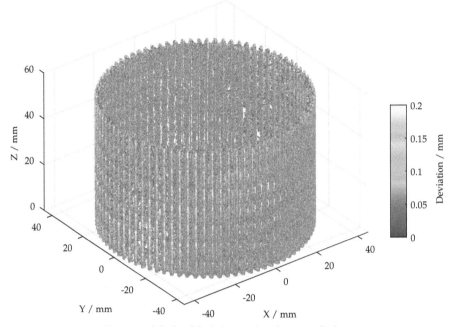

Figure 20. Calculated deviations projected onto a cylinder.

Figure 21b shows the histogram of all deviations. The arithmetic average is 0.0741 mm with a standard deviation of 0.0371 mm. In comparison to this, Figure 21a shows the histogram of one single measurement.

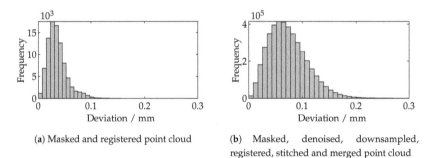

(a) Masked and registered point cloud

(b) Masked, denoised, downsampled, registered, stitched and merged point cloud

Figure 21. Comparison of the deviation histograms of a single measurement and the final point cloud of all 588 measurements.

The arithmetic average is 0.0356 mm with a standard deviation of 0.0188 mm. The greater measurement uncertainty is presumably caused by averaging (due to merging) all minor misalignments throughout all 588 measurements.

9. Discussion

A robust system for time-effective and automated full-field 3D measurements at high resolution for industrial applications has been introduced. The compact and flexible design of the 3D measuring endoscope, based on the endoscopic fringe projection approach, can be used to measure the shape of geometries which remain out of reach for traditional tactile or other measurement principles.

In this application, several hundred single measurements were performed to scan all features of a rotationally symmetric forming tool. Much better stitching can be achieved by using a CAD based registration approach. This is especially relevant when the rotational axes are slightly misaligned and the kinematics and the hand-eye transformation remain uncalibrated and therefore the positing data of the stages cannot be used to precisely stitch all point clouds. Against the background of rapid inspection and heavy and unwieldy specimen, requiring extremely precise alignment, this approach can be considered reasonable. By utilizing and resampling CAD data, the registration can successfully stitch all measured data and calculate the overall deviation. The registration approach shifts the focus in favor of a very detailed examination of local features like gear geometries while reducing the sensitivity to detect shape deviations which should be taken into consideration.

To reduce fragmentation at the back edges, the number of vertical measuring positions should be enlarged. The fragmentation at the side edges is probably caused by the arrangement of the camera and projector optics at the measuring head. A clear improvement could be achieved by adding another projector into the system, arranged on the other side of the camera.

Furthermore, this setup is supposed to enable a much better characterization of abrasion and wear. This can be used to optimize the threshold level, introduced in Section 7.2, setting the ratio between measurement density and measurement uncertainty. This ensures that no defects are skipped while the system is sufficiently sensitive to detect all possible types of defects. Beyond that, the deeper process of understanding can be used to optimize certain process parameters.

Since the original geometrical properties of the specimen are not known exactly, the measurement uncertainty cannot be concluded. A potentially significant source of deviation is given by the layer of applied anti-glossy spray which presumably leads to a more or less constant deviation offset. Nevertheless, despite this first approach, it is expected that future potential damage of the specimen can be identified and measured reliably. A detailed assessment of the systems performance and sensitivity to identify various defects should be made shortly after enough data has been collected and the system has proven to be is suitable for permanent use in various industrial applications.

10. Future Improvements

This initial approach is providing scope for further improvements. It is supposed, that higher uncertainty in contrast to one ideal single measurement appears to be connected to the registration and the merging operation. Whereas the error susceptibility of the registration seems difficult to reduce, merging could be significantly improved. Instead of combining the points in each voxel the filter should keep only the points of good condition in future developments. Therefore, the same map which was originally used to mask each point cloud (see Section 7.2) could be used. Furthermore, the local quality of calibration could be a considerable factor since distortion is a major error source at the edges of the camera image. A selection could be done by comparing the estimated local reprojection error of all points within one voxel. The reprojection error is an important criterion to estimate the calibration quality for each calibrated feature, based on the camera calibration by [18,19].

Measuring and all data processing took about 3 h, which can be significantly accelerated. At first by parallelizing point cloud computation and measurement. Despite GPU support, the most time-consuming step is calculating the final point-wise deviation based on the smallest Euclidean distance of the alpha and measured point cloud. To increase performance, additional voxel filters shall be applied. An alternative approach would be the transformation of both point clouds into cylinder coordinates and the rearrangement of every axial layer in order of ascending angle. Each layer can then be resampled by interpolation according to a given grid. Finally, all points can be projected onto a plane. By subtracting both images (alpha data and measured data), the deviation can be calculated. According to a first assessment, a deviation can be calculated within seconds while harboring minor interpolation errors.

A more precise approach would be the calculation of the deviation based on a point-plane metric by using the polygon data stored in the alpha mesh. This metric could also be used for the registration algorithm superseding the alpha point cloud resampling step.

Author Contributions: Conceptualization, L.H. and M.K.; methodology, L.H.; software, L.H.; validation, L.H., M.K. and E.R.; formal analysis, L.H.; investigation, L.H.; resources, L.H.; data curation, L.H.; writing—original draft preparation, L.H.; writing—review and editing, L.H. and M.K.; visualization, L.H.; supervision, M.K.; project administration, E.R.; funding acquisition, E.R.

Funding: This work was funded by the German Research Foundation (DFG), project B6 "Endoscopic geometry inspection" within the Collaborative Research Center (CRC) / TR 73.

Conflicts of Interest: The authors declare no conflict of interest.

Abbreviations

The following abbreviations are used in this manuscript:

CAD	Computer aided design
CMOS	Complementary metal-oxide-semiconductor
DOF	Depth of field
GPU	Graphics processing unit
GRIN	Gradient index
HDR	High dynamic range
ICP	Iterative closest point
LED	Light-emitting diode
RMS	Root mean square
WD	Working distance

References

1. Weckenmann, A.; Krämer, P.; Hoffmann, J. Manufacturing Metrology—State of the Art and Prospects. In Proceedings of the 9th International Symposium on Measurement and Quality Control, Madras, India, 21–24 November 2007; pp. 1–8.
2. Frankowski, G.; Hainich, R. DLP-Based 3D Metrology by Structured Light or Projected Fringe Technology for Life Sciences and Industrial Metrology. In Proceedings of the SPIE MOEMS-MEMS: Micro- and Nanofabrication, San Jose, CA, USA, 24–29 January 2009.
3. Merklein, M.; Allwood, J.M.; Behrens, B.-A.; Brosius, A.; Hagenah, H.; Kuzman, K.; Mori, K.; Tekkaya, A.E.; Weckenmann, A. Bulk forming of sheet metal. *CIRP Ann.* **2012**, *61*, 725–745. [CrossRef]
4. Gröbel, D.; Schulte, R.; Hildenbrand, P.; Lechner, M.; Engel, U.; Sieczkarek, P.; Wernicke, S.; Gies, S.; Tekkaya, A.E.; Behrens, B.A.; et al. Manufacturing of functional elements by sheet-bulk metal forming processes. *Prod. Eng.* **2016**, , 63–80. [CrossRef]
5. Geng, J.; Xie, J. Review of 3-D endoscopic surface imaging techniques. *IEEE Sens. J.* **2013**, *14*, 945–960. [CrossRef]
6. Sansoni, G.; Trebeschi, M.; Docchio, F. State-of-The-Art and Applications of 3D Imaging Sensors in Industry, Cultural Heritage, Medicine, and Criminal Investigation. *Sensors* **2009**, *9*, 568–601. [CrossRef] [PubMed]
7. Matthias, S.; Loderer, A.; Koch, S.; Gröne, M.; Kästner, M.; Hübner, S.; Krimm, R.; Reithmeier, E.; Hausotte, T.; Behrens, B.-A. Metrological solutions for an adapted inspection of parts and tools of a sheet-bulk metal forming process. *Prod. Eng.* **2016**, *10*, 51–61. [CrossRef]
8. Matthias, S.; Kästner, M.; Reithmeier, E. Comparison of LASER and LED illumination for fiber optic fringe projection. In Proceedings of the SPIE Photonics Europe, Brussels, Belgium, 4–7 April 2016.
9. Matthias, S.; Schlobohm, J.; Kästner, M.; Reithmeier, E. Fringe projection profilometry using rigid and flexible endoscopes. *tm Tech. Mess.* **2017**, *84*, 123–129. [CrossRef]
10. Peng, T.; Gupta, S.K. Algorithms for Generating Adaptive Projection Patterns for 3D Shape Measurement. *J. Comput. Inf. Sci. Eng.* **2008**, *8*, 031009. [CrossRef]
11. Rusu, R.B.; Marton, Z.C.; Blodow, N.; Dolha, M.; Beetz, M. Towards 3D Point cloud based object maps for household environments. *Robot. Auton. Syst.* **2008**, *56*, 927–941. [CrossRef]

12. Besl, P.J.; McKay, N.D. A method for registration of 3-D shapes. In Proceedings of the 1991 Robotics, Boston, MA, USA, 14–15 November 1991; pp. 586–607.
13. Kabsch, W. A solution for the best rotation to relate two sets of vectors. *Acta Crystallogr.* **1976**, *32*, 922–923. [CrossRef]
14. Osada, R.; Funkhouser, T.; Chazelle, B.; Dobkin, D. Shape Distributions. *ACM Trans. Graph.* **2002**, *21*, 807–832. [CrossRef]
15. Zhang, Z. Iterative Point Matching for Registration of Free-Form Curves and Surfaces. *Int. J. Comput. Vis.* **1994**, *13*, 119–152. [CrossRef]
16. Faugeras, O.D.; Herbert, M. The Representation, Recognition, and Locating of 3-D Objects. *Int. J. Robot. Res.* **1986**, *5*, 27–52. [CrossRef]
17. Horn, B. Closed-form solution of absolute orientation using unit quaternions. *Int. J. Opt. Soc. Am. A* **1994**, *4*, 629–642. [CrossRef]
18. Zhang, Z. A Flexible New Technique for Camera Calibration. *IEEE Trans. Pattern Anal. Mach. Intell.* **2000**, *22*, 1330–1334. [CrossRef]
19. Heikkilä, J.; Silvén, O. A four-step camera calibration procedure with implicit image correction. In Proceedings of the IEEE International Conference on Computer Vision and Pattern Recognition, San Juan, Puerto Rico, 17–19 June 1997.

Article

A High Sensitivity Temperature Sensing Probe Based on Microfiber Fabry-Perot Interference

Zhoubing Li [1], Yue Zhang [1], Chunqiao Ren [1], Zhengqi Sui [1] and Jin Li [1,2,*]

[1] College of Information Science and Engineering, Northeastern University, Shenyang 110819, China; lizhoubing50@163.com (Z.L.); 20174113@stu.neu.edu.cn (Y.Z.); Rencq12138@163.com (C.R.); 15942415413@163.com (Z.S.)

[2] State Key Laboratory of Synthetical Automation for Process Industries, Northeastern University, Shenyang 110819, China

* Correspondence: lijin@ise.neu.edu.cn

Received: 27 March 2019; Accepted: 15 April 2019; Published: 16 April 2019

Abstract: In this paper, a miniature Fabry-Perot temperature probe was designed by using polydimethylsiloxane (PDMS) to encapsulate a microfiber in one cut of hollow core fiber (HCF). The microfiber tip and a common single mode fiber (SMF) end were used as the two reflectors of the Fabry-Perot interferometer. The temperature sensing performance was experimentally demonstrated with a sensitivity of 11.86 nm/°C and an excellent linear fitting in the range of 43–50 °C. This high sensitivity depends on the large thermal-expansion coefficient of PDMS. This temperature sensor can operate no higher than 200 °C limiting by the physicochemical properties of PDMS. The low cost, fast fabrication process, compact structure and outstanding resolution of less than 10^{-4} °C enable it being as a promising candidate for exploring the temperature monitor or controller with ultra-high sensitivity and precision.

Keywords: fiber sensors; temperature sensors; Fabry-Perot interferometer; microfiber; PDMS; integrated optics

1. Introduction

As a typical physical parameter, the temperature must be carefully controlled and monitored in many fields, such as clinical medicine, biochemical reactions, industrial production, aviation safety and so on [1–3]. In recent years, optical fiber temperature sensors have aroused widespread research interest, because of their unique advantages compared with electrical ones, such as remote monitoring capability, high sensitivity, anti-electromagnetic interference properties, and intrinsic safety [4,5]. By combining the resonance enhancement effect of the optical coupling technique, multi-modes interference, optical evanescent field, optical time domain reflecting and optical ring-down technology produced by different special optical fiber structures, various optical fiber temperature sensors were realized [6–9]. Multi-modes interference is carried out by splicing together different kinds of fibers to excite the modes' interference. The splicing joints are fragile and the length for each section must be carefully controlled during the fabrication process; Optical evanescent fields can be obtained around micro/nanofibers with diameters comparable to the wavelength of the incident light. Although micro/nanofibers offer excellent performance, the sensor probes based on them are difficult to fabricate because of their thin diameter and environmentally sensitive properties. The optical time domain reflection technique was used to sense temperature and strain based on Raman or Brillouin scattering [10]. The sensitivity of the temperature sensor based on optical ring-down technology only can be increased by extending the fiber length.

In addition to the basic sensing mechanism, the sensing performance was further improved by means of temperature sensitive materials [11]. Many materials, such as polymers and metal oxides, have been reported to be elaborated by surface or inner coating, and used to encapsulate the whole fiber structure [12–14]. In addition to the temperature dependence, the effect of humidity, strain and other related parameters on the sensing performance must be determined and eliminated. At present, the most common commercial optical fiber temperature sensor is the fiber Bragg grating (FBG) having good repeatability and stable sensing characteristics [15]. It can be prepared by ultraviolet exposure or nano-etching technology to meet the working requirements of different temperature ranges [16]. However, high sensitivity or precision is difficult to obtain for FBG temperature sensors, which seriously hinders their commercial application [17].

The optical fiber temperature sensors based on multi-wavelength interference mainly include Mach–Zehnder and Fabry-Perot interferometers. The former typically perform as transmission structures, which separate and transmit an independent signal light and reference light by using different special optical fibers or structures, such as micro/nano fibers, photonic crystal fibers (PCFs), dislocation fusion fibers and multi-core fibers. A corresponding sensitivity of up to 6.5 nm/°C was observed [18]. However, the structures of the Mach-Zehnder interferometers are complex due to their dual-optical-paths system [19–22]. To simplify the structures, the two optical paths can be revealed in single fiber, named the in-line Mach-Zehnder interferometer, such as C-typed PCFs [23], side-hole PCFs [24], D-shaped-hole fibers [25] and muti-core fibers [26]. These compact structures were precisely machined using femtosecond lasers, focused ion beams and chemical vapor deposition, and display excellent stability and sensing performance. However, these are hard to manufacture in batches due to the high cost and technical requirements. In addition to the above complex optical fiber structures, single polymer optical fibers have been demonstrated with a temperature sensitivity of ~10^{-3} °C [27], where the temperature performance were revealed by the transmission power and the effect of relative and twist have been experimentally obtained [28,29]. Furthermore, their packaging size is hard to reduce further depending on the bending loss of the optical fiber [30], which will seriously limit their application in a narrow space; the latter ones are carried out as reflective structures, where the temperature sensitive cavity was constructed at the end of the optical fiber by laser or ion beam processing, chemical etching or film forming and special fiber splicing technologies [31–37]. Among them, femtosecond laser processing can machine a refractive index turning point with good repeatability in the optical fiber, which was used as a Fabry-Perot cavity and can work at high temperatures up to 1000 °C [31]; focused ion beams can etch an air cavity at the tip of an optical fiber, based which a Fabry-Perot temperature sensor with a sensitivity of −654 pm/°C has been experimentally demonstrated [32]. However, the expensive and complex preparation processing, as well as the high technique requirements for engineers have become huge obstacles for commercial production [33]. The Fabry-Perot interferometer probe can be obtained conveniently and quickly by chemical etching or film forming technology [34], however, the fabrication repeatability is low, and the structural parameters are difficult to control accurately [35]. By using the special hollow-core photonic bandgap fiber (HC-PBF) or PCF, the temperature working range and sensitivity of cascaded splicing fiber based Fabry-Perot interferometer has been experimentally verified as high as 1200 °C and 17 nm/°C, respectively, but their structures are relatively fragile [36,37].

Compared with conventional temperature sensors, the proposed Fabry-Perot interferometer temperature sensor costs less and is easier and faster to prepare. This compact Fabry-Perot temperature probe was proposed by encapsulating a microfiber and a single mode fiber (SMF) tip in a hollow core fiber (HCF), between which temperature sensitive polydimethylsiloxane (PDMS) was filled and cured. The microfiber was prepared by the one-step heating-stretching technique from a normal SMF. The microfiber and SMF can be easily aligned due to the comparable inner diameter of HCFs. The high transparency and low refractive index of PDMS causes little impact on the incident light. Furthermore, a sensitivity of higher than 11 nm/°C has been experimentally demonstrated due to its high thermal expansion coefficient. This temperature sensor will be a promising candidate for monitoring temperature fluctuations in small spaces due to its high sensitivity and tiny scale (200 μm in diameter and <5 mm in length).

2. Materials and Methods

To fabricate the Fabry-Perot interferometer, a cut of transparent HCF was prepared firstly, as shown in Figure 1. The coating layer of a HCF (TSP150200, inner diameter: ~150 μm, outer diameter: ~200 μm, coating layer: polyimide, Polymicro Technologies, Inc., Phoenix, AZ, USA) was removed by a Bunsen burner (Dragon 200, fuel: butane, max-temperature 1300 °C, Rocker Scientific Co., Ltd., New Taipei, Taiwan), as shown in inset (a) of Figure 1. The cavity length of Fabry-Perot interferometer can be observed through its transparent wall. Both the microfiber and SMF can be inserted and aligned easily due to their small diameter difference. The microfiber was obtained from the SMF (Coating removed diameter: 125 μm, SMF-28, Corning Inc., Corning, NY, USA) using the scanning flame heating-stretching technique (inset (b) of Figure 1). Where, the diameter and length of microfiber were precisely controlled by optimizing the fabrication process of a fiber melting-drawing system (IPCS-5000-ST, Idealphotonics Inc., Hong Kong, China). This system uses the high-purity hydrogen and oxygen as the fuel to obtain a high heating temperature of up to 2500–3000 °C. When SMF reaches a melting state at high temperature, its two ends were fixed onto two motorized displacement platforms and stretched in opposite directions. By carefully adjusting the speed and scanning region of the flame, the microfiber with uniform diameter can be obtained in the heating zone. Different diameters were easily achieved by controlling the stretching velocity. Fabry-Perot interferometer was finally fabricated by assistance of a homemade micromanipulation system (inset (c) of Figure 1). A cut of transparent HCF was fixed on a slide glass substrate with UV glue. One end of SMF and microfiber were cut with a flat-face and acted as two reflecting surfaces of Fabry-Perot interferometer. The other tail-ends of SMF and microfiber were clamped by two fiber claps and fixed onto two three-dimensional (3-D) optical fiber adjusting frames (APFP-XYZ, adjusting precision <2 μm, Zolix Instruments Co., Ltd., Beijing, China).

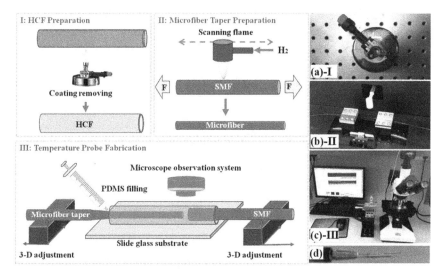

Figure 1. Fabrication process of the microfiber and PDMS based Fabry-Perot temperature probe. I: Coating layer of HCF was removed to prepare the transparent HCF (inset (**a**)); II: MF taper was prepared by scanning flame stretching technique (inset (**b**)); III: Fabry-Perot temperature probe was fabricated by assistance of the micromanipulation method under a microscope (insets (**c**) & (**d**)).

In this case, the Fabry-Perot structure can be timely observed and measured by a microscope system (DMM-300C, Shanghai Caikon Optical Instrument Co., Ltd., Shanghai, China) on a computer and its cavity length was also timely precisely manipulated according to the reflected spectrum. The basic component and curing agent were mixed with a weight ratio of 10:1 to obtain the PDMS sol, which was filled into the HCF using a syringe (inset (d) of Figure 1) and cured in ~20 min. The experimental schematic was illustrated in Figure 2.

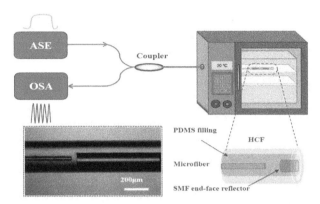

Figure 2. Experimental schematic of the microfiber and PDMS-based Fabry-Perot interferometer for sensing temperature. The light source, temperature sensor and spectrometer were contacted by a 1×2 coupler. The enlarged schematic and microscope picture of the temperature probe were illustrated.

An amplified stimulated emission (ASE, ASE-C light source, 1520–1610 nm, Shenzhen Golight Technology Co., Ltd., Shenzhen, China) was used as the light source. The light was launched into the Fabry-Perot interferometer temperature probe through a 1×2 coupler (with the splitter ratio of 50:50). The reflection optical signal was collected by an optical spectrum analyzer (OSA, AQ6370, 600–1700 nm, resolution 20 pm, Yokogawa Electric Corp., Tokyo, Japan). In this work, the optical polarization direction does not affect the sensing performance due to the circularly polarized light output of ASE source and the cylindrical structure of microfiber. The temperature probe was placed in a thermostat (25–250 °C, resolution 0.1 °C, Shanghai Boxun Medical Biological Instruments Co., Ltd., Shanghai, China). The inset is a micrograph of the proposed temperature probe. The microfiber has a uniform diameter of ~63 μm and a length of ~2 cm. This length should be carefully controlled depending on the cone angle of the microfiber taper. Too long and thin microfiber will be easily adsorbed on the inner surface of HCF because of van der Waals force. In this case, it will be difficult to parallel its end-face with the reflecting surface of SMF to construct the two reflectors of Fabry-Perot interferometer. The cavity length was finally determined as ~34 μm.

3. Results

In the experiment, the Fabry-Perot interferometer temperature probe was placed in a thermostat. The temperature was increased from room temperature to 100 °C with steps of 1 °C. The reflection spectra of the Fabry-Perot temperature probe were recorded by a spectrometer. The printed pictures of the spectrometer screen at different temperature (40 °C and 41 °C) are illustrated in Figure 3. The spectrum curve refers to the original spectrum reflected from the microfiber. The free spectral range (FSR) is ~21 nm, during which the wavelength values of the resonance dips were determined with demodulation equipment with the resolution of 1 pm. When temperature changed from 40 °C (Figure 3a) to 41 °C (Figure 3b), the resonance dip moved with a wavelength location shift of ~10.5 nm, indicating a resolution of lower than 10^{-4} °C. Due to the limitation of one period of FSR, the ultra-sensitive temperature fluctuation monitoring can be achieved in the maximum range of 0 °C to ~2 °C (fluctuation level: ±1 °C). In order to achieve a commercial low-cost device, a photodiode can be used to monitor the change in intensity of a single wavelength to determine the direction and magnitude of temperature fluctuations.

Sensors **2019**, *19*, 1819

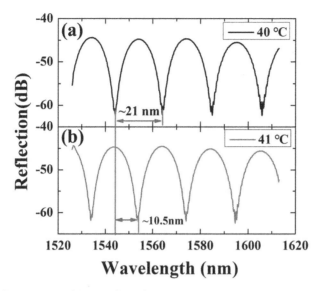

Figure 3. Reflection spectra of the microfiber Fabry-Perot temperature probe at temperature of (**a**) 40 °C and (**b**) 41 °C, respectively.

The sensitivity of Fabry-Perot interferometer is dependent on the resonance shift as a function of temperature [38]:

$$s = \frac{\Delta \lambda_T}{\Delta T} = \lambda_m \alpha_{FP} = \lambda_m \left(\frac{L_{PDMS}\alpha_{PDMS,T} - L_{fiber}\alpha_{fiber,T}}{L_{PDMS}} \right) \quad (1)$$

where, α_{FP} refers to the relative change for the cavity length of the Fabry-Perot interferometer. In this work, it is depended on both the thermal expansion of PDMS and silica fibers. The FSR can be expressed as:

$$FSR = \frac{\lambda^2}{2n_{PDMS}L_{PDMS}} \quad (2)$$

In addition to the incident wavelength, FSR is inversely proportional to the change in refractive index and length of PDMS, which are determined by its thermo-optic coefficient and thermal expansion coefficient, respectively. Here, the thermal expansion coefficient plays a dominant role in the temperature change process, since the bulk expansion of PDMS is limited by the HCF wall and transferred into a change in cavity length to improve the sensitivity of the sensor. For the proposed Fabry-Perot interferometer, FSR is inversely proportional to the spacing between the microfiber and SMF tips, which was demonstrated experimentally when we continuously moved the microfiber towards the SMF in the HCF.

To demonstrate the temperature sensing performance in a wider range, the wavelength movement of one resonance dip was marked and traced in the whole spectrum range of the light source from 1520 nm to 1610 nm, as shown in Figure 4. When the temperature increased from 43 °C to 50 °C with steps of 1 °C, a resonance dip was marked to trace its shift amount. The inset of Figure 4 illustrates eight reflection spectra recorded at the different temperature values, where the resonance dip is red-shifted for almost a half cycle of the FSR. This resonance dip was independently selected to clearly display the temperature sensing characteristics from 43 °C to 50 °C. The resonance dip red-shifted continuously from 1534.8 nm (43 °C) to 1607.3 nm (50 °C).

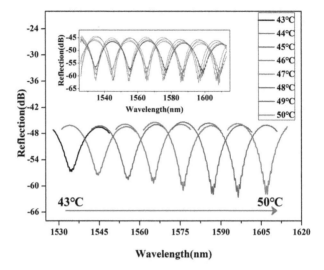

Figure 4. Reflection spectra of the microfiber Fabry-Perot temperature probe for the temperature increased from 43 °C to 50 °C with a step of 1 °C.

The corresponding temperature sensing characteristic curve was shown in Figure 5. Black (circle) and red (pentagon) experimental data points and corresponding linear fitting represent the temperature response results for heating and cooling process, respectively.

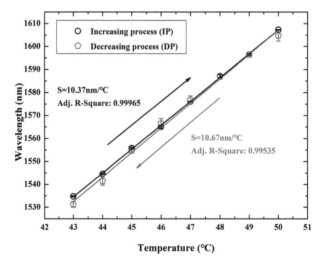

Figure 5. Location of resonance dip changed as a function of temperature with an excellent linear fitting during the increasing and decreasing process of 43–50 °C. The corresponding sensitivities were determined to be 10.37 nm/°C and 10.67 nm/°C, respectively.

During the heating process, a sensitivity of up to 10.37 nm/°C for the temperature sensing was experimentally demonstrated with a linearity of 0.99965. To verify the recovery characteristics of the temperature sensor, the movement of resonance dip was recorded through the cooling process in the same temperature range (from 50 °C to 43 °C in steps of 1 °C). By linearly fitting the experimental data points, a sensitivity of up to 10.67 nm/°C was obtained with a linearity of 0.99535. The performance curve illustrates the relationship between resonance dip and temperature, which will be stable for a temperature probe with fixed structure parameters. When a new sensor is used, the temperature can be determined by referring to calibration curve.

To reveal the repeatability and stability of the proposed temperature sensor, three-cycle experiments for a sensing probe with the cavity length of 31 μm and the microfiber diameter of 61 μm were performed, where the corresponding wavelength shift values depending on the temperature increasing/decreasing were recorded and illustrated in Figure 6. The highest sensitivity of 11.86 nm/°C was experimentally demonstrated for the temperature increase process in the first round, which was higher than the probe in Figure 5 mainly due to the shorter cavity, which matches well with the theoretical analysis. Equation (1) indicates that the sensitivity is proportional to the relative change in cavity length. A shorter cavity will result in a more significant change than that of the longer one. Furthermore, its larger FSR enables high-precise temperature fluctuation monitoring in a wider range (see the analysis of Equation (2) and Figure 3).

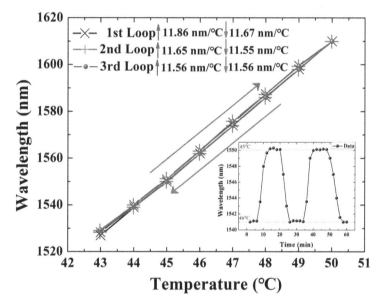

Figure 6. Three-cycle experiments for the temperature-dependence curves for a Fabry-Perot interferometer with the cavity length of 31 μm and the microfiber diameter of 61 μm. Inset: Stability of the wavelength locations for the temperature changing between 45 °C and 46 °C.

The maximum wavelength backlash was determined as ~1.3 nm during the three-cycle measurement process. On the one hand, this is related to the thermal expansion relaxation time of PDMS; on the other hand, it is also limited by the temperature control accuracy of the oven, which is also indicated in the stability measurement of the proposed temperature probe (inset of Figure 6). When the temperature fluctuates between 45 °C and 46 °C, the positional fluctuation of the resonance wavelength was less than ~0.2 nm, and the corresponding response time (stabilization time) was ~3 min. The above fluctuations fall within the performance range of the thermostatic oven. In order to calibrate the sensing characteristics of this temperature probe in a larger working range, the specific resonance dips should be dynamically selected in different temperature ranges. Thereafter, the temperature sensing characteristic curve can be obtained by using the relative shift of the labeled resonance dips. In addition to the microfiber and SMF, the final working range of this temperature probe will be limited by the sensitive materials. PDMS in solid status has the stable physicochemical property in the temperature range of −55–200 °C. Therefore, this temperature probe can work in a wider range, not limited to the results reported in this work.

4. Discussion

The optical fiber temperature sensor proposed in this work is compact and easily prepared. Its sensitivity is significantly higher than most of other fiber temperature sensors reported in recent years, as compared in Table 1.

Table 1. Sensing performance comparison for typical temperature probes based on optical fibers.

Mechanism	Structure	Sensitivity	Range	Reference
Grating interference	Copper tube/FBG	27.6 pm/°C	0–35 °C	[16]
	FBG	18.8 pm/°C	20–90 °C	[17]
Mach-Zehnder interference	SMS/Microfiber	6.5 nm/°C	51–65 °C	[18]
	Micro-bend fiber	1.92×10^{-3}/°C	29–52 °C	[19]
	SMS/Liquid	−1.88 nm/°C	0–80 °C	[21]
	Liquid cored PCF	−2.15 nm/°C	20–80 °C	[6]
	Liquid-filled PCF	−1.83 nm/°C	23–58 °C	[20]
	C-typed PCF	−7.609 nm/°C	15–30 °C	[23]
	NOA 73/PMMA	−431 pm/°C	25–75 °C	[22]
	PMMA	1.04×10^{-3}/°C	25–120 °C	[27]
	Abrupt tapered fiber	0.0833 dBm/°C	30–50 °C	[30]
Fabry-Perot interference	Single RI turning dot	13.9 pm/°C	100–500 °C	[31]
		18.6 pm/°C	500–1000 °C	
	Open microcavity	−654 pm/°C	30–120 °C	[32]
	HC-PBF/HCF splicing	17 nm/°C	100–800 °C	[36]
	SMF/PCF splicing	15.61 pm/°C	300–1200 °C	[37]
	LOCTITE 3493 film	~5.2 nm/°C	15–22 °C	[35]
	Microfiber taper	1.97 pm/°C	50–150 °C	[38]
	Nafion film	2.71 nm/°C	−15–65 °C	[39]
	Microfiber/SMF/PDMS	10.67 nm/°C	43–50 °C	This work

SMS: Single-muti-single mode fiber; NOA 73: Norland optical adhesive 73; RI: Refractive index; PMMA: poly(methyl methacrylate; LOCTITE 3493: Light cure adhesive 3493.

As can be seen from Table 1, the temperature sensitivity of FBGs is low, and the encapsulation technology and demodulation optical path are complex [16]. The dual-arms system of Mach-Zehnder interferometers are commonly built using special optical fibers (for example PCF [6,20] or microfibers [18,19,30]) or by splicing different optical fibers [18,21], where the sensitive liquid or polymer were introduced to create a temperature-sensitive probe [6,20,21]. In contrast, the Fabry-Perot fiber interferometer can be easily fabricated on a single fiber. It has a more compact structure for developing high-performance temperature microprobes. Femtosecond laser [31] or ion beam etching technology [32], as well as high-precision fiber-splicing technology [36,37], can improve its temperature detection limit to as high as 1200 °C, making it suitable for extreme high temperature environments; furthermore, sol coating [35] or temperature-sensitive polymer encapsulation technology [39] can be used for enhance normal temperature microprobes, which will be a promising candidate for implantable microsensors for health or environmental monitoring under 200 °C.

Compared with the polymer film reflector, in this work, the smooth end-faces of SMF and microfiber were used as the two reflectors of the Fabry-Perot interferometer. PDMS is used to fix the two reflectors and realize a highly sensitive response to temperature changing. The temperature response properties can be revealed by the contribution of the negative thermal-optics coefficient (α_{to}: -450×10^{-6}/°C) and the thermal-expansion coefficient (α_{te}: 960×10^{-6}/°C) effects of PDMS. When the temperature increases, a smaller effective refractive index and a longer cavity length will be obtained, respectively. In view of their contributions to the effective optical path between the two reflectors of the Fabry-Perot interferometer, they have the opposite impact on the cavity length when the temperature changes.

5. Conclusions

In this paper, a compact and miniature Fabry-Perot interferometer based on a microfiber and SMF in a cut of HCF was proposed and experimentally demonstrated. The morphology parameters, such as microfiber diameter and cavity length, can be precisely controlled by the microfiber fabrication (scanning flame stretching technique) and micromanipulation processes (microscope- assised micromanipulation method), respectively. By filling PDMS into this Fabry-Perot interferometer with the microfiber diameter of ~63 μm and cavity length of ~34 μm, a temperature sensitivity of higher than 10 nm/°C was experimentally obtained. When the cavity length was reduced to ~31 μm, a highest sensitivity of 11.86 nm/°C has been experimentally demonstrated with an excellent repeatability and stability. Due to its high sensitivity and easily adjustable morphology, this Fabry-Perot temperature sensor has promising applications for precisely monitoring temperature fluctuations in biochemical reaction processes, industrial production and food storage.

Author Contributions: Z.L. and Y.Z. conceived and performed the experiments; they and C.R. analyzed the results and was involved in the paper writing; Z.S. was involved in the paper writing; J.L. was the supervisor and involved in the paper review and editing.

Funding: This research was funded by Fundamental Research Funds for the Central Universities (N170405003 and N170407005), Liaoning Province Natural Science Foundation (20180510015) and National Undergraduate Innovation and Entrepreneurship Program (190158).

Conflicts of Interest: The authors declare no conflict of interest.

References

1. Fajkus, M.; Nedoma, J.; Martinek, R.; Vasinek, V.; Nazeran, H.; Siska, P. A non-invasive multichannel hybrid fiber-optic sensor system for vital sign monitoring. *Sensors* **2017**, *17*, 111. [CrossRef] [PubMed]
2. Fan, R.; Mu, Z.; Li, J. Miniature temperature sensor based on polymer-packaged silica microfiber resonator. *J. Phys. Chem. Solids* **2019**, *129*, 307–311. [CrossRef]
3. Mamidi, V.R.; Kamineni, S.; Ravinuthala, L.S.P.; Thumu, V.; Pachava, V.R. Characterization of encapsulating materials for fiber Bragg grating-based temperature sensors. *Fiber Integr. Opt.* **2014**, *33*, 325–335. [CrossRef]
4. Ghazanfari, A.; Li, W.; Leu, M.C.; Zhuang, Y.; Huang, J. Advanced ceramic components with embedded sapphire optical fiber sensors for high temperature applications. *Mater. Des.* **2016**, *112*, 197–206. [CrossRef]
5. Joe, H.E.; Yun, H.; Jo, S.H.; Jun, M.B.; Min, B.K. A review on optical fiber sensors for environmental monitoring. *Int. J. Precis. Eng. Mannuf.-Green Technol.* **2018**, *5*, 173–191. [CrossRef]
6. Liu, Q.; Li, S.G.; Chen, H.L.; Li, J.S.; Fan, Z.K. High-sensitivity plasmonic temperature sensor based on photonic crystal fiber coated with nanoscale gold film. *Appl. Phys. Express* **2015**, *8*, 046701. [CrossRef]
7. Lou, J.Y.; Wang, Y.P.; Tong, L.M. Microfiber optical sensors: A review. *Sensors* **2014**, *14*, 5823–5844. [CrossRef] [PubMed]
8. Castano, L.M.; Flatau, A.B. Smart fabric sensors and e-textile technologies: A review. *Smart Mater. Struct.* **2014**, *23*, 053001. [CrossRef]
9. Bao, Y.; Huang, Y.; Hoehler, M.S.; Chen, G. Review of fiber optic sensors for structural fire engineering. *Sensors* **2019**, *19*, 877. [CrossRef]
10. Liu, Y.; Ma, L.; Yang, C.; Tong, W.; He, Z. Long-range Raman distributed temperature sensor with high spatial and temperature resolution using graded-index few-mode fiber. *Opt. Express* **2018**, *26*, 20562–20571. [CrossRef]
11. Urrutia, A.; Goicoechea, J.; Ricchiuti, A.L.; Barrera, D.; Sales, S.; Arregui, F.J. Simultaneous measurement of humidity and temperature based on a partially coated optical fiber long period grating. *Sens. Actuator B Chem.* **2018**, *227*, 135–141. [CrossRef]
12. Hernández-Romano, I.; Cruz-Garcia, M.A.; Moreno-Hernández, C.; Monzón-Hernández, D.; López-Figueroa, E.O.; Paredes-Gallardo, O.E.; Torres-Cisneros, M.; Villatoro, J. Optical fiber temperature sensor based on a microcavity with polymer overlay. *Opt. Express* **2016**, *24*, 5654–5661. [CrossRef]

13. Sun, H.; Luo, H.; Wu, X.; Liang, L.; Wang, Y.; Ma, X.; Zhang, J.; Hu, M.; Qiao, X. Spectrum ameliorative optical fiber temperature sensor based on hollow-core fiber and inner zinc oxide film. *Sens. Actuators B Chem.* **2017**, *245*, 423–427. [CrossRef]

14. Ramakrishnan, M.; Rajan, G.; Semenova, Y.; Farrell, G. Overview of fiber optic sensor technologies for strain/temperature sensing applications in composite materials. *Sensors* **2016**, *16*, 99. [CrossRef]

15. Hong, C.Y.; Zhang, Y.F.; Zhang, M.X.; Leung, L.M.G.; Liu, L.Q. Application of FBG sensors for geotechnical health monitoring, a review of sensor design, implementation methods and packaging techniques. *Sens. Actuators A Phys.* **2016**, *244*, 184–197. [CrossRef]

16. Zhang, D.P.; Wang, J.; Wang, Y.J.; Dai, X. A fast response temperature sensor based on fiber Bragg grating. *Meas. Sci. Technol.* **2014**, *25*, 075105. [CrossRef]

17. Gonzalez-Reyna, M.A.; Alvarado-Mendez, E.; Estudillo-Ayala, J.M.; Vargas-Rodriguez, E.; Sosa-Morales, M.E.; Sierra-Hernandez, J.M.; Jauregui-Vazquez, D.; Rojas-Laguna, R. Laser temperature sensor based on a fiber Bragg grating. *IEEE Photonics Technol. Lett.* **2015**, *27*, 1141–1144. [CrossRef]

18. Zhang, Y.J.; Tian, X.J.; Xue, L.L.; Zhang, Q.J.; Yang, L.; Zhu, B. Super-high sensitivity of fiber temperature sensor based on leaky-mode bent SMS structure. *IEEE Photonic Technol. Lett.* **2013**, *25*, 560–563. [CrossRef]

19. Moraleda, A.T.; García, C.V.; Zaballa, J.Z.; Arrue, J. A temperature sensor based on a polymer optical fiber macro-bend. *Sensors* **2013**, *13*, 13076–13089. [CrossRef]

20. Geng, Y.F.; Li, X.J.; Tan, X.L.; Deng, Y.L.; Hong, X.M. Compact and ultrasensitive temperature sensor with a fully liquid-filled photonic crystal fiber Mach–Zehnder interferometer. *IEEE Sens. J.* **2014**, *14*, 167–170. [CrossRef]

21. Silva, S.; Pachon, E.G.; Franco, M.A.; Hayashi, J.G.; Malcata, F.X.; Frazão, O.; Jorge, P.; Cordeiro, C.M. Ultrahigh-sensitivity temperature fiber sensor based on multimode interference. *Appl. Opt.* **2012**, *51*, 3236–3242. [CrossRef] [PubMed]

22. Niu, D.H.; Wang, X.B.; Sun, S.Q.; Jiang, M.H.; Xu, Q.; Wang, F.; Wu, Y.D.; Zhang, D.M. Polymer/silica hybrid waveguide temperature sensor based on asymmetric Mach–Zehnder interferometer. *J. Opt.* **2018**, *20*, 045803. [CrossRef]

23. Zhao, Y.; Wu, Q.L.; Zhang, Y.N. Theoretical analysis of high-sensitive seawater temperature and salinity measurement based on C-type micro-structured fiber. *Sens. Actuators B Chem.* **2018**, *258*, 822–828. [CrossRef]

24. Reyes-Vera, E.; Cordeiro, C.M.; Torres, P. Highly sensitive temperature sensor using a Sagnac loop interferometer based on a side-hole photonic crystal fiber filled with metal. *Appl. Opt.* **2017**, *56*, 156–162. [CrossRef]

25. Weng, S.; Pei, L.; Wang, J.; Ning, T.; Li, J. High sensitivity D-shaped hole fiber temperature sensor based on surface plasmon resonance with liquid filling. *Photonics Res.* **2017**, *5*, 103–107. [CrossRef]

26. Wales, M.D.; Clark, P.; Thompson, K.; Wilson, Z.; Wilson, J.; Adams, C. Multicore fiber temperature sensor with fast response times. *OSA Contin.* **2018**, *1*, 764–771. [CrossRef]

27. Leal-Junior, A.; Frizera-Neto, A.; Marques, C.; Pontes, M.J. A polymer optical fiber temperature sensor based on material features. *Sensors* **2018**, *18*, 301. [CrossRef]

28. Leal-Junior, A.; Frizera-Neto, A.; Marques, C.; Pontes, M. Measurement of temperature and relative humidity with polymer optical fiber sensors based on the induced stress-optic effect. *Sensors* **2018**, *18*, 916. [CrossRef]

29. Leal-Junior, A.; Frizera, A.; Marques, C.; Pontes, M.J. Polymer-optical-fiber-based sensor system for simultaneous measurement of angle and temperature. *Appl. Opt.* **2018**, *57*, 1717–1723. [CrossRef] [PubMed]

30. Raji, Y.M.; Lin, H.S.; Ibrahim, S.A.; Mokhtar, M.R.; Yusoff, Z. Intensity-modulated abrupt tapered fiber Mach-Zehnder interferometer for the simultaneous sensing of temperature and curvature. *Opt. Laser Technol.* **2016**, *86*, 8–13. [CrossRef]

31. Chen, P.C.; Shu, X.W. Refractive-index-modified-dot Fabry-Perot fiber probe fabricated by femtosecond laser for high-temperature sensing. *Opt. Express* **2018**, *26*, 5292–5299. [CrossRef]

32. Gomes, A.D.; Becker, M.; Dellith, J.; Zibaii, M.I.; Latifi, H.; Rothhardt, M.; Bartelt, H.; Frazão, O. Multimode Fabry–Perot interferometer probe based on Vernier effect for enhanced temperature sensing. *Sensors* **2019**, *19*, 453. [CrossRef] [PubMed]

33. Poeggel, S.; Duraibabu, D.; Lacraz, A.; Kalli, K.; Tosi, D.; Leen, G.; Lewis, E. Femtosecond-laser-based inscription technique for post-fiber-Bragg grating inscription in an extrinsic Fabry–Perot interferometer pressure sensor. *IEEE Sens. J.* **2016**, *16*, 3396–3402. [CrossRef]

34. Wang, R.H.; Qiao, X.G. Intrinsic Fabry-Pérot interferometer based on concave well on fiber end. *IEEE Photonic Technol. Lett.* **2014**, *26*, 1430–1433. [CrossRef]

35. Zhang, G.L.; Yang, M.H.; Wang, M. Large temperature sensitivity of fiber-optic extrinsic Fabry–Perot interferometer based on polymer-filled glass capillary. *Opt. Fiber Technol.* **2013**, *19*, 618–622. [CrossRef]

36. Zhang, Z.; He, J.; Du, B.; Zhang, F.; Guo, K.; Wang, Y. Measurement of high pressure and high temperature using a dual-cavity Fabry–Perot interferometer created in cascade hollow-core fibers. *Opt. Lett.* **2018**, *43*, 6009–6012. [CrossRef]

37. Yu, H.H.; Wang, Y.; Ma, J.; Zheng, Z.; Luo, Z.Z.; Zheng, Y. Fabry-Perot interferometric high-temperature sensing up to 1200 °C based on a silica glass photonic crystal fiber. *Sensors* **2018**, *18*, 273.

38. Zhang, X.; Peng, W.; Zhang, Y. Fiber Fabry–Perot interferometer with controllable temperature sensitivity. *Opt. Lett.* **2015**, *40*, 5658–5661. [CrossRef]

39. Liu, S.Q.; Ji, Y.K.; Yang, J.; Sun, W.M.; Li, H.Y. Nafion film temperature/humidity sensing based on optical fiber Fabry-Perot interference. *Sens. Actuators A Phys.* **2018**, *269*, 313–321. [CrossRef]

Review

Hybrid Plasmonic Fiber-Optic Sensors

Miao Qi [1,†], Nancy Meng Ying Zhang [1,†], Kaiwei Li [2], Swee Chuan Tjin [1,*] and Lei Wei [1,*]

[1] School of Electrical and Electronic Engineering and the Photonics Institute, Nanyang Technological University, 50 Nanyang Avenue, Singapore 639798, Singapore; miao001@e.ntu.edu.sg (M.Q.); mzhang018@e.ntu.edu.sg (N.M.Y.Z.)

[2] Institute of Photonics Technology, Jinan University, Guangzhou 510632, China; likaiwei11@163.com

* Correspondence: esctjin@ntu.edu.sg (S.C.T.); wei.lei@ntu.edu.sg (L.W.)

† These authors contributed equally to this work.

Received: 20 April 2020; Accepted: 6 June 2020; Published: 8 June 2020

Abstract: With the increasing demand of achieving comprehensive perception in every aspect of life, optical fibers have shown great potential in various applications due to their highly-sensitive, highly-integrated, flexible and real-time sensing capabilities. Among various sensing mechanisms, plasmonics based fiber-optic sensors provide remarkable sensitivity benefiting from their outstanding plasmon–matter interaction. Therefore, surface plasmon resonance (SPR) and localized SPR (LSPR)-based hybrid fiber-optic sensors have captured intensive research attention. Conventionally, SPR- or LSPR-based hybrid fiber-optic sensors rely on the resonant electron oscillations of thin metallic films or metallic nanoparticles functionalized on fiber surfaces. Coupled with the new advances in functional nanomaterials as well as fiber structure design and fabrication in recent years, new solutions continue to emerge to further improve the fiber-optic plasmonic sensors' performances in terms of sensitivity, specificity and biocompatibility. For instance, 2D materials like graphene can enhance the surface plasmon intensity at the metallic film surface due to the plasmon–matter interaction. Two-dimensional (2D) morphology of transition metal oxides can be doped with abundant free electrons to facilitate intrinsic plasmonics in visible or near-infrared frequencies, realizing exceptional field confinement and high sensitivity detection of analyte molecules. Gold nanoparticles capped with macrocyclic supramolecules show excellent selectivity to target biomolecules and ultralow limits of detection. Moreover, specially designed microstructured optical fibers are able to achieve high birefringence that can suppress the output inaccuracy induced by polarization crosstalk and meanwhile deliver promising sensitivity. This review aims to reveal and explore the frontiers of such hybrid plasmonic fiber-optic platforms in various sensing applications.

Keywords: optical fibers; hybrid plasmonic sensors; surface plasmon resonance; localized surface plasmon resonance; 2D materials; graphene; transition metal oxides; gold nanoparticles; cyclodextrin

---

1. Introduction

Plasmonic fiber-optic sensors have captured intensive research attention in recent years due to their high degree of integration, high sensitivity, flexibility and remote-sensing capability [1–3]. Fiber-optic plasmonic sensors can be generally classified into two categories, surface plasmon resonance (SPR)-based sensors and localized surface plasmon resonance (LSPR)-based sensors. Conventionally, SPR- and LSPR-based hybrid fiber-optic sensors are realized by depositing thin metal films and metallic nanoparticles on various fiber structures (e.g., fiber gratings, side-polished fiber, microfiber, etc.) respectively to contribute strong plasmon–matter interaction. To further improve the measurement accuracy, sensitivity and selectivity to analyte molecules, specially designed fiber structures or functional materials are normally applied to strengthen the intensity of surface plasmon or the adsorption to target molecules. Specialty optical fibers like microstructured optical fibers (MOFs) can achieve high-level integration so that the very small dimension waveguide and the microfluidic channels are able to be integrated within a single fiber with only micrometer-scale diameter, leading to effective plasmon–matter interaction. The recent breakthroughs in 2D materials such as graphene, transition metal dichalcogenides (MX_2), transition metal oxides (TMOs), etc. reveal new opportunities in plasmon–matter enhancement by constructing 2D material/metal hybrid plasmonic structures or heavily doping-free carriers in 2D TMOs to realize intrinsic strong plasmonics in frequently used visible or near-infrared (NIR) optical windows. In addition, macrocyclic supramolecules have been recently proven to be excellent surface functionalization candidates for metallic nanoparticles, contributing to a simple functionalization process, selective target molecules recognition and improved biocompatibility.

In this review, the background and the state-of-the-art of SPR/LSPR fiber-optic sensors will be reviewed. More importantly, the aforementioned emerging hybrid fiber-optic plasmonic sensing solutions will be illustrated in detail. For instance, the exploration of how the highly birefringent MOF based SPR sensor can suppress polarization crosstalk and improve sensitivity in the meantime; how to integrate graphene-on-gold hybrid structure on the fiber-optic platform to strengthen the surface plasmon intensity and to effectively adsorb biomolecules; how to dope abundant free electrons in 2D MoO_3 and achieve highly integrated microfiber based plasmonic sensing in NIR optical frequencies; how to synthesize cyclodextrin-capped gold nanoparticles (AuNPs) in a one-step process and realize microfiber based highly selective detection of cholesterol in human serum, etc. will be demonstrated.

2. Fiber-Optic Surface Plasmon Resonance Sensors

A surface plasmon polariton (SPP) is an electromagnetic wave that propagates in parallel at the interface between the metal film and dielectric medium. The SPP is TM-polarized. As illustrated in Figure 1a, the polarization direction of SPP is perpendicular to the metal–dielectric interface [4]. The SPP has the evanescent nature, which is strongest at the surface of the metal film and exponentially decays into the dielectric material. Conventionally, SPR is realized on a Kretschmann–Raether silica prism of which the base is coated with a nanometer-scale thin metal film (Figure 1b) [5]. The ambient medium of the thin metal film is dielectric and considered as semi-infinite, by solving the Maxwell equations, the propagation constant of SPP is given by:

$$K_{SP} = \frac{\omega}{c} \sqrt{\frac{\varepsilon_{metal}\varepsilon_{dielectric}}{\varepsilon_{metal} + \varepsilon_{dielectric}}} \tag{1}$$

where ω, c and ε are the frequency of the TM-polarized incident light, light velocity and dielectric constant, independently. The propagation constant of the evanescent wave at the interface is:

$$K_{ev} = \frac{\omega}{c} \sqrt{\varepsilon_{prism}} \sin \theta \tag{2}$$

where θ is the angle of light incidence. SPR can be excited by the TM-polarized total reflected light at the silica-metal interface when the phase matches, that the propagation constant of reflected light equals to the propagation constant of SPP:

$$\frac{\omega}{c} \sqrt{\varepsilon_{prism}} \sin(\theta_{res}) = \frac{\omega}{c} \sqrt{\frac{\varepsilon_{metal}\varepsilon_{dielectric}}{\varepsilon_{metal} + \varepsilon_{dielectric}}} + \Delta\beta \tag{3}$$

The $\Delta\beta$ of the right expression denotes the effects of finite metal layer thickness and high refractive index of prism in real situation. In most fiber-optic SPR sensors, the wavelength interrogation method is employed. The refractive index sensitivity is defined as:

$$S_n = \delta\lambda_{res}/\delta n_s \tag{4}$$

where δn_s is the change in the refractive index of the analyte and $\delta\lambda_{res}$ is the shift of resonant wavelength [6].

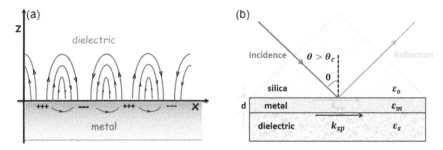

Figure 1. (**a**) The propagating surface plasmon polariton (SPP) at the metal–dielectric interface (Figure adapted with permission from reference [4]). (**b**) The schematic illustration of conventional Kretschmann-Raether prism configuration (Figure adapted with permission from reference [5]).

Along with the increasing demand for compact, highly-integrated, flexible and even in situ sensing devices, optical fiber-based SPR sensors receive more and more attention [7–11]. Various fiber-optic SPR configurations have been investigated to achieve highly sensitive SPR sensors. The key point in the design of the fiber-optic SPR sensor is to realize the phase-matching between the guided mode in fiber and the SPP at the metal-dielectric interface. Hence it is essential to coat the thin metal film at the surface of fiber structure where a strong evanescent field of guide mode can be exposed, leading to the strong SPP at the metal surface for effective light–matter interaction. Fiber gratings like long-period fiber grating (LPG) [12–14] and tilted fiber Bragg grating (TFBG) [15–17], tapered fiber [18–23], side-polished fiber [18,24], etc. have been demonstrated to be feasible for SPR sensing (Figure 2).

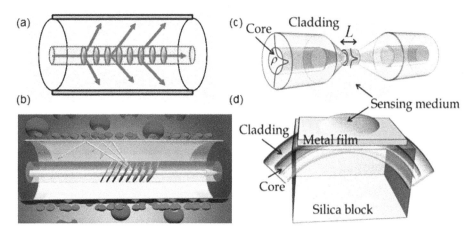

Figure 2. Fiber-optic surface plasmon resonance (SPR) sensor based on (**a**) long-period fiber grating (LPG) (Figure adapted with permission from reference [12]). (**b**) tilted fiber Bragg grating (TFBG) (Figure adapted with permission from reference [15]). (**c**) tapered fiber. (**d**) side-polished fiber (Figure (**c**) and (**d**) adapted with permission from reference [18]).

In recent years, MOFs are favored due to higher degree of integration, longer interaction distance and improved robustness, that the cladding air holes can function as microfluidic channels for liquid or gas analyte infiltration [25–27]. With the distinctive design of core dimension and cladding air holes arrangement, the thin metal films coated on the inner surface of air holes can effectively interact with the evanescent field of the core mode, which grants access to infiltrated analyte to the strong SPP. Numerous MOF structures have been proposed, including hexagonal MOFs [28], semicircular- channel MOFs [29], exposed-core MOFs [30], etc. In most MOF-based SPR designs, the prime consideration is to facilitate simple analyte infiltration and large interaction area. Hence, birefringence commonly exists in MOF-based SPRs. Based on Equation (1), birefringence leads to the offset between SPR wavelengths corresponding to two orthogonal polarizations of core mode. When external perturbations such as fiber bending, twisting and pressure are applied on the fiber, the coupling from the desired mode polarization to undesired mode polarization will occur. Therefore, the overall SPR peak, which is the superposition of SPR of two orthogonal polarizations, will be unstable, leading to inaccurate sensing results.

To address the issue of birefringence induced measurement instability, polarization-maintaining MOF-based SPR sensor with high birefringence could be a promising solution. A large birefringence can be realized in a near-panda MOF with the two central air holes of the photonic-crystal arranged cladding holes enlarged (Figure 3a). The material of the MOF is fused silica. The enlarged two central holes can facilitate easier thin noble metal film deposition and analyte infiltration. Strong surface plasmons can be excited by the x-polarized fundamental core mode with the thin gold film deposited on the inner walls of central holes. As discussed earlier, SPP can only be excited by the TM-polarized incident light (i.e., the polarization perpendicular to the metal film surface), and y-polarized core mode corresponds to a much weaker SPP compared with that of x-polarized mode (Figure 3b). This indicates the SPR sensing output is predominated by the plasmonic behaviors of x-polarized code mode. For a low-birefringent MOF, of which the diameter of central holes (d2) is comparable to that of other cladding holes (d1) (e.g., d1/d2 = 0.95), both x- and y-polarized mode can excite relatively strong SPP. As a result, the existence of unwanted polarization could induce an offset of overall resonant wavelength as high as 0.67 nm from that of the desired polarization, which means the SPR sensing accuracy is considerably compromised (Figure 3c). On the contrary, even though a highly-birefringent MOF consists of two modal polarizations corresponding to even larger resonant wavelength difference, the immensely suppressed SPP of unwanted polarization has a bare influence on the overall resonant wavelength and the sensing accuracy. For instance, the wavelength offset of the proposed highly birefringent near-panda MOF with d1/d2 = 0.4 is as small as 0.06 nm (Figure 3d).

Based on the finite element method (FEM) simulation of photonic-crystal arranged MOFs with different d1/d2 ratios, the relation between phase birefringence and sensing inaccuracy can be deduced. As shown in Figure 3e, the resonant wavelength offset could increase to be as large as 18.89 nm when the phase birefringence increases from $\sim 4 \times 10^{-5}$ to $\sim 1 \times 10^{-4}$. When the phase birefringence exceeds beyond a threshold ($\sim 1 \times 10^{-4)}$, the wavelength offset effectively reduces and even tends toward 0 after 4×10^{-4} phase birefringence. The investigation indicates that small birefringence that commonly exists in MOF-based SPR sensors could induce non-negligible undesired resonant wavelength offset, which affects sensing accuracy. The proposed highly-birefringent MOF with intentionally introduced large phase birefringence $\sim 4.2 \times 10^{-4}$ can effectively suppress such impact of polarization crosstalk to be extremely small. In addition, more expanded central holes enhance the plasmon–matter interaction, thereby providing higher sensitivity. Figure 3f compares the sensitivities when d1/d2 = 0.4, 0.5, 0.6 and 1.0. It is clear that the sensitivity is improved when the central holes expand. At a high analyte refractive index range of 1.37–1.38, the proposed highly-birefringent MOF SPR sensor can achieve a sensitivity as high as 3000 nm/RIU.

Besides optimizing the design of fiber structure, integrating functional nanomaterials with a fiber-optic platform can also effectively promote the light-matter interaction. In the past decade, 2D materials have drawn extensive attention in various research fields including the highly integrated sensors. The extremely large surface-to-volume ratio, in situ plasmonic properties tunability and near field confinement are the great advantages of 2D materials in sensing applications [33–36]. The plasmonics of most common 2D materials such as graphene and MX_2 fall in MIR or terahertz regions, which are not compatible with the well-developed optical communication window even though they can achieve superior plasmonic sensing performance [37,38]. Therefore, numerous research efforts focus on enhancing the plasmon–matter interaction by applying 2D material/metal film hybrid structures to SPR configurations. For instance, the thin gold film in conventional Kretschmann configuration has

Sensors **2020**, *20*, 3266

been upgraded to graphene/gold [6,39–41], graphene oxide/gold [42–45], graphene-MoS$_2$/gold [46], etc. hybrid film-like architectures (Figure 4). It is proven that the intensity of SPP on the gold film surface can be effectively strengthened by the seamlessly integrated graphene layer. When graphene and gold are in contact, the work function difference between the two materials (4.5 eV for graphene and 5.54 eV for gold) causes electrons to flow from graphene to gold to equilibrate the Fermi levels [47,48]. As a result, the electron density at the gold film surface increases as the graphene becomes p-type doped. Therefore, a stronger SPP so a higher sensitivity can be achieved.

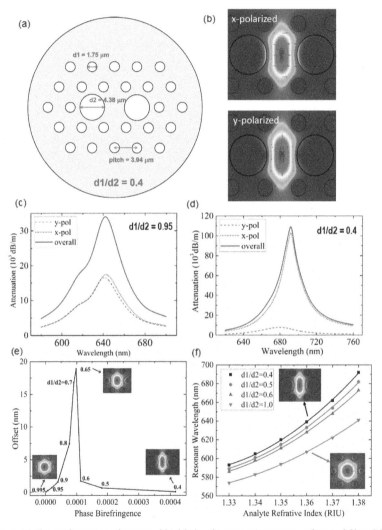

Figure 3. (**a**) The configuration of proposed highly birefringent microstructured optical fiber (MOF). (**b**) x-polarized and y-polarized core mode pattern of the SPR sensor (d1/d2 = 0.4) (Figure adapted with permission from reference [31]). (**c**) Attenuation spectra of highly birefringent MOF when d1/d2 = 0.95 and (**d**) d1/d2 = 0.4. (**e**) The variation of wavelength offset along with phase birefringence. (Inset) The x-polarized core mode pattern. (**f**) The SPR sensitivities when d1/d2 = 1.0, 0.6, 0.5 and 0.4 respectively. (Inset) The x-polarized core mode (Figure adapted with permission from reference [32]).

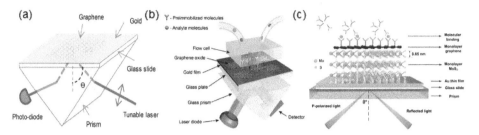

Figure 4. The prism based SPR configurations with hybrid plasmonic structures of (**a**) single-layer graphene/gold (Figure adapted with permission from reference [39]); (**b**) multilayer graphene/Py/gold (Figure adapted with permission from reference [42]); (**c**) graphene-MoS2/gold (Figure adapted with permission from reference [46]).

Even though the 2D material/metal film hybrid structure had been widely proposed on the prism-based SPR configuration, systematic analysis and experimental demonstration of integrating such hybrid plasmonic structure with flexible waveguides such as optical fibers were rare. As a proof of concept, a graphene-on-gold hybrid structure is proposed to be seamlessly integrated with a side-polished optical fiber, purposing to demonstrate that the 2D material/metal hybrid structures could achieve enhanced plasmonic biosensing performance on flexible waveguide platforms. As illustrated in Figure 5a, the exposed evanescent field of guided core mode interacts with the graphene-on-gold structure deposited at the surface of polished facet of optical fiber, leading to strong SPP-biomolecules interaction. Meanwhile, the single graphene layer functions as excellent surface functionalization of the thin gold film. Since the SPP at the gold film surface exponentially decays with the penetration depth, the thickness of surface functionalization is a crucial factor that affects sensitivity. The graphene layer, as thin as 0.34 nm, could hardly compromise the SPR sensitivity [49]. Moreover, the carbon atoms of graphene arranged in honeycomb format can easily form π-stacking interaction with the aromatic rings commonly existed in biomolecules [50]. Hence it facilitates effective adsorption of target biomolecules such as ssDNA, providing high sensitivity and low limit of detection (LOD).

Simulation can verify the SPP enhancement capability of the additional graphene sheet on conventional gold film coated side-polished fiber. The inset of Figure 5b plots the whole electrical field distribution of guided core mode in the fiber as well as the SPP on the side-polished facet. The magnified field distribution of the SPP at the gold/graphene surface is shown in Figure 5b. As expected, introducing single or multiple graphene layers can effectively enhance the SPP intensity on the thin gold film surface, which benefited from the electrons transfer, as explained above. Another interesting finding in the simulation is that bilayer or multi-layer graphene slightly compromises the SPP intensity compared with the single-layer graphene. This is due to the electrons' energy loss induced by the increase of graphene layers [49]. Therefore, with the SPP intensity boosted by ~30.2%, a single graphene layer most enhances the plasmonic sensing behavior. The experimental results further verify that the graphene-on-gold hybrid structure can effectively improve the plasmonic sensing behavior. Figure 5c compares the resonant peaks of the conventional thin gold film coated side-polished fiber and the graphene-on-gold hybrid structure integrated side-polished fiber when both sensing configurations are immersed in deionized (DI) water. The inset of Figure 5c shows the microscopic view of the boundary of transferred graphene on the thin gold film coated side-polished fiber facet. The graphene-on-gold hybrid structure corresponds to a deeper resonant peak, indicating a stronger SPP intensity, which matches well with the simulation.

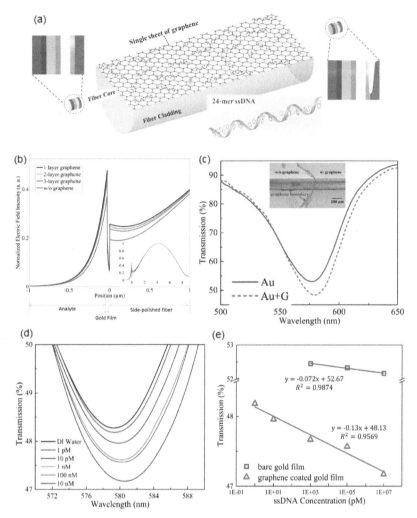

Figure 5. (a) The schematic illustration of proposed hybrid graphene-on-gold SPR sensor. The nucleobases of target ssDNA molecules can form stable π-stacking interaction with the honeycomb arrange carbon atoms of graphene. (b) The comparison of electric field intensities when the SPP is excited by bare thin gold film, single layer graphene/gold, 2-layer graphene/gold and 3-layer graphene/gold. (Inset) The electric field distribution over the entire fiber-optic graphene/gold hybrid structure. (c) The comparison of fiber transmission spectra with and without single graphene layer (Inset) The microscopic view of the single graphene layer transferred on the side-polished fiber. (d) The transmission spectra variation of proposed hybrid plasmonic sensor along with the increase of ssDNA concentration. (e) The comparison of sensitivities to ssDNA concentration of fiber-optic plasmonic sensors with and without graphene layer (Figure adapted with permission from reference [51]).

The biosensing capability of the proposed plasmonic hybrid SPR configuration can be validated by detecting ssDNA concentration. ssDNA quantization provides biomedical significance in gene expression, DNA sequencing and polymerase chain reaction (PCR) [52]. Figure 5d shows the magnified SPR peaks of the biosensing platform with the incrementing ssDNA concentration. This can be explained by Equation (1) that the surrounding refractive index of the plasmonic architecture is increased due to the efficient adsorption of ssDNA molecules on the graphene surface via π-stacking interaction. Also, the SPP evanescent field is scattered by the bonding of ssDNA molecules, which further induces transmission loss, thereby a deeper SPR peak. The LOD of the biosensor to ssDNA molecules is as small as 1 pM based on the distinguishable enhancement of the SPR peak (the red curve of Figure 5d). To experimentally verify that the biosensing performance is improved by the additional graphene layer, a conventional thin gold film based side-polished fiber-optic SPR sensor is prepared and applied to measure the same ssDNA solutions. The comparison of sensitivities corresponding to the two structures in Figure 5e can obviously indicate that the graphene-on-gold hybrid structure can effectively improve the sensitivity almost two-fold.

Wei et al. also compared the theoretically and experimentally performances of the fiber-optic SPR sensors with and without graphene in evaluating bovine serum albumin (BSA) concentration [6]. As shown in Figure 6a,b, the unclad portion of a plastic optical fiber is deposited with a gold film, and the graphene monolayer is transferred to the gold surface by PMMA. Figure 6d shows the variation of the reflection spectra of the graphene/Au fiber-optic SPR sensor with BSA concentration ranging from 0 to 2 mg/mL, which displays a 13.8-nm redshift compared with 6.1 nm for the Au fiber-optic SPR sensor (Figure 6c). After linearly fitting the resonant wavelengths and BSA concentration, the sensitivity of the graphene/Au hybrid sensor is 7.01 nm/(mg/mL), while the sensor without graphene is only 2.98 nm/(mg/mL). Additionally, regarding the full width half maximum (FWHM) of the two sensors, the graphene/Au hybrid sensor also possesses a more obvious variation tendency.

The finite element analysis (FEA) method based on COMSOL Multiphysics is established to clarify the improved sensing capability by graphene. The calculated electric field mode diagrams of fiber-optic SPR sensors with and without graphene are displayed in Figure 6e,f, respectively. On the sensing medium/Au interface, both sensors exhibit the similar confined electric field distributions, while the presence of graphene can strengthen the confined electric field with a maximum intensity of 6.4×10^4 V/m. Furthermore, in Figure 6g, the electric field intensity of the two structures perpendicular to the sensing interface (white dashed line) is extracted and compared. As can be seen, both electric field intensity exponentially decays along with the distance from Au film, and graphene/Au hybrid structure reveals a more considerable penetration depth of 256 nm, thus improving the sensitivity to the surrounding medium.

Based on the same mechanism, Wang et al. developed an SPR immunosensor employing graphene oxide (GO)-modified photonic crystal fiber (PCF) for human IgG detection [44]. PCF is a type of MOF consisting of a honeycomb structure with air holes, infiltrating liquid crystals into these air holes enables PCF tunable optical characteristics [53–60]. As shown in Figure 7a, the PCF with five layers of air holes is spliced between two multimode fibers (MMFs). After being deposited with Au film, the fiber is cleaned with piranha and then modified with Mercapto ethylamine (MEA) to enrich amine (-NH$_2$) groups for further reaction with epoxy groups on GO. Subsequently, the EDC/NHS system is used to activate the carboxyl of GO, and anti-IgG is directionally linked by staphylococcal protein A (SPA) orientation. Finally, BSA is introduced to block the free SPA surface, and the sensor is ready for human IgG sensing.

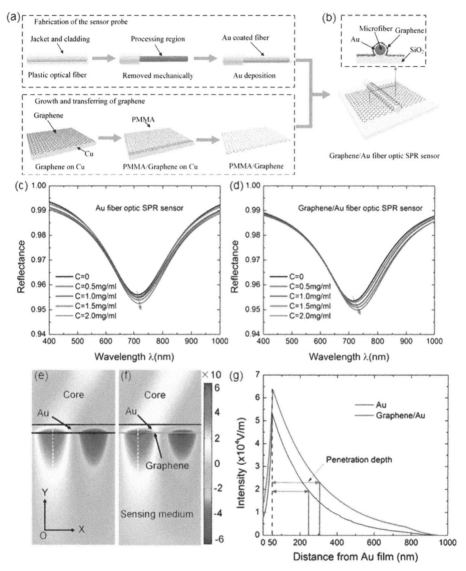

Figure 6. (**a**) Fabrication of Au coated fiber and preparation of graphene monolayer. (**b**) The schematic illustration of cross-section view of the proposed graphene/Au fiber-optic SPR sensor. (**c**) Reflection spectra of Au (**c**) and graphene/Au (**d**) fiber-optic SPR sensor with varying bovine serum albumin (BSA) concentration. (**e**) Finite element analysis (FEA) simulation of electric field distribution of fiber-optic SPR sensors with (**e**) and without (**f**) graphene. (**g**) Electric field decaying along Y-direction (Figure adapted with permission from reference [6]).

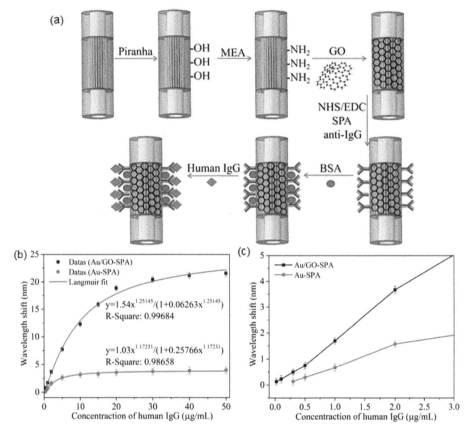

Figure 7. (**a**) Fabrication of graphene oxide (GO)-modified SPR immunosensor. (**b**) Fitting curve of wavelength shift versus human IgG concentration. (**c**) Local enlarged drawing (Figure adapted with permission from reference [44]).

The immune reaction between anti-IgG and human IgG will cause wavelength redshift, as shown in Figure 7b, the wavelength shift and human IgG concentration can be fitted using the Langmuir equation. Compared with the Au-SPA sensor, Au/GO-SPA sensor exhibits a distinct redshift of 0.02 nm to 21.57 nm. After zooming in (Figure 7c), it can be observed that the LOD of Au/GO-SPA sensor (0.01 µg/mL) is 30 times lower than the Au-SPA sensor (0.3 µg/mL), which indicated GO significantly enhanced the immunosensor sensitivity.

Hu et al. also incorporate the graphene monolayer on a gold-coated TFBG (Figure 8a), the TBFG is further functionalized with ssDNAs by π–π stacking for dopamine detection. As shown in Figure 8d, there is an obvious differential amplitude increase when dopamine concentration raises from 10^{-14} M to 10^{v13} M, and a quite linear correlation ($R^2 = 99\%$) is observed over dopamine concentration from 10^{-13} M to 10^{v8} M. TFBG enables the optic-fiber sensor with high RI sensitivity, narrow cladding and innate insensitivity to temperature and optical power fluctuations, which is feasible for biomedical sensing.

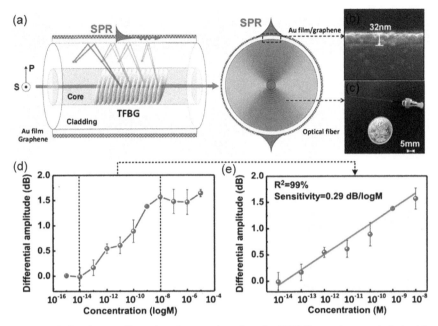

Figure 8. (a) The schematic illustration of proposed graphene/Au TFBG fiber-optic sensor (polarimetric sensing characteristic of TFBG and the energy distribution along fiber cross section). (b) Scanning electron microscope (SEM) image of graphene monolayer coated on Au surface. (c) photograph of the whole fiber-optic probe. (d) The differential amplitude output versus dopamine concentration. (e) The linear relationship between differential amplitude and dopamine concentration. (Figure adapted with permission from reference [41]).

Although the 2D material/metal hybrid structures facilitate remarkable light-matter interaction in plasmonic sensing, the intrinsic SPP of most common 2D materials (e.g., graphene and MX_2) located at the MIR range is almost impossible for practical applications. Therefore, an alternative class of 2D plasmonic material, heavily doped ultrathin TMOs, have captured research attention in recent years aiming for manipulating the intrinsic plasmonics of 2D materials with exceptional field confinement and in situ plasmonic tunability in the frequently used visible and NIR optical window [61–64]. To realize SPP in visible or NIR frequencies, sufficient free carrier concentration must be achieved in 2D materials. The unique character of outer-d valence electrons enables TMOs to achieve sufficient free carrier doping via ionic intercalation. Taking the most representative TMOs, molybdenum trioxide (MoO_3) and tungsten oxide (WO_3), as examples, free electrons can be abundantly doped by introducing oxygen vacancies in the TMO lattice [65,66]. Therefore, the plasmonic behavior of 2D TMOs can be easily tuned by manipulating the oxygen vacancies. So far, the tunable plasmonics of heavily doped MoO_3 nanoflakes in visible or NIR region has been most widely studied, yet the exploration on integrating such emerging 2D materials with optical devices especially the highly-integrated waveguide based sensing devices is very limited.

Driven by the purpose of investigating the potential of 2D TMOs on highly-integrated plasmonic devices, a biosensor based on a microfiber functionalized with α-MoO_3 nanoflakes is developed and validated by BSA molecules detection. As shown in Figure 9a, few-layer α-MoO_3 nanoflakes are synthesized by the liquid-phase exfoliation method [67] and then heavily doped with free electrons via an H^+ intercalation process [68]. After doping, a sub-stoichiometric α-MoO_{3-x} nanoflakes solution with strong SPP at the NIR region is formed. Pristine MoO_3 only introduces absorption at ultraviolet (UV) wavelengths, which is due to the large bandgap of 3.2 eV [69]. After electrons are increasingly doped, a distinct absorption peak appears and enhances at 700–800 nm range, in the meantime, undergoes a blueshift. This phenomenon can be explained by Drude model that the plasma frequency is inversely correlated to electron density [70,71].

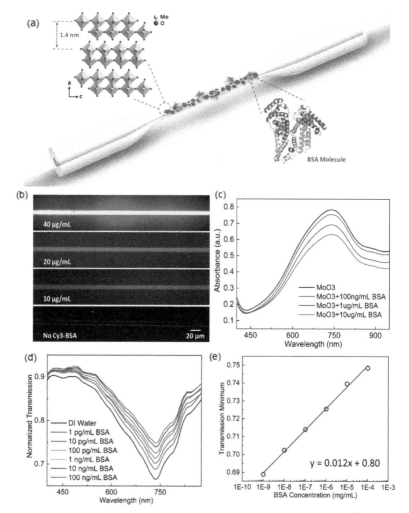

Figure 9. (a) The schematic illustration of heavily-doped MoO_{3-x} nanoflakes based hybrid fiber-optic plasmonic biosensor. (Inset 1) The crystal structure of α-MoO_3 lattice. (Inset 2) Molecular structure of BSA. (b) The fluorescent microscopic view of MoO_{3-x} nanoflakes functionalized microfibers coated with different concentrations of dye labelled BSA molecules. (c) The absorption spectra of MoO_{3-x} nanoflakes solutions mixed with different BSA concentrations. (d) Transmission spectrum variation of proposed hybrid plasmonic biosensor along with increasing BSA concentration. (e) The linear increase of transmission minimum against log-scale BSA concentration (Figure adapted with permission from reference [68]).

The MoO_{3-x} nanoflakes can be stably immobilized on the microfiber surface via electrostatic interaction. Since MoO_{3-x} is positively charged, the microfiber surface functionalized with evenly distributed negative charges (e.g., self-assembled poly(allylamine) (PAA)/poly(styrene sulfonate) (PSS) bilayer) applies strong attraction to the nanoflakes. Similarly, the immobilized positively charged MoO_{3-x} nanoflakes on the microfiber surface can effectively attract negatively charged target molecules, such as BSA [67]. Dye-labeled BSA molecules are adopted to verify the effectiveness of electrostatic interaction-based target molecule adsorption as well as fiber surface functionalization. Figure 9b shows the fluorescent microscope views of four MoO_{3-x} nanoflake-deposited microfibers after immersing in different concentrations of dye-labeled BSA solutions. It is evident that the fiber brightens as the BSA concentration increases. Also, the even brightness on the fiber surface implies the uniformity of adsorbed BSA molecules so as the MoO_{3-x} nanoflakes.

The binding of negatively charged BSA molecules on the MoO_{3-x} nanoflakes surface impacts the plasmonic behavior. When MoO_{3-x} nanoflakes suspensions mix with different concentrations of BSA solution, the absorption peak of MoO_{3-x} weakens as the BSA concentration increases (Figure 9c). This is due to the free electrons at the MoO_{3-x} surface being repelled by the negatively charged BSA molecules, resulting in the reduced free electron density involved in the plasmonic resonance [61,64,72]. Therefore, a fiber-optic sensor based on MoO_{3-x} nanoflakes shows a unique characteristic that the resonance peak on the fiber transmission spectrum gradually shallows along with the increasing concentration of target BSA (Figure 9d). Profited from the full utilization of a high aspect ratio of 2D MoO_{3-x}, a LOD of BSA as low as 1 pg/mL is achieved. Moreover, the transmission minimum of the plasmonic resonance peak provides a linear response to the log-scale BSA concentration (Figure 9e).

With the vigorous development of material science, there is abundant research to introduce diverse materials into fiber-optic SPR sensors [73–77]. For instance, Santos et al. propose a refractive index sensor by combining Al_2O_3-Ag metamaterial film with D-type PCF fiber, in which the sensor performance can be adjusted by the thickness and component of metamaterial [78]. Semwal et al. wrapped Ag-coated optical fiber with the enzyme (ADH) and coenzyme (NAD)–containing hydrogel to establish an ethanol sensor. These will surely boost the advancement of fiber-optic fiber sensors [79].

3. Fiber-Optic Localized Surface Plasmon Resonance Sensors

By contrast with the propagating SPP at a thin metal film surface, the resonant electron oscillation induced by light interacting with a metallic nanoparticle is non-propagating due to the particle size restriction. Therefore, it is called localized SPR (LSPR). LSPR can be excited when the oscillation frequency of nanoparticle electron cloud matches with the frequency of incident light (Figure 10) [80,81]. A proper model for understanding how incident light is scattered and absorbed by a nanoparticle with a diameter much smaller than the wavelength is the Mie theory. The Mie theory constructs a model to deduce the extinction cross-section of nanoparticle based on the assumption that the nanoparticle is a homogeneous conducting sphere:

$$\sigma_{ext} = 9\left(\frac{\omega}{c}\right)(\varepsilon_{dielectric})^{\frac{3}{2}} V \frac{\varepsilon''_{metal}}{\left(\varepsilon'_{metal} + 2\varepsilon_{dielectric}\right)^2 + \left(\varepsilon''_{metal}\right)^2} \tag{5}$$

where V is the volume of nanoparticle, and ε'_{metal} and ε''_{metal} are the real and the imaginary parts of the metal-dielectric function, respectively, in the Drude model [82]:

$$\varepsilon'_{metal} = 1 - \frac{\omega_p^2}{(\omega^2 + \gamma^2)} \tag{6}$$

$$\varepsilon''_{metal} = \frac{\omega_p^2 \gamma}{(\omega^2 + \gamma^2)\omega} \tag{7}$$

where γ is the damping of electron oscillation and ω_p is the bulk plasma frequency. More detailed definitions of γ and ω_p can be found in [83]. Since LSPR operation frequencies are generally within the visible and NIR optical windows where $\gamma \ll \omega_p$, Equation (7) can be simplified as:

$$\varepsilon'_{metal} = 1 - \frac{\omega_p}{\omega^2} \tag{8}$$

Based on Equation (5), the resonance is satisfied (i.e., the extinction cross-section is maximum) when $\varepsilon'_{metal} = -2\varepsilon_{dielectric}$. The LSPR resonant frequency is thereby expressed as:

$$\omega_{LSPR} = \frac{\omega_p}{\sqrt{2\varepsilon_{dielectric} + 1}} \tag{9}$$

Furthermore, for dielectric medium, $\varepsilon_{dielectric} = n^2_{dielectric}$. Therefore, the refractive index of the ambient dielectric medium of the nanoparticle impacts the LSPR resonant wavelength:

$$\lambda_{LSPR} = \lambda_p \sqrt{2n^2_{dielectric} + 1} \tag{10}$$

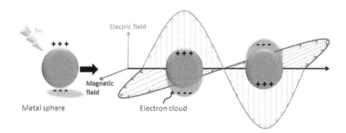

Figure 10. The schematic diagram of localized surface plasmon resonance (Figure adapted with permission from Ref. [81]).

Similar to SPR-based fiber-optic sensing platforms, LSPR-based fiber-optic devices have also captured intensive research attention [84–89]. Various fiber structures such as microfiber, cascaded unclad fiber, fiber endface, etc. have been integrated with silver or gold nanoparticles and shown promising plasmonic sensing performance (Figure 11). To achieve efficient selectivity to analyte molecules, it is necessary to apply surface functionalization on metallic nanoparticles. Taking the most widely employed gold nanoparticle as an example, many effective functionalization strategies have been proven, such as biomolecule coating [90–92], ligand substitution [93], polymer deposition [94,95], etc. However, when sensitivity is the crucial factor of plasmonic sensors, it is critical to keep the surface functionalization as thin as possible. Due to the evanescent nature of surface plasmon, thick surface functionalization considerably compromises the plasmon–matter interaction. For instance, a study has compared the LODs and sensitivities of two functionalization strategies with thicknesses of 4.24 nm and 0.96 nm respectively and shown that surface functionalization thinner than 1 nm significantly improves the sensing performance [96]. In such cases, macrocyclic supramolecules have shown the potential to meet the challenges of achieving both sub-nanometer functionalization thickness and target molecules recognition.

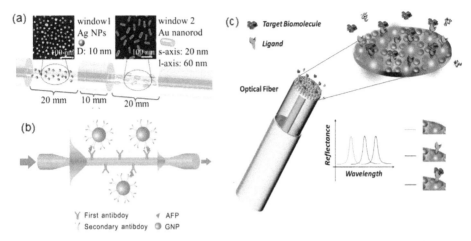

Figure 11. The LSPR devices based on various optical fiber structures such as (**a**) cascaded unclad fiber (Figure adapted with permission from reference [97]); (**b**) microfiber (Figure adapted with permission from reference [98]); (**c**) optical fiber endface (Figure adapted with permission from reference [99]).

Benefiting from their macrocyclic cavities, macrocyclic supramolecules like cyclodextrins (CDs), cucurbiturils, pillararenes, calixarenes, etc. show excellent molecular recognition capability by the host-guest interaction. The host–guest interaction is noncovalent interaction between macrocyclic supramolecules and corresponding guest molecules to form inclusion complexations [100]. Encouragingly, it is proven that host–guest interaction is a more effective target molecule recognition and adsorption mechanism compared with the conventional biomolecule-ligand binding [101]. Another advantage of macrocyclic supramolecules being surface functionalization of metallic nanoparticles is the heights of their macrocyclic cavities being normally less than 1 nm [102–104], which facilitates sensitive molecular detection as discussed above. In addition, the macrocyclic supramolecules also eliminate the cytotoxicity of nanoparticles that is often favored in bio-medical sensing applications. In recent years, studies have been carried out to achieve functionalizing metallic nanoparticles with macrocyclic supramolecules during the synthesis of nanoparticles instead of through post-processing surface modification. In most of these attempts (Figure 12), however, harsh reducing reagents such as thiols, NaBH$_4$, NaOH have to be introduced, which violates the purpose of achieving biocompatibility in many LSPR biosensing applications [105–107]. Therefore, inspired by the method proposed by Zhao et al. [108], where CDs act as both reducing and capping agent for AuNPs synthesis, a microfiber based LSPR biosensor is developed to comprehensively investigate the plasmonic sensing potential of one-step synthesized macrocyclic supramolecules decorated metallic nanoparticles.

Figure 13 illustrates the configuration of LSPR biosensor based on a microfiber integrated with one-step synthesized β-CD-capped AuNPs [109]. Functioning as both the reducing and capping agent, β-CD facilitates the AuNPs formation and forms biocompatible functionalization layer via the conjunction of carboxyl groups and gold surface. The evanescent field of guide mode in microfiber leaks out at the tapered portion and interacts with immobilized AuNPs at the fiber surface, leading to a strong LSPR peak on the transmission spectrum. Cholesterol, the guest molecule of β-CD [108,110], is employed to validate the biosensing performance of the proposed LSPR device. The sterol groups of cholesterol molecules can tightly fit into the β-CD macrocyclic cavity, meanwhile forming stable host–guest interaction through the hydrophobic associations [111,112].

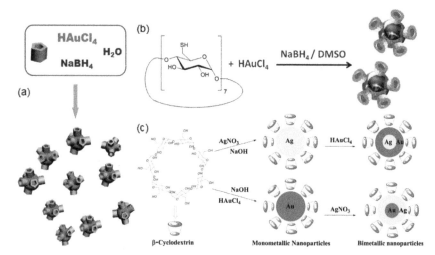

Figure 12. (a) The synthesis formulas of (a) carboxylatopillar[5]arene capped gold nanoparticles (AuNPs) ((Figure adapted with permission from reference [105]). (b) Cyclodextrin (CD)-capped AuNPs (Figure adapted with permission from Ref. [106]). (c) CD capped silver nanoparticles, AuNPs and Ag_{core}-Au_{shell}/Au_{core}-Ag_{shell} bimetallic nanoparticles (Figure adapted with permission from reference [107]).

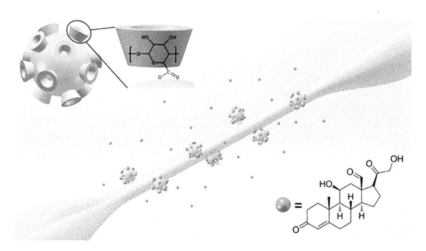

Figure 13. The schematic illustration of the β-CD-capped AuNPs based fiber-optic biosensor. (Inset1): The conjunction between β-CD molecule and AuNP surface. (Inset2): The molecular structure of cholesterol (Figure adapted with permission from reference [109]).

The as-synthesized β-CD-capped AuNPs show good uniformity in particle size with diameters range from ~18 nm to ~21 nm. The dynamic light scattering (DLS) characterization of the AuNPs further verifies the observation [109]. The resonance band of the absorption peak of β-CD-capped AuNPs solution with linewidth as narrow as 47 nm also indicates the monodispersity of the particles, which is comparable with that of conventionally synthesized AuNPs. Proton nuclear magnetic resonance (^{1}H NMR) and Fourier transform infrared (FTIR) spectra are performed to further illustrate that the hydroxyl groups in β-CDs mainly contribute to reducing Au^{3+} ions to Au^0 atoms [113].

The as-synthesized AuNPs are negatively charged. Hence it can be stably immobilized on positively charged microfiber surface (e.g., functionalize the fiber surface with homogeneous PAA layer) via electrostatic attraction. As shown in Figure 14a,b the prepared microfiber is 4 μm in diameter and decorated with evenly distributed AuNPs. The attached AuNPs induce a deep resonance band centered at 530.7 nm on the microfiber transmission spectrum. When the fiber-optic sensing device is sequentially immersed in cholesterol solutions with concentrations ranging from 5 aM to 0.5 μM, the LSPR resonance band gradually deepens along with the increasing cholesterol concentration meanwhile the resonant wavelength shifts from 530.7 nm to 531.4 nm (Figure 14c). Such an ultra-low LOD of 5 aM is profited from the highly efficient host–guest interaction between β-CD and cholesterol. The transmission minimum of the resonance band can be taken as the sensing parameter and provides a linear response to the log-scale cholesterol concentration (Figure 14d).

The selectivity of the proposed biosensor to cholesterol is validated by an interference study, where common interfering substances in human serum such as glutamic acid, cysteine, ascorbic acid, dopamine and human serum albumin (HSA) are introduced. Figure 14e shows the real-time average transmission intensity within 530–535 nm of microfiber when the interfering substances are introduced during the detection of cholesterol. It is clear that the β-CD-capped AuNPs based fiber-optic sensor only responses to cholesterol molecules but not interfered by other substances. To further validate the cholesterol recognition capability of the proposed sensor, recovery experiments are also carried out to evaluate the accuracy of detecting real human serum samples diluted by a factor of 1014 and spiked with different cholesterol concentrations. As summarized in Table 1, the measurement of cholesterol concentration in the unspiked human serum sample is 4.23 mM. The measurement of the same sample using commercial blood cholesterol monitor is 4.35 mM, which indicates the proposed fiber-optic biosensor is reliable. In addition, the recoveries of the spiked samples are 105.2–112.2%, which is also within a satisfactory range, further verifies the accuracy of the proposed sensor. Therefore, it indicates the tremendous plasmonic sensing potential of highly integrated fiber-optic sensors based on the macrocyclic supramolecules modified metallic nanoparticles.

Table 1. Recovery results of detecting cholesterol in human serum samples.

Sample	Added (aM)	Found * (aM)	Recovery (%)
Human serum (male)	0.0	42.3 ± 2.8	-
	50.0	94.9 ± 6.7	105.2
	100.0	154.5 ± 16.5	112.2

* The values are mean of 4 independent experiments ± standard deviation (Reprinted with permission from reference [110]).

Another polysaccharide, chitosan, has also been used for AuNPs synthesis as a reducing and stabilizing agent [114]. Sadani et al. immobilize the synthesized chitosan-capped AuNPs (ChGnP) with a diameter of 20 nm on U-bent fiber for mercury (Hg(II)) detection [115]. The U-bent fiber is firstly incubating in (3-Aminopropyl)triethoxysilane (APTES) solution to enrich amine on the fiber surface, with glutaraldehyde crosslinking followed. Thereafter, BSA is linked to glutaraldehyde for further AuNPs' immobilization (Figure 15a,b)

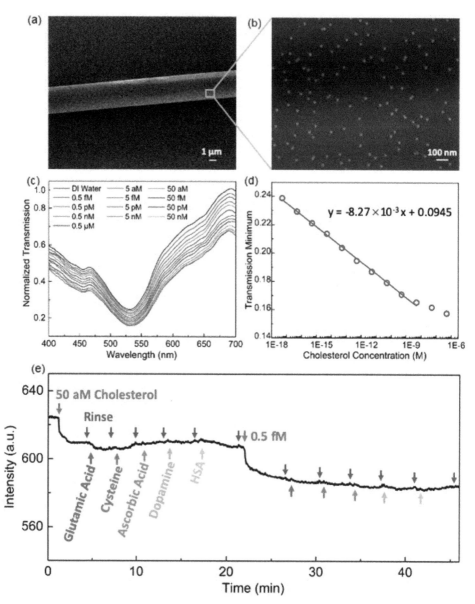

Figure 14. (**a**) The SEM image of 4-μm-diameter microfiber. (**b**) The distribution of β-CD-capped AuNPs on the microfiber surface. (**c**) Transmission spectrum variation of microfiber based hybrid plasmonic biosensor along with increasing cholesterol concentration. (**d**) The linear decrease of transmission minimum against log-scale cholesterol concentration. (**e**) The real-time average transmission intensity within 530–535 nm of microfiber when the interfering substances are introduced during cholesterol detection (Figure adapted with permission from reference [109]).

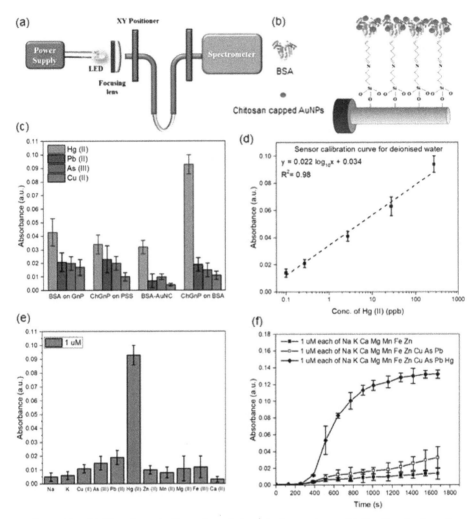

Figure 15. (a) The schematic illustration of the detection system. (b) Functionalization of chitosan-capped AuNPs on U-bent fiber. (c) Selection of optimal receptor for Hg(II) detection. (d) The linear increase of absorbance against Hg(II) concentration. (e) Absorbance at 520 nm for 1 μM individual metal ions detection. (f) Absorbance increasement against time for 1 μM metal ions mixture detection (Figure adapted with permission from reference [115]).

The sensitivity and selectivity of four sensors: BSA attaching to citrate capped AuNPs (BSA on GnP), polyanionic poly(sodium 4-styrenesulfonate) (PSS) immobilized ChGnP (ChGnP on PSS), fluorescent BSA-Au nanoclusters (BSA-AuNC) and BSA immobilized ChGnP (ChGnP on BSA) are compared. As shown in Figure 15c, compared to ChGnP on the BSA system, the first three show more deficient absorbance at the same Hg(II) concentration, and only the BSA-AuNC exists insignificant selectivity. The proposed ChGnP on the BSA LSPR sensor shows a linear calibration curve from Hg(II) concentration 0.1 ppb to 540 ppb (Figure 15d). 1 μM of different metal ions are dissolved in DI water separately, the absorbance for Hg(II) is greater than 0.9 a.u. while all other control ions are less than 0.2 a.u. (Figure 15e). Also, as shown in Figure 15f, when 1 μM of diverse metal ions mixtures with and without Hg(II) are detected, the change of absorbance at 520 nm over time shows that only mixture with the presence of Hg (II) reveals significant enhancement, further proving the excellent selectivity. The chemisorbed of Hg(II) on lone pair electrons of N, O and S atoms in chitosan and BSA, the hydrophobic interaction and the Van-der-Waals interaction with thiol groups in BSA are hypothesized the dominant factors of sensitivity and selectivity towards Hg(II).

Lee et al. fabricate a fiber-optic LSPR sensor for the detection of ochratoxin A (OTA) utilizing aptamer-modified gold nanorods (GNRs) [116]. The GNRs are immobilized on the optical fiber by Au–S interaction, after being dipped into OTA solution, the LSPR spectrum is monitored exploiting the light reflection of a silver mirror at the end of the fiber. The aptamer's specific recognition of OTA induces an LSPR peak shift (Figure 16c), and OTA can be specifically and quantitatively detected with a LOD of 12.0 pM and excellent linear response. This fiber-optic LSPR sensor possesses superior simplicity, which only demands to dip into OTA solution. The methods and performances of the hybrid fiber-optic sensors referred to are summarized is Table 2.

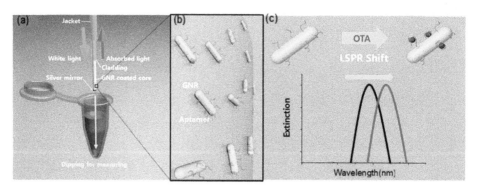

Figure 16. (a) The schematic illustration of the fiber-optic LSPR aptasensor. (b) Localized magnification of the fiber surface with GNPs immbolized on. (c) LSPR shift after ochratoxin A (OTA) recognization (Figure adapted with permission from reference [116]).

Table 2. Summary of hybrid fiber-optic sensors in this review.

Material	Mode	Analyst	LOD	Remark	Ref
Graphene/Au	SPR	ssDNA	1 pM	π-Stacking with graphene	[51]
Graphene/Au	SPR	BSA	NA	Significance of graphene	[6]
GO/Au	SPR	Human IgG	0.01 μg/mL	Anti-IgG/IgG interaction	[44]
Graphene/Au	SPR	Dopamine	10^{-13} M	π-Stacking with ss-DNA	[41]
MoO_{3-x} nanoflake	SPR	BSA	1 pg/mL	Electrostatic interaction	[68]
β-CD/AuNPs	LSPR	Cholesterol	5 aM	Host-guest interaction	[109]
Chitosan/AuNPs	LSPR	Hg(II)	0.1 ppb	Chemisorbed	[115]
Aptamer/GNRs	LSPR	OTA	12.0 pM	Aptamer's specific recognition	[116]

4. Conclusions

As discussed in this review, the proper design of MOFs with high birefringence provides wide possibilities in highly integrated microfluidic sensing devices with improved measurement accuracy and stability. Profiting from 2D material-based hybrid plasmonic structures, fiber-optic plasmonic sensors can deliver more promising sensing capability. The exceptional surface to volume ratio, near-field confinement and in situ plasmonic properties tunability of 2D materials facilitate the further enhancement of plasmon-matter interaction so as to enhance the sensitivity and LOD. In addition,

Sensors **2020**, *20*, 3266

the development of supramolecular chemistry brings new solutions in LSPR nanoparticles surface functionalization, leading to excellent target molecule selectivity via host–guest interaction. Given the numerous possibilities in optical design and hybrid plasmonic architectures construction, fiber-optic hybrid plasmonic sensors possess vast potential in various sensing scenarios with distinct advantages of high sensitivity, flexibility, miniaturization and high degree of integration. To further integrate the advances of specialty fibers and various nanomaterials, one major direction in the future is to achieve advanced functional fiber-based sensing by integrating multi-functional materials inside a single fiber or large-scale fabrics [117–135].

Funding: This work was supported in part by the Singapore Ministry of Education Academic Research Fund Tier 2 (MOE2019-T2-2-127), the Singapore Ministry of Education Academic Research Fund Tier 1 (MOE2019-T1-001-103 and MOE2019-T1-001-111) and the Singapore National Research Foundation Competitive Research Program (NRF-CRP18-2017-02). This work was also supported in part by Nanyang Technological University.

Conflicts of Interest: The authors declare no conflict of interest.

References

1. Gandhi, M.S.; Chu, S.; Senthilnathan, K.; Babu, P.R.; Nakkeeran, K.; Li, Q. Recent advances in plasmonic sensor-based fiber optic probes for biological applications. *Appl. Sci.* **2019**, *9*, 949. [CrossRef]
2. Allsop, T.; Neal, R. A Review: Evolution and Diversity of Optical Fibre Plasmonic Sensors. *Sensors* **2019**, *19*, 4874. [CrossRef] [PubMed]
3. Kim, H. Plasmonics on optical fiber platforms. In *Plasmonics*; Tatjana Gric; IntechOpen: London, UK, 2018; pp. 181–200.
4. Jana, J.; Ganguly, M.; Pal, T. Enlightening surface plasmon resonance effect of metal nanoparticles for practical spectroscopic application. *RSC Adv.* **2016**, *6*, 86174–86211. [CrossRef]
5. Zhao, Y.; Deng, Z.; Li, J. Photonic crystal fiber based surface plasmon resonance chemical sensors. *Sens. Actuators B Chem.* **2014**, *202*, 557–567. [CrossRef]
6. Wei, W.; Nong, J.; Zhu, Y.; Zhang, G.; Wang, N.; Luo, S.; Chen, N.; Lan, G.; Chuang, C.-J.; Huang, Y. Graphene/Au-enhanced plastic clad silica fiber optic surface plasmon resonance sensor. *Plasmonics* **2018**, *13*, 483–491. [CrossRef]
7. Shi, S.; Wang, L.; Su, R.; Liu, B.; Huang, R.; Qi, W.; He, Z. A polydopamine-modified optical fiber SPR biosensor using electroless-plated gold films for immunoassays. *Biosens. Bioelectron.* **2015**, *74*, 454–460. [CrossRef]
8. Wang, W.; Mai, Z.; Chen, Y.; Wang, J.; Li, L.; Su, Q.; Li, X.; Hong, X. A label-free fiber optic SPR biosensor for specific detection of C-reactive protein. *Sci. Rep.* **2017**, *7*, 1–8. [CrossRef]
9. Tabassum, R.; Gupta, B.D. Fiber optic manganese ions sensor using SPR and nanocomposite of ZnO–polypyrrole. *Sens. Actuators B Chem.* **2015**, *220*, 903–909. [CrossRef]
10. Shi, S.; Wang, L.; Wang, A.; Huang, R.; Ding, L.; Su, R.; Qi, W.; He, Z. Bioinspired fabrication of optical fiber SPR sensors for immunoassays using polydopamine-accelerated electroless plating. *J. Mater. Chem. C* **2016**, *4*, 7554–7562. [CrossRef]
11. Gao, D.; Guan, C.; Wen, Y.; Zhong, X.; Yuan, L. Multi-hole fiber based surface plasmon resonance sensor operated at near-infrared wavelengths. *Opt. Commun.* **2014**, *313*, 94–98. [CrossRef]
12. Caucheteur, C.; Guo, T.; Albert, J. Review of plasmonic fiber optic biochemical sensors: Improving the limit of detection. *Anal. Bioanal. Chem.* **2015**, *407*, 3883–3897. [CrossRef] [PubMed]
13. Schuster, T.; Herschel, R.; Neumann, N.; Schaffer, C.G. Miniaturized Long-Period Fiber Grating Assisted Surface Plasmon Resonance Sensor. *J. Light. Technol.* **2012**, *30*, 1003–1008. [CrossRef]
14. Zhang, N.M.Y.; Dong, X.; Shum, P.P.; Hu, D.J.J.; Su, H.; Lew, W.S.; Wei, L. Magnetic field sensor based on magnetic-fluid-coated long-period fiber grating. *J. Opt.* **2015**, *17*, 065402. [CrossRef]
15. Shevchenko, Y.; Francis, T.J.; Blair, D.A.D.; Walsh, R.; DeRosa, M.C.; Albert, J. In Situ Biosensing with a Surface Plasmon Resonance Fiber Grating Aptasensor. *Anal. Chem.* **2011**, *83*, 7027–7034. [CrossRef] [PubMed]
16. Caucheteur, C.; Guo, T.; Liu, F.; Guan, B.-O.; Albert, J. Ultrasensitive plasmonic sensing in air using optical fibre spectral combs. *Nat. Commun.* **2016**, *7*, 13371. [CrossRef] [PubMed]

17. Wang, Q.; Jing, J.-Y.; Wang, B.-T. Highly Sensitive SPR Biosensor Based on Graphene Oxide and Staphylococcal Protein A Co-Modified TFBG for Human IgG Detection. *IEEE Trans. Instrum. Meas.* **2018**, *68*, 3350–3357. [CrossRef] .

18. Roh, S.; Chung, T.; Lee, B. Overview of the characteristics of micro-and nano-structured surface plasmon resonance sensors. *Sensors* **2011**, *11*, 1565–1588. [CrossRef]

19. Lin, H.-Y.; Huang, C.-H.; Cheng, G.-L.; Chen, N.-K.; Chui, H.-C. Tapered optical fiber sensor based on localized surface plasmon resonance. *Opt. Express* **2012**, *20*, 21693–21701. [CrossRef]

20. Li, K.; Zhang, N.; Ying Zhang, N.M.; Zhou, W.; Zhang, T.; Chen, M.; Wei, L. Birefringence induced Vernier effect in optical fiber modal interferometers for enhanced sensing. *Sens. Actuators B Chem.* **2018**, *275*, 16–24. [CrossRef]

21. Li, K.; Zhang, N.M.Y.; Zhang, N.; Zhang, T.; Liu, G.; Wei, L. Spectral characteristics and ultrahigh sensitivity near the dispersion turning point of optical microfiber couplers. *J. Light. Technol.* **2018**, *36*, 2409–2415. [CrossRef]

22. Li, K.; Zhang, N.; Zhang, N.M.Y.; Liu, G.; Zhang, T.; Wei, L. Ultra-sensitive measurement of gas refractive index using an optical nanofiber coupler. *Opt. Lett.* **2018**, *43*, 679–682. [CrossRef] [PubMed]

23. Li, K.; Zhang, T.; Liu, G.; Zhang, N.; Zhang, M.; Wei, L. Ultrasensitive optical microfiber coupler based sensors operating near the turning point of effective group index difference. *Appl. Phys. Lett.* **2016**, *109*, 101101. [CrossRef]

24. Zhang, Y.; Chen, Y.; Yang, F.; Hu, S.; Luo, Y.; Dong, J.; Zhu, W.; Lu, H.; Guan, H.; Zhong, Y.; et al. Sensitivity-Enhanced Fiber Plasmonic Sensor Utilizing Molybdenum Disulfide Nanosheets. *J. Phys. Chem. C* **2019**, *123*, 10536–10543. [CrossRef]

25. Yu, X.; Zhang, Y.; Pan, S.; Shum, P.; Yan, M.; Leviatan, Y.; Li, C. A selectively coated photonic crystal fiber based surface plasmon resonance sensor. *J. Opt.* **2009**, *12*, 015005. [CrossRef]

26. Zhang, N.; Li, K.; Cui, Y.; Wu, Z.; Shum, P.P.; Auguste, J.-L.; Dinh, X.Q.; Humbert, G.; Wei, L. Ultra-sensitive chemical and biological analysis via specialty fibers with built-in microstructured optofluidic channels. *Lab Chip* **2018**, *18*, 655–661. [CrossRef]

27. Zhang, N.; Humbert, G.; Gong, T.; Shum, P.P.; Li, K.; Auguste, J.-L.; Wu, Z.; Hu, D.J.J.; Luan, F.; Dinh, Q.X.; et al. Side-channel photonic crystal fiber for surface enhanced Raman scattering sensing. *Sens. Actuators B Chem.* **2016**, *223*, 195–201. [CrossRef]

28. Peng, Y.; Hou, J.; Huang, Z.; Lu, Q. Temperature sensor based on surface plasmon resonance within selectively coated photonic crystal fiber. *Appl. Opt.* **2012**, *51*, 6361–6367. [CrossRef]

29. Hassani, A.; Skorobogatiy, M. Design criteria for microstructured-optical-fiber-based surface-plasmon-resonance sensors. *J. Opt. Soc. Am. B* **2007**, *24*, 1423–1429. [CrossRef]

30. Yang, X.; Lu, Y.; Wang, M.; Yao, J. An exposed-core grapefruit fibers based surface plasmon resonance sensor. *Sensors* **2015**, *15*, 17106–17114. [CrossRef]

31. Zhang, N.M.Y.; Hu, D.J.J.; Shum, P.P.; Wu, Z.; Li, K.; Huang, T.; Wei, L. Design and analysis of surface plasmon resonance sensor based on high-birefringent microstructured optical fiber. *J Opt.* **2016**, *18*, 065005. [CrossRef]

32. Zhang, N.M.Y.; Hu, D.J.J.; Shum, P.P.; Wu, Z.; Li, K.; Huang, T.; Wei, L. High-birefringent microstructured optical fiber based surface plasmon resonance sensor. In Proceedings of the CLEO: Applications and Technology, San Jose, CA, USA, 5–10 June 2016; p. JTu5A.116.

33. Luo, X.; Qiu, T.; Lu, W.; Ni, Z. Plasmons in graphene: Recent progress and applications. *Mater. Sci. Eng. R Rep.* **2013**, *74*, 351–376. [CrossRef]

34. Rodrigo, D.; Limaj, O.; Janner, D.; Etezadi, D.; Garcia de Abajo, F.J.; Pruneri, V.; Altug, H. Mid-infrared plasmonic biosensing with graphene. *Science* **2015**, *349*, 165–168. [CrossRef]

35. Sharma, A.K.; Kaur, B. Fiber optic SPR sensing enhancement in NIR via optimum radiation damping catalyzed by 2D materials. *IEEE Photon. Technol. Lett.* **2018**, *30*, 2021–2024. [CrossRef]

36. Wang, Q.; Jiang, X.; Niu, L.-Y.; Fan, X.-C. Enhanced sensitivity of bimetallic optical fiber SPR sensor based on MoS$_2$ nanosheets. *Opt. Laser Eng.* **2020**, *128*, 105997. [CrossRef]

37. Ju, L.; Geng, B.G.; Horng, J.; Girit, C.; Martin, M.; Hao, Z.; Bechtel, H.A.; Liang, X.; Zettl, A.; Shen, Y.R.; et al. Graphene plasmonics for tunable terahertz metamaterials. *Nat. Nanotechnol.* **2011**, *6*, 630. [CrossRef]

38. Koppens, F.H.; Chang, D.E.; Garcia de Abajo, F.J. Graphene plasmonics: A platform for strong light–matter interactions. *Nano Lett.* **2011**, *11*, 3370–3377. [CrossRef]

39. Salihoglu, O.; Balci, S.; Kocabas, C. Plasmon-polaritons on graphene-metal surface and their use in biosensors. *Appl. Phys. Lett.* **2012**, *100*, 213110. [CrossRef]

40. Zhang, C.; Li, Z.; Jiang, S.Z.; Li, C.H.; Xu, S.C.; Yu, J.; Li, Z.; Wang, M.H.; Liu, A.H.; Man, B.Y. U-bent fiber optic SPR sensor based on graphene/AgNPs. *Sens. Actuators B Chem.* **2017**, *251*, 127–133. [CrossRef]

41. Hu, W.; Huang, Y.; Chen, C.; Liu, Y.; Guo, T.; Guan, B.-O. Highly sensitive detection of dopamine using a graphene functionalized plasmonic fiber-optic sensor with aptamer conformational amplification. *Sens. Actuators B Chem.* **2018**, *264*, 440–447. [CrossRef]

42. Stebunov, Y.V.; Aftenieva, O.A.; Arsenin, A.V.; Volkov, V.S. Highly sensitive and selective sensor chips with graphene-oxide linking layer. *ACS Appl. Mater. Interfaces* **2015**, *7*, 21727–21734. [CrossRef] [PubMed]

43. Wang, Q.; Wang, B.-T. Surface plasmon resonance biosensor based on graphene oxide/silver coated polymer cladding silica fiber. *Sens. Actuators B Chem.* **2018**, *275*, 332–338. [CrossRef]

44. Wang, Q.; Wang, B. Sensitivity enhanced SPR immunosensor based on graphene oxide and SPA co-modified photonic crystal fiber. *Opt. Laser Technol.* **2018**, *107*, 210–215. [CrossRef]

45. Kaushik, S.; Tiwari, U.K.; Pal, S.S.; Sinha, R.K. Rapid detection of Escherichia coli using fiber optic surface plasmon resonance immunosensor based on biofunctionalized Molybdenum disulfide (MoS$_2$) nanosheets. *Biosens. Bioelectron.* **2019**, *126*, 501–509. [CrossRef] [PubMed]

46. Zeng, S.; Hu, S.; Xia, J.; Anderson, T.; Dinh, X.-Q.; Meng, X.-M.; Coquet, P.; Yong, K.-T. Graphene–MoS$_2$ hybrid nanostructures enhanced surface plasmon resonance biosensors. *Sens. Actuators B Chem.* **2015**, *207*, 801–810. [CrossRef]

47. Giovannetti, G.; Khomyakov, P.A.; Brocks, G.; Karpan, V.M.; van den Brink, J.; Kelly, P.J. Doping graphene with metal contacts. *Phys. Rev. Lett.* **2008**, *101*, 026803. [CrossRef]

48. Khomyakov, P.A.; Giovannetti, G.; Rusu, P.C.; Brocks, G.; van den Brink, J.; Kelly, P.J. First-principles study of the interaction and charge transfer between graphene and metals. *Phys. Rev. B* **2009**, *79*, 195425. [CrossRef]

49. Kim, J.A.; Hwang, T.; Dugasani, S.R.; Amin, R.; Kulkarni, A.; Park, S.H.; Kim, T. Graphene based fiber optic surface plasmon resonance for bio-chemical sensor applications. *Sens. Actuators B Chem.* **2013**, *187*, 426–433. [CrossRef]

50. Song, B.; Li, D.; Qi, W.; Elstner, M.; Fan, C.; Fang, H. Graphene on Au (111): A highly conductive material with excellent adsorption properties for high-resolution bio/nanodetection and identification. *Chem. Phys. Chem.* **2010**, *11*, 585–589. [CrossRef]

51. Zhang, N.M.Y.; Li, K.; Shum, P.P.; Yu, X.; Zeng, S.; Wu, Z.; Wang, Q.J.; Yong, K.T.; Wei, L. Hybrid Graphene/Gold Plasmonic Fiber-Optic Biosensor. *Adv. Mater. Technol.* **2017**, *2*, 1600185. [CrossRef]

52. Bhat, S.; Curach, N.; Mostyn, T.; Bains, G.S.; Griffiths, K.R.; Emslie, K.R. Comparison of methods for accurate quantification of DNA mass concentration with traceability to the international system of units. *Anal. Chem.* **2010**, *82*, 7185–7192. [CrossRef]

53. Wei, L.; Alkeskjold, T.T.; Bjarklev, A. Tunable and rotatable polarization controller using photonic crystal fiber filled with liquid crystal. *Appl. Phys. Lett.* **2010**, *96*, 241104. [CrossRef]

54. Wei, L.; Alkeskjold, T.T.; Bjarklev, A. Electrically tunable bandpass filter using solid-core photonic crystal fibers filled with multiple liquid crystals. *Opt. Lett.* **2010**, *35*, 1608–1610. [CrossRef] [PubMed]

55. Wei, L.; Weirich, J.; Alkeskjold, T.T.; Bjarklev, A. On-chip tunable long-period grating devices based on liquid crystal photonic bandgap fibers. *Opt. Lett.* **2009**, *34*, 3818–3820. [CrossRef] [PubMed]

56. Wei, L.; Alkeskjold, T.T.; Bjarklev, A. Compact design of an electrically tunable and rotatable polarizer based on a liquid crystal photonic bandgap fiber. *IEEE Photon. Technol. Lett.* **2009**, *21*, 1633–1635. [CrossRef]

57. Wei, L.; Xue, W.; Chen, Y.; Alkeskjold, T.T.; Bjarklev, A. Optically fed microwave true-time delay based on a compact liquid-crystal photonic-bandgap-fiber device. *Opt. Lett.* **2009**, *34*, 2757–2759. [CrossRef]

58. Wei, L.; Eskildsen, L.; Weirich, J.; Scolari, L.; Alkeskjold, T.T.; Bjarklev, A. Continuously tunable all-in-fiber devices based on thermal and electrical control of negative dielectric anisotropy liquid crystal photonic bandgap fibers. *Appl. Opt.* **2009**, *48*, 497–503. [CrossRef]

59. Li, K.; Zhang, T.; Zhang, N.; Zhang, M.; Zhang, J.; Wu, T.; Ma, S.; Wu, J.; Chen, M.; He, Y.; et al. Integrated liquid crystal photonic bandgap fiber devices. *Front. Optoelectron.* **2016**, *9*, 466–482. [CrossRef]

60. Zhang, N.; Humbert, G.; Wu, Z.; Li, K.; Shum, P.P.; Zhang, N.M.Y.; Cui, Y.; Auguste, J.-L.; Dinh, X.Q.; Wei, L. In-line optofluidic refractive index sensing in a side-channel photonic crystal fiber. *Opt. Express* **2016**, *24*, 27674–27682. [CrossRef]

61. Alsaif, M.M.Y.A.; Latham, K.; Field, M.R.; Yao, D.D.; Medhekar, N.V.; Beane, G.A.; Kaner, R.B.; Russo, S.P.; Ou, J.Z.; Kalantar-zadeh, K. Tunable plasmon resonances in two-dimensional molybdenum oxide nanoflakes. *Adv. Mater.* **2014**, *26*, 3931–3937. [CrossRef]

62. Cheng, H.; Kamegawa, T.; Mori, K.; Yamashita, H. Surfactant-Free Nonaqueous Synthesis of Plasmonic Molybdenum Oxide Nanosheets with Enhanced Catalytic Activity for Hydrogen Generation from Ammonia Borane under Visible Light. *Angew. Chem. Int. Ed.* **2014**, *53*, 2910–2914. [CrossRef]

63. Cheng, H.; Wen, M.; Ma, X.; Kuwahara, Y.; Mori, K.; Dai, Y.; Huang, B.; Yamashita, H. Hydrogen doped metal oxide semiconductors with exceptional and tunable localized surface plasmon resonances. *J. Am. Chem. Soc.* **2016**, *138*, 9316–9324. [CrossRef] [PubMed]

64. Liu, W.; Xu, Q.; Cui, W.; Zhu, C.; Qi, Y. CO$_2$-Assisted Fabrication of Two-Dimensional Amorphous Molybdenum Oxide Nanosheets for Enhanced Plasmon Resonances. *Angew. Chem. Int. Ed.* **2017**, *56*, 1600–1604. [CrossRef] [PubMed]

65. Manthiram, K.; Alivisatos, A.P. Tunable localized surface plasmon resonances in tungsten oxide nanocrystals. *J. Am. Chem. Soc.* **2012**, *134*, 3995–3998. [CrossRef] [PubMed]

66. Lounis, S.D.; Runnerstrom, E.L.; Llordés, A.; Milliron, D.J. Defect chemistry and plasmon physics of colloidal metal oxide nanocrystals. *J. Phys. Chem. Lett.* **2014**, *5*, 1564–1574. [CrossRef]

67. Balendhran, S.; Walia, S.; Alsaif, M.; Nguyen, E.P.; Ou, J.Z.; Zhuiykov, S.; Sriram, S.; Bhaskaran, M.; Kalantar-zadeh, K. Field effect biosensing platform based on 2D α-MoO$_3$. *ACS Nano* **2013**, *7*, 9753–9760. [CrossRef]

68. Zhang, N.M.Y.; Li, K.; Zhang, T.; Shum, P.; Wang, Z.; Wang, Z.; Zhang, N.; Zhang, J.; Wu, T.; Wei, L. Electron-rich two-dimensional molybdenum trioxides for highly integrated plasmonic biosensing. *ACS Photonics* **2018**, *5*, 347–352. [CrossRef]

69. Cheng, H.; Qian, X.; Kuwahara, Y.; Mori, K.; Yamashita, H. A Plasmonic Molybdenum Oxide Hybrid with Reversible Tunability for Visible-Light-Enhanced Catalytic Reactions. *Adv. Mater.* **2015**, *27*, 4616–4621. [CrossRef]

70. Liu, X.; Kang, J.-H.; Yuan, H.; Park, J.; Kim, S.J.; Cui, Y.; Hwang, H.Y.; Brongersma, M.L. Electrical tuning of a quantum plasmonic resonance. *Nat. Nanotechnol.* **2017**, *12*, 866–870. [CrossRef]

71. Prabhakaran, V.; Mehdi, B.L.; Ditto, J.J.; Engelhard, M.H.; Wang, B.; Gunaratne, K.D.D.; Johnson, D.C.; Browning, N.D.; Johnson, G.E.; Laskin, J. Rational design of efficient electrode–electrolyte interfaces for solid-state energy storage using ion soft landing. *Nat. Commun.* **2016**, *7*, 1–10. [CrossRef]

72. Alsaif, M.M.Y.A.; Field, M.R.; Daeneke, T.; Chrimes, A.F.; Zhang, W.; Carey, B.J.; Berean, K.J.; Walia, S.; Embden, J.; Zhang, B.; et al. Exfoliation solvent dependent plasmon resonances in two-dimensional sub-stoichiometric molybdenum oxide nanoflakes. *ACS Appl. Mater. Interfaces* **2016**, *8*, 3482–3493. [CrossRef]

73. Li, K.; Zhang, N.; Zhang, T.; Wang, Z.; Chen, M.; Wu, T.; Ma, S.; Zhang, M.; Zhang, J.; Dinish, U.S.; et al. Formation of ultra-flexible, conformal, and nano-patterned photonic surfaces via polymer cold-drawing. *J. Mater. Chem. C* **2018**, *6*, 4649–4657. [CrossRef]

74. Zhang, N.; Liu, H.; Stolyarov, A.M.; Zhang, T.; Li, K.; Shum, P.P.; Fink, Y.; Sun, X.W.; Wei, L. Azimuthally polarized radial emission from a quantum dot fiber laser. *ACS Photonics* **2016**, *3*, 2275–2279. [CrossRef]

75. Polley, N.; Basak, S.; Hass, R.; Pacholski, C. Fiber optic plasmonic sensors: Providing sensitive biosensor platforms with minimal lab equipment. *Biosens. Bioelectron.* **2019**, *132*, 368–374. [CrossRef] [PubMed]

76. Liang, Y.; Yu, Z.; Li, L.; Xu, T. A self-assembled plasmonic optical fiber nanoprobe for label-free biosensing. *Sci. Rep.* **2019**, *9*, 1–7. [CrossRef]

77. Wei, Y.; Hu, J.; Wu, P.; Su, Y.; Liu, C.; Wang, S.; Nie, X.; Liu, L. Optical Fiber Cladding SPR Sensor Based on Core-Shift Welding Technology. *Sensors* **2019**, *19*, 1202. [CrossRef]

78. Santos, D.F.; Guerreiro, A.; Baptista, J.M. SPR optimization using metamaterials in a D-type PCF refractive index sensor. *Opt. Fiber Technol.* **2017**, *33*, 83–88. [CrossRef]

79. Semwal, V.; Shrivastav, A.M.; Verma, R.; Gupta, B.D. Surface plasmon resonance based fiber optic ethanol sensor using layers of silver/silicon/hydrogel entrapped with ADH/NAD. *Sens. Actuators B Chem.* **2016**, *230*, 485–492. [CrossRef]

80. Mayer, K.M.; Hafner, J.H. Localized surface plasmon resonance sensors. *Chem. Rev.* **2011**, *111*, 3828–3857. [CrossRef]

81. Peiris, S.; McMurtrie, J.; Zhu, H.-Y. Metal nanoparticle photocatalysts: Emerging processes for green organic synthesis. *Catal. Sci. Technol.* **2016**, *6*, 320–338. [CrossRef]

82. Parkins, G.R.; Lawrence, W.E.; Christy, R.W. Intraband optical conductivity σ (ω, T) of Cu, Ag, and Au: Contribution from electron-electron scattering. *Phys. Rev. B* **1981**, *23*, 6408. [CrossRef]

83. Luther, J.M.; Jain, P.K.; Ewers, T.; Alivisatos, A.P. Localized surface plasmon resonances arising from free carriers in doped quantum dots. *Nat. Mater.* **2011**, *10*, 361–366. [CrossRef] [PubMed]

84. Cao, J.; Tu, M.H.; Sun, T.; Grattan, K.T.V. Wavelength-based localized surface plasmon resonance optical fiber biosensor. *Sens. Actuators B Chem.* **2013**, *181*, 611–619. [CrossRef]

85. Tu, M.H.; Sun, T.; Grattan, K.T.V. LSPR optical fibre sensors based on hollow gold nanostructures. *Sens. Actuators B Chem.* **2014**, *191*, 37–44. [CrossRef]

86. Urrutia, A.; Goicoechea, J.; Rivero, P.J.; Pildain, A.; Arregui, F.J. Optical fiber sensors based on gold nanorods embedded in polymeric thin films. *Sens. Actuators B Chem.* **2018**, *255*, 2105–2112. [CrossRef]

87. Paul, D.; Biswas, R. Highly sensitive LSPR based photonic crystal fiber sensor with embodiment of nanospheres in different material domain. *Opt. Laser Technol.* **2018**, *101*, 379–387. [CrossRef]

88. Shukla, G.M.; Punjabi, N.; Kundu, T.; Mukherji, S. Optimization of plasmonic U-shaped optical fiber sensor for mercury ions detection using glucose capped silver nanoparticles. *IEEE Sens. J.* **2019**, *19*, 3224–3231. [CrossRef]

89. Kim, J.; Oh, S.Y.; Shukla, S.; Hong, S.B.; Heo, N.S.; Bajpai, V.K.; Chun, H.S.; Jo, C.-H.; Choi, B.G.; Huh, Y.S.; et al. Heteroassembled gold nanoparticles with sandwich-immunoassay LSPR chip format for rapid and sensitive detection of hepatitis B virus surface antigen (HBsAg). *Biosens. Bioelectron.* **2018**, *107*, 118–122. [CrossRef]

90. Sciacca, B.; Monro, T.M. Dip biosensor based on localized surface plasmon resonance at the tip of an optical fiber. *Langmuir* **2014**, *30*, 946–954. [CrossRef]

91. Jia, S.; Bian, C.; Sun, J.; Tong, J.; Xia, S. A wavelength-modulated localized surface plasmon resonance (LSPR) optical fiber sensor for sensitive detection of mercury (II) ion by gold nanoparticles-DNA conjugates. *Biosens. Bioelectron.* **2018**, *114*, 15–21. [CrossRef]

92. Liang, G.; Zhao, Z.; Wei, Y.; Liu, K.; Hou, W.; Duan, Y. Plasma enhanced label-free immunoassay for alpha-fetoprotein based on a U-bend fiber-optic LSPR biosensor. *RSC Adv.* **2015**, *5*, 23990–23998. [CrossRef]

93. Oliverio, M.; Perotto, S.; Messina, G.C.; Lovato, L.; De Angelis, F. Chemical functionalization of plasmonic surface biosensors: A tutorial review on issues, strategies, and costs. *ACS Appl. Mater. Interfaces* **2017**, *9*, 29394–29411. [CrossRef] [PubMed]

94. Bolduc, O.R.; Masson, J.-F. Advances in surface plasmon resonance sensing with nanoparticles and thin films: Nanomaterials, surface chemistry, and hybrid plasmonic techniques. *Anal. Chem.* **2011**, *83*, 8057–8062. [CrossRef] [PubMed]

95. Mieszawska, A.J.; Mulder, W.J.M.; Fayad, Z.A.; Cormode, D.P. Multifunctional gold nanoparticles for diagnosis and therapy of disease. *Mol. Pharm.* **2013**, *10*, 831–847. [CrossRef]

96. Tadepalli, S.; Kuang, Z.; Jiang, Q.; Liu, K.-K.; Fisher, M.A.; Morrissey, J.J.; Kharasch, E.D.; Slocik, J.M.; Naik, R.R.; Singamaneni, S. Peptide functionalized gold nanorods for the sensitive detection of a cardiac biomarker using plasmonic paper devices. *Sci. Rep.* **2015**, *5*, 16206. [CrossRef]

97. Lin, H.-Y.; Huang, C.-H.; Huang, C.-C.; Liu, Y.-C.; Chau, L.-K. Multiple resonance fiber-optic sensor with time division multiplexing for multianalyte detection. *Opt. Lett.* **2012**, *37*, 3969–3971. [CrossRef]

98. Li, K.; Liu, G.; Wu, Y.; Hao, P.; Zhou, W.; Zhang, Z. Gold nanoparticle amplified optical microfiber evanescent wave absorption biosensor for cancer biomarker detection in serum. *Talanta* **2014**, *120*, 419–424. [CrossRef] [PubMed]

99. Ricciardi, A.; Crescitelli, A.; Vaiano, P.; Quero, G.; Consales, M.; Pisco, M.; Esposito, E.; Cusano, A. Lab-on-fiber technology: A new vision for chemical and biological sensing. *Analyst* **2015**, *140*, 8068–8079. [CrossRef] [PubMed]

100. Qu, D.-H.; Wang, Q.-C.; Zhang, Q.-W.; Ma, X.; Tian, H. Photoresponsive host–guest functional systems. *Chem. Rev.* **2015**, *115*, 7543–7588. [CrossRef] [PubMed]

101. Yang, Y.-W.; Sun, Y.-L.; Song, N. Switchable host–guest systems on surfaces. *Acc. Chem. Res.* **2014**, *47*, 1950–1960. [CrossRef]

102. Lagona, J.; Mukhopadhyay, P.; Chakrabarti, S.; Isaacs, L. The cucurbit[n]uril family. *Angew. Chem. Int. Ed.* **2005**, *44*, 4844–4870. [CrossRef]

103. Yu, G.; Jie, K.; Huang, F. Supramolecular amphiphiles based on host–guest molecular recognition motifs. *Chem. Rev.* **2015**, *115*, 7240–7303. [CrossRef] [PubMed]

104. Xie, J.; Zuo, T.; Huang, Z.; Huan, L.; Gu, Q.; Gao, C.; Shao, J. Theoretical study of a novel imino bridged pillar[5]arene derivative. *Chem. Phys. Lett.* **2016**, *662*, 25–30. [CrossRef]

105. Li, H.; Chen, D.-X.; Sun, Y.-L.; Zheng, Y.B.; Tan, L.-L.; Weiss, P.S.; Yang, Y.-W. Viologen-mediated assembly of and sensing with carboxylatopillar[5]arene-modified gold nanoparticles. *J. Am. Chem. Soc.* **2013**, *135*, 1570–1576. [CrossRef]

106. Sánchez, A.; Díez, P.; Villalonga, R.; Martínez-Ruiz, P.; Eguílaz, M.; Fernández, I.; Pingarrón, J.M. Seed-mediated growth of jack-shaped gold nanoparticles from cyclodextrin-coated gold nanospheres. *Dalton Trans.* **2013**, *42*, 14309–14314. [CrossRef] [PubMed]

107. Bhoi, V.I.; Kumar, S.; Murthy, C.N. Cyclodextrin encapsulated monometallic and inverted core–shell bimetallic nanoparticles as efficient free radical scavengers. *New J. Chem.* **2016**, *40*, 1396–1402. [CrossRef]

108. Zhao, Y.; Huang, Y.; Zhu, H.; Zhu, Q.; Xia, Y. Three-in-one: Sensing, self-assembly, and cascade catalysis of cyclodextrin modified gold nanoparticles. *J. Am. Chem. Soc.* **2016**, *138*, 16645–16654. [CrossRef]

109. Zhang, N.M.Y.; Qi, M.; Wang, Z.; Wang, Z.; Chen, M.; Li, K.; Shum, P.; Wei, L. One-step synthesis of cyclodextrin-capped gold nanoparticles for ultra-sensitive and highly-integrated plasmonic biosensors. *Sens. Actuators B Chem.* **2019**, *286*, 429–436. [CrossRef]

110. Lu, Y.; Li, H.; Qian, X.; Zheng, W.; Sun, Y.; Shi, B.; Zhang, Y. Beta-cyclodextrin based reflective fiber-optic SPR sensor for highly-sensitive detection of cholesterol concentration. *Opt. Fiber Technol.* **2020**, *56*, 102187. [CrossRef]

111. Yu, Y.; Chipot, C.; Cai, W.; Shao, X. Molecular dynamics study of the inclusion of cholesterol into cyclodextrins. *J. Phys. Chem. B* **2006**, *110*, 6372–6378. [CrossRef]

112. Christoforides, E.; Papaioannou, A.; Bethanis, K. Crystal structure of the inclusion complex of cholesterol in β-cyclodextrin and molecular dynamics studies. *Beilstein J. Org. Chem.* **2018**, *14*, 838–848. [CrossRef]

113. Schneider, H.-J.; Hacket, F.; Rüdiger, V.; Ikeda, H. NMR studies of cyclodextrins and cyclodextrin complexes. *Chem. Rev.* **1998**, *98*, 1755–1786. [CrossRef] [PubMed]

114. Huang, H.; Yang, X. Synthesis of polysaccharide-stabilized gold and silver nanoparticles: A green method. *Carbohydr. Res.* **2004**, *339*, 2627–2631. [CrossRef] [PubMed]

115. Sadani, K.; Nag, P.; Mukherji, S. LSPR based optical fiber sensor with chitosan capped gold nanoparticles on BSA for trace detection of Hg (II) in water, soil and food samples. *Biosens. Bioelectron.* **2019**, *134*, 90–96. [CrossRef] [PubMed]

116. Lee, B.; Park, J.-H.; Byun, J.-Y.; Kim, J.H.; Kim, M.-G. An optical fiber-based LSPR aptasensor for simple and rapid in-situ detection of ochratoxin A. *Biosens. Bioelectron.* **2018**, *102*, 504–509. [CrossRef] [PubMed]

117. Xu, B.; Ma, S.; Xiang, Y.; Zhang, J.; Zhu, M.; Wei, L.; Tao, G.; Deng, D. In-fiber structured particles and filament array from the perspective of fluid instabilities. *Adv. Opt. Mater.* **2020**, *2*, 1–12. [CrossRef]

118. Zhang, Q.; Man, P.; He, B.; Li, C.; Li, Q.; Pan, Z.; Wang, Z.; Yang, J.; Wang, Z.; Zhou, Z.; et al. Binder-free NaTi₂(PO₄)₃ anodes for high-performance coaxial-fiber aqueous rechargeable sodium-ion batteries. *Nano Energy* **2020**, *67*, 104212. [CrossRef]

119. Zhang, J.; Wang, Z.; Wang, Z.; Zhang, T.; Wei, L. In-fiber production of laser-structured stress-mediated semiconductor particles. *ACS Appl. Mater. Interfaces* **2019**, *11*, 45330–45337. [CrossRef]

120. Zhang, J.; Wang, Z.; Wang, Z.; Zhang, T.; Wei, L. In-fibre particle manipulation and device assembly via laser induced thermocapillary convection. *Nat. Commun.* **2019**, *10*, 5206. [CrossRef]

121. Zhang, Q.; Li, C.; Li, Q.; Pan, Z.; Sun, J.; Zhou, Z.; He, B.; Man, P.; Xie, L.; Kang, L.; et al. Flexible and high-voltage coaxial-fiber aqueous rechargeable zinc-ion battery. *Nano Lett.* **2019**, *19*, 4035–4042. [CrossRef]

122. Zhang, Q.; Li, L.; Li, H.; Tang, L.; He, B.; Li, C.; Pan, Z.; Zhou, Z.; Lia, Q.; Sun, J.; et al. Ultra-endurance coaxial-fiber stretchable sensing systems fully powered by sunlight. *Nano Energy* **2019**, *60*, 267–274. [CrossRef]

123. Zhang, T.; Wang, Z.; Srinivasan, B.; Wang, Z.; Zhang, J.; Li, K.; Boussard-Pledel, C.; Troles, J.; Bureau, B.; Wei, L. Ultra-flexible glassy semiconductor fibers for thermal sensing and positioning. *ACS Appl. Mater. Interfaces* **2019**, *11*, 2441–2447. [CrossRef]

124. Yan, W.; Page, A.; Nguyen-Dang, T.; Qu, Y.; Sordo, F.; Wei, L.; Sorin, F. Advanced multi-material electronic and optoelectronic fibers and textiles. *Adv. Mater.* **2018**, *31*, 1802348. [CrossRef] [PubMed]

125. Ma, S.; Ye, T.; Zhang, T.; Wang, Z.; Li, K.; Chen, M.; Zhang, J.; Wang, Z.; Ramakrishna, S.; Wei, L. Highly oriented electrospun P(VDF-TrFE) fibers via mechanical stretching for wearable motion sensing. *Adv. Mater. Technol.* **2018**, *3*, 1800033. [CrossRef]

Sensors **2020**, *20*, 3266

126. Zhang, T.; Li, K.; Zhang, J.; Chen, M.; Wang, Z.; Ma, S.; Zhang, N.; Wei, L. High-performance, flexible, and ultralong crystalline thermoelectric fibers. *Nano Energy* **2017**, *41*, 35–42. [CrossRef]

127. Zhang, J.; Li, K.; Zhang, T.; Buenconsejo, P.J.S.; Chen, M.; Wang, Z.; Zhang, M.; Wang, Z.; Wei, L. Laser induced in-fiber fluid dynamical instabilities for precise and scalable fabrication of spherical particles. *Adv. Funct. Mater.* **2017**, *27*, 1703245. [CrossRef]

128. Wang, S.; Zhang, T.; Li, K.; Ma, S.; Chen, M.; Lu, P.; Wei, L. Flexible piezoelectric fibers for acoustic sensing and positioning. *Adv. Electron. Mater.* **2017**, *3*, 1600449. [CrossRef]

129. Wei, L.; Hou, C.; Levy, E.; Lestoquoy, G.; Gumennik, A.; Abouraddy, A.F.; Joannopoulos, D.; Fink, Y. Optoelectronic fibers via selective amplification of in-fiber capillary instabilities. *Adv. Mater.* **2017**, *29*, 1603033. [CrossRef]

130. Shabahang, S.; Tao, G.; Kaufman, J.J.; Qiao, Y.; Wei, L.; Bouchenot, T.; Gordon, A.; Fink, Y.; Bai, Y.; Hoy, R.S.; et al. Controlled fragmentation of multimaterial fibres and films via polymer cold-drawing. *Nature* **2016**, *534*, 529–533. [CrossRef]

131. Hou, C.; Jia, X.; Wei, L.; Tan, S.-C.; Zhao, X.; Joannopoulos, J.D.; Fink, Y. Crystalline silicon core fibres from aluminium core preforms. *Nat. Commun.* **2015**, *6*, 6248. [CrossRef]

132. Gumennik, A.; Wei, L.; Lestoquoy, G.; Stolyarov, A.M.; Jia, X.; Rekemeyer, P.H.; Smith, M.J.; Liang, X.; Grena, B.; Johnson, S.G.; et al. Silicon-in-silica spheres via axial thermal gradient in-fibre capillary instabilities. *Nat. Commun.* **2013**, *4*, 2216. [CrossRef]

133. Hou, C.; Jia, X.; Wei, L.; Stolyarov, A.M.; Shapira, O.; Joannopoulos, J.D.; Fink, Y. Direct atomic-level observation and chemical analysis of ZnSe synthesized by in situ high-throughput reactive fiber drawing. *Nano Lett.* **2013**, *13*, 975–979. [CrossRef] [PubMed]

134. Stolyarov, A.M.; Wei, L.; Shapira, O.; Sorin, F.; Chua, S.L.; Joannopoulos, J.D.; Fink, Y. Microfluidic directional emission control of an azimuthally polarized radial fibre laser. *Nat. Photonics* **2012**, *6*, 229–233. [CrossRef]

135. Stolyarov, A.M.; Wei, L.; Sorin, F.; Lestoquoy, G.; Joannopoulos, J.D.; Fink, Y. Fabrication and characterization of fibers with built-in liquid crystal channels and electrodes for transverse incident-light modulation. *Appl. Phys. Lett.* **2012**, *101*, 011108. [CrossRef]

Review

Carbon Allotrope-Based Optical Fibers for Environmental and Biological Sensing: A Review

Stephanie Hui Kit Yap [†], Kok Ken Chan [†], Swee Chuan Tjin * and Ken-Tye Yong *

School of Electrical and Electronic Engineering, Nanyang Technological University, 50 Nanyang Avenue, Singapore 639798, Singapore; step0031@e.ntu.edu.sg (S.H.K.Y.); kchan019@e.ntu.edu.sg (K.K.C.)
* Correspondence: esctjin@ntu.edu.sg (S.C.T.); ktyong@ntu.edu.sg (K.-T.Y.)
† S.H.K.Y. and K.K.C. contributed equally in this work.

Received: 3 March 2020; Accepted: 31 March 2020; Published: 5 April 2020

Abstract: Recently, carbon allotropes have received tremendous research interest and paved a new avenue for optical fiber sensing technology. Carbon allotropes exhibit unique sensing properties such as large surface to volume ratios, biocompatibility, and they can serve as molecule enrichers. Meanwhile, optical fibers possess a high degree of surface modification versatility that enables the incorporation of carbon allotropes as the functional coating for a wide range of detection tasks. Moreover, the combination of carbon allotropes and optical fibers also yields high sensitivity and specificity to monitor target molecules in the vicinity of the nanocoating surface. In this review, the development of carbon allotropes-based optical fiber sensors is studied. The first section provides an overview of four different types of carbon allotropes, including carbon nanotubes, carbon dots, graphene, and nanodiamonds. The second section discusses the synthesis approaches used to prepare these carbon allotropes, followed by some deposition techniques to functionalize the surface of the optical fiber, and the associated sensing mechanisms. Numerous applications that have benefitted from carbon allotrope-based optical fiber sensors such as temperature, strain, volatile organic compounds and biosensing applications are reviewed and summarized. Finally, a concluding section highlighting the technological deficiencies, challenges, and suggestions to overcome them is presented.

Keywords: graphene; carbon nanotubes; carbon dots; nanodiamonds; nanomaterials; optical fiber; sensors; nanocoating

1. Introduction

Carbon allotropes have been extensively used in many sensing applications for targets such as temperature, pressure, magnetics, environmental pollutants, and biomolecules, either on their own or via other host-supports such as optical fibers, electrodes, and field-effect transistors [1–4]. Among these technologies, optical fiber-based sensors have attracted significant interest due to the surface versatility of silica or plastic optical fibers that allows a wide range of surface modifications. Other interesting properties, such as small dimensions and lightweight features that enable compact system design, real-time monitoring, multiplexing capabilities, and resistance to harsh environments, also offer significant advantages for the development of practical carbon allotrope-based optical fiber sensors (OFS) [5]. Carbon allotrope-based OFS exist in various system designs due to the variety of fiber structures, optical interrogation methods, and deposition techniques that can be adopted to achieve highly sensitive and selective detection of target molecules. Nevertheless, these sensors usually share a common sensing scheme, where a small portion of the guided wave energy, known as the evanescence wave, penetrates the fiber cladding and interacts with the nanocarbon coating [6]. The bindings of molecules or physical changes in the surrounding environment can be detected by the evanescence wave and contribute to a refractive index and/or optical property change that enables quantification of target molecules or physical parameters. In some instances, surface-modified carbon allotropes also permit multi-parameter detection capabilities of carbon allotrope-based OFS [7]. As such, an insight into the recent advances of carbon allotrope-based OFS can serve as a guideline to develop an effective detection approach for emerging real-world applications. Many review articles on carbon allotrope-based OFS mainly focus on a single type of carbon allotrope and the fundamental theory of carbon allotrope-based OFS [8,9]. Thus, this comprehensive review article encompasses the properties, preparation, sensing mechanisms, nanocoating deposition techniques, physical and biochemical sensing applications. This review aims to highlight the recent research work on carbon allotrope-based OFS that will serve as a reference guide for researchers to develop optimal detection approaches for physical parameters or trace level monitoring of chemical and biomolecules. Four groups of carbon allotropes, including carbon nanotubes (CNT), carbon dots (CDs), graphene, and nanodiamonds (NDs), will be studied. Commonly adopted synthesis approaches, classification of various deposition techniques for the integration of carbon allotropes as the thin film coating of OFS as well as numerous examples of these carbon allotrope-based OFS will also be outlined.

2. Classifications of Carbon Allotropes

2.1. Carbon Nanotubes

CNTs are one of the most well-known members of the nanocarbon family. CNTs can be divided into single-walled CNTs (SWCNTs) and multi-walled CNTs (MWCNTs), both of which were discovered by Iijima in 1991 and 1993, respectively [10,11]. Since then, there has been a great interest in these species due to their outstanding structural, mechanical, and electronic properties. Generally, SWCNTs comprise one layer of sp^2-hybridized carbon atoms rolled up into a seamless cylinder with diameter and length in the nanometer and micrometer range, respectively. On the other hand, MWCNTs consist of multiple concentric CNTs with an interlayer spacing of 3.4 Å [12]. Physically, CNTs have high length-to-diameter aspect ratios, often exceeding 10,000, and therefore are one of the most anisotropic nanomaterials ever produced. Mechanically, CNTs are among the strongest and stiffest fibers, attributed to the strong sp^2 bonds between the individual carbon atoms.

Typically, the surface properties of CNTs are the main reason for the inability of CNTs to disperse in organic or polar solvents. Even though the two ends of the CNTs exhibit oxygen-containing moieties that are generally hydrophilic, the wall that constitutes a major portion of the CNT's surface area is hydrophobic [13,14]. Thus, CNTs are often solidly held together in bundles due to the strong van der Waals interaction. The dispersion and modification of hydrophobic CNTs are often a major challenge during the functionalization of CNTs onto the optical fibers [15]. Since CNTs in aqueous or polar solvents tend to aggregate swiftly, they are often dispersed in non-polar organic solvents such as dimethylformamide (DMF) or by modifying the CNTs with polymers or surfactants. Nevertheless, the dispersity obstacle in aqueous solution can be indirectly viewed as an advantage since optical fibers functionalized with CNTs dispersed in organic solvents enable the solvent to be easily removed by evaporation [15].

The electronic properties of CNTs are highly dependent on the physical structure, such as the atomic arrangement of the carbon atoms (chirality), length, and diameter of CNTs (Figure 1). The chirality of SWCNT indicates the angle at which a graphene sheet is being rolled up, as well as the alignment of the π-orbitals. The atomic structure of CNTs can be defined in the form of a chiral vector: $\vec{c_h} = n\vec{a_1} + m\vec{a_2}$, where n and m can be termed as the number of steps along with the zigzag carbon bonds of the honeycomb lattice and $\vec{a_1}$ and $\vec{a_2}$ are the unit vectors. There are three different types of CNTs, namely armchair, zigzag, and chiral tubes. Armchair tubes have an equal n and m values and have a chiral angle of 30°. On the other hand, zigzag tubes have m=0, and exhibit a chiral angle of 0° while chiral tubes can exhibit any other values. The chirality is vital in determining the conductivity property of the CNTs. Conducting CNTs are in achiral and armchair configuration (n, n). Alternatively, chiral (n, m) and achiral zigzag $(n, 0)$ CNTs are semiconductors with the exception when the $\frac{n-m}{3}$ results in a whole number. As the diameter of CNTs increases, the band gap tends to reduce and can result in a zero-bandgap semiconductor [15]. Besides chirality, the properties of CNTs can also be influenced by catalytic particles and dopants [16,17], as well as functionalization of the side walls [18,19]. The physical and electronic properties have led CNTs to be integrated into sensors with various types of sensing modalities, exhibiting outstanding adaptive and sensory capabilities.

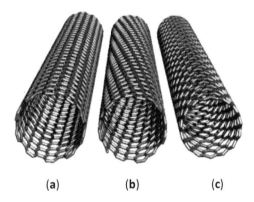

(a)   **(b)**   **(c)**

Figure 1. The idealized illustration of three distinctive SWCNTs with open ends. (**a**) an achiral metallic conductive armchair (10,10) SWCNT, (**b**) a chiral semi-conductive (12,7) SWCNT, (**c**) an achiral conductive zigzag (15,0) SWCNT. SWCNT in (c) is conductive as $\frac{n-m}{3}$ results in a whole number [18]. Copyright © 2002, John Wiley and sons.

2.2. Carbon Dots

Fluorescent CDs were discovered accidentally during a typical gel electrophoresis purification of SWCNTs prepared via an arc-discharge procedure [20]. The fluorescent carbon nanoparticles, which later became known as CDs, are identified and separated from the carbon soot as a by-product of the arc-discharge process. CDs, also known as carbon quantum dots (CQDs) or carbon nanodots (CNDs), is another category of carbonaceous material. CDs are sometimes used interchangeably with graphene quantum dots (GQDs). However, there is some obvious physical distinctions between CDs and GQDs. GQDs refer to graphene monolayers that are fragmentized into nanosized pieces, comprising mainly of sp²-hybridized carbon atoms [21]. On the other hand, CDs are quasi-spherical carbon nanoparticles with a dimension below 10 nm [22–24]. CDs typically consist of amorphous or crystalline cores with sp²-hybridized carbon configuration. Some studies have also described CDs to exhibit diamond-like sp³ carbon configuration [25].

Surface moieties on CDs are typically introduced during the synthesis process. Different types of surface functional groups such as C=O, C–O, C–OH, and C=C can exist on the CDs surface and is highly dependent on the types of precursors used. For instance, Dong et al. [26] synthesized fluorescent CDs using citric acid and branched polyethyleneimine (BPEI) and low-temperature heating. The resultant product was found to be covered with amino-rich BPEI. These functional groups are vital for sensing applications since they can form certain coordination bonds with specific molecules and trigger some optical properties changes. The precursors used for the preparation of CDs can also introduce various types of dopant ions into CDs that can modulate the optical and sensing properties of CDs [27,28]. Shan et al. [29] employed boron-doped CDs prepared via a one-pot solvothermal synthesis using boron tribromide as the boron source and hydroquinone as the carbon precursor to sensitively detect hydrogen peroxide and glucose. Post functionalization with specific chelating groups or biomolecules is another method to endow CDs with sensitive and specific targeting capabilities [30]. In many CDs-based sensing schemes, there will be changes in fluorescence emission intensity that can be attributed to three main mechanisms, namely the inner filter effect, photo-induced electron transfer, and Forster resonance energy transfer [3].

The fluorescence property of CDs is a unique feature that is widely used in sensing applications, and the emission of CDs can be either excitation-dependent or excitation-independent [31–33]. For excitation-dependent CDs, the fluorescence emission wavelength can be tuned from 400 to 750 nm with a gradual increase in excitation wavelength [34]. The fluorescence intensity of the CDs can also be influenced by environmental factors such as types of solvents [35], Ph [36], temperature [37], and the concentration of CDs [38]. Interestingly, CDs also exhibit up-conversion fluorescence emission that is highly beneficial for in vivo biological applications [24,33,39,40]. Even though it is still a matter of intensive discussion, the origin of the fluorescence property is generally attributed to the quantum confinement effect, various surface functional groups, and existence of fluorophore species on the CDs surface [3].

2.3. Graphene

Graphene consists of two-dimensional covalently bonded monolayer carbon atoms arranged in a hexagonal network. Graphene has been long believed as a hypothetical structure until proven experimentally by the ground-breaking work of Novoselov et al. [41], which earned them the Nobel Prize in 2010. Novoselov et al. [41] developed a strategy to isolate single-layer graphene from the highly oriented pyrolytic graphite via repeated peeling using scotch tape. Since then, there has been an exponential increase in research employing graphene in optoelectronics [42], energy storage [43], energy conversion [44], and biomedical applications [45,46], due to its many unique virtues. Graphene is also an interesting candidate for sensing applications owing to its high sensitivity towards external stimuli since each carbon atom is a surface atom, thus having an extremely high surface to volume ratio [47]. Due to the delocalized π-electrons on its surface, the physical properties of graphene can also be tuned and modified to allow specific interaction with certain molecules [48].

The electrical properties of graphene also play an important role in sensing applications. Specifically, the electrical conductivity of graphene will change after the absorption of molecules on the surface of graphene because the molecules may act as electron donors or acceptors and thus, affecting the carrier concentration [49,50]. Moreover, graphene is highly conductive with low Johnson noise therefore a small variation in carrier concentration can result in a notable change in electrical conductivity. Besides, graphene has also been reported to possess surface-enhanced Raman scattering (SERS) property, enabling trace-level of target molecules to be detected by amplifying the characteristic Raman signals [51–53]. Serving as an alternative to noble metals such as gold or silver, graphene exhibits tunable surface plasmons at infrared and THz frequencies [54,55]. The electronic band configuration of graphene is determined by a combination of linear dispersion relation and vanishing density of states at the Fermi level in its neutral state [56,57]. As the Fermi energy deviates from the neutrality point, graphene exhibits metallic optical response, leading to the existence of plasmons. Many studies also described the advantages of graphene plasmons that include lengthy lifetime, large spatial confinement and field enhancement, as well as tunability via electrostatic grating [58].

Apart from the electronic properties, graphene presents distinctive optical properties that are widely used for sensing applications. Despite being a zero-bandgap material at pristine condition, graphene oxide (GO) that has heterogeneous functional moieties exhibits strong emission from the UV to near-infrared range [47]. The strong emission is attributed to the electronic transition between the pristine sp^2 carbon domain and the functional moieties located at the boundaries of the GO sheets [59]. The fluorescence emission of GO can be enhanced or quenched, depending on the presence and concentration of the target molecules [60]. GO is often applied as an active material that is functionalized on an optical fiber. In many cases, the optical fiber served as the transmission medium to send the excitation signal as well as to capture the fluorescence emission to the photodetector. Therefore, an effective setup configuration that can couple maximum fluorescence emission intensity back to the fiber is critical, particularly for microstructured optical fibers [61].

2.4. Nanodiamonds

Diamond is long-known for its outstanding properties such as superior thermal conductivity and extreme hardness. However, nano-scaled diamonds, also known as NDs, were only discovered by Soviet scientists in the 1960s [62]. The NDs were detected in the soot after the detonation of oxygen-deficient TNT/hexogen composition in an inert environment with no additional carbon supply [63]. NDs continued to be relatively unknown until the end of the 1980s [64]. Unlike CNTs, graphene, and CDs that consist of sp^2 graphitic carbon, NDs comprise only of sp^3 carbon atoms and have a diamonoid-like morphology. Generally, the dimension of NDs ranges from 2 to 20 nm, which are considerably smaller than bulk diamond and diamond abrasive powders but are bigger than organic diamonoid molecules [47]. NDs tend to form aggregates, and even commercial NDs contain large NDs clusters that cannot be dispersed via ultrasonication treatment [65]. Thus, many methods have been developed to de-aggregate the NDs. Osawa et al. [66] developed two methods to break up NDs aggregate in various non-aqueous mediums. The first method is by using stirred-media milling with zirconia microbeads, capable of reducing the diameter of the NDs from 200 nm to 4–5 nm within 100 min. Despite being able to break up large aggregates, this method also wears out the beads, blades, and vessels that introduces zirconia contamination to the NDs solution. Thus, further treatment with strong acids is required to remove the zirconia nanoparticles. The second method is by using high-power ultrasonication with the assistance of zirconia beads. This bead-assisted sonic disintegration (BASD) method can reduce the size of the aggregates to a similar dimension as compared to the first method without requiring any post-treatment. Dry milling is another economical and facile approach that does not introduce any contaminant species and reduces the size of NDs aggregates from micrometer to nanometer range [67]. Pentecost employed water-soluble compounds such as sodium chloride and sucrose during the milling process that can then be removed by rinsing the NDs with water. Ultracentrifugation is another contaminant-free strategy to separate NDs into different sizes by mass and dimension, but the yield of this method to obtain single-digit NDs is very low [68].

The most unique property of NDs which distinguishes them from other carbon allotropes is the presence of the fluorescent defect center, known as nitrogen-vacancy (N-V) center. N-V centers have been an important characteristic for sensing applications as their PL is strong and resistant to photobleaching with an obvious zero-phonon line even at ambient temperature, electron spin triplet nature of the electronic ground state as well as the dependence of PL emission intensity on the strength of spin projection on the symmetry axis of the N-V center [69]. The N-V center is a defect in the crystal structure of NDs, as shown in Figure 2. One out of the two neighboring carbon atoms in the NDs crystalline lattice is substituted with a nitrogen atom while the other is a vacancy without any replacement atom. The two unpaired nitrogen electrons form the spin triplet ground and excited states $m_s = 0, \pm1$. These states can be optically initialized, manipulated and determined at ambient temperature. After being optically excited, the N-V centers can transit between the ground and electronically excited states. The N-V centers can relax to the ground state via radiative and nonradiative pathways. Practically, the radiative transition can results in a wide PL spectrum with a zero-phonon line at 637 nm that can be employed for accurate sensing measurement [70,71]. On the other hand, the nonradiative pathway is the intersystem crossing (ISC) to the singlet states located below the excited triplet state [72]. A resonant microwave frequency can also be used as the excitation source to excite the population to $m_s = \pm1$ spin level. Subsequently, non-radiative decay can take place as a result of ISC, and a decrease in PL emission can be detected. For magnetometry, the N-V center under an unknown magnetic field will produce a split between the $m_s = \pm1$ spin sub-level. The electron spin resonance transition between the sublevels can be employed to determine the strength and direction of the magnetic field. In summary, Table 1 shows the properties comparison for different carbon allotropes.

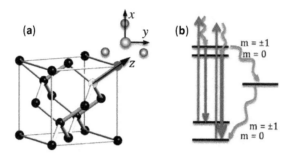

Figure 2. (**a**) N-V center in NDs. N and C are colored in red and black, respectively while vacancy is represented by yellow. The x and y-axes are shown on top of the z-axis. (**b**) The electronic states of NDs at room temperature. As a result of the double degeneracy of the molecular orbitals of the excited spin-triplet state, there are two orthogonal transition dipole moments. Spin-preserving PL and optical excitation are illustrated by green and red arrows, respectively [69]. Copyright © 2017, Elsevier.

Table 1. Comparison of carbon allotropes properties.

Type of Carbon Allotrope	Properties
Carbon Nanotubes	• High length to diameter aspect ratios • Requires functionalization to reduce the hydrophobicity • Tends to aggregate due to strong van der Waals interaction • Electronic properties are dependent on the chirality, length and diameter of CNTs.
Carbon Dots	• Have tunable fluorescence • Can be excitation dependent or excitation independent • Have numerous surface moieties on the surface • Doping and functionalization can improve sensing capabilities • Flexibility in selecting the starting precursors
Graphene	• High surface to volume ratio • Conductivity of graphene can be influenced by the attachment of analyte • Can be employed for various sensing modalities based on their electronic, SERS and fluorescence characteristics
Nanodiamonds	• Fluorescence based on the nitrogen-vacancy center • Resistant to photobleaching • Fluorescence is sensitive to magnetic field changes

3. Synthesis Approaches of Carbon Allotropes

Generally, carbon allotropes are prepared from carbon precursors such as graphite, organic gases, green organic compound, or volatile organic compounds (VOCs) by using various synthetic approaches to reorganize the carbon atoms. In this section, the preparation methods of CNTs, CDs, graphene, and NDs will be comprehensively discussed and reviewed.

3.1. Carbon Nanotubes

CNTs can be prepared via three main techniques as follow: arc-discharge method, chemical vapor deposition (CVD), and laser ablation method. The arc-discharge procedure is carried out in a vacuum chamber using two carbon electrodes as the precursor [73,74]. The chamber is filled with inert gas to expedite the carbon deposition to form MWCNTs with near-perfect morphology. A similar condition is necessary for the formation of SWCNTs but with the addition of catalysts such as Ni, Fe, Co, Pt, and Rh. On the other hand, the laser ablation technique employs an intense laser pulse to hit a carbon target in a furnace filled with inert air with the assistance of a catalyst. The bombardment of the laser beam will vaporize the carbon precursor and form a graphene film on the substrate. Despite being able to produce high-quality CNTs, both of these methods require high preparation temperatures of about 3000–4000 °C to evaporate the carbon atoms from the carbon precursor.

In contrast, the CVD technique can be used to prepare CNTs at a much lower temperature [75]. Typically, hydrocarbon gases such as methane or ethylene are channeled into a reaction chamber and will break down into reactive species at a temperature between 500–1000 °C. In the presence of metallic particles such as Ni, Fe, or Co that serve as the catalyst, the reactive species will be coated on the substrate, leading to the formation of CNTs. By adjusting the synthesis parameters and catalysts, different types of CNTs can be prepared. As the preparation requirements are lower, the CVD technique has the potential to be employed for large scale synthesis processes. Nevertheless, CNTs prepared using the CVD technique suffers from relatively high defect densities in MWCNTs, which can be due to insufficient thermal energy. Regardless of the preparation methods, the resultant CNTs are usually contaminated with carbonaceous and metallic impurities arising from the reaction process that can adversely affect the properties of CNTs. To eliminate the carbonaceous contaminants gas phase and liquid phase purification methods are introduced [76]. Gas-phase purification uses high temperature while liquid phase purification involves washing the CNTs with acidic solutions such as nitric acid or sulfuric acid. On the other hand, metallic contaminants can be removed by heating the CNTs to the evaporation temperature of the contaminant. As such, CNTs with a purity of up to 99.6% can be achieved [77].

3.2. Carbon Dots

In general, there are two main routes to prepare CDs, known as the top-down and bottom-up. These synthesis routes can be performed via optical, chemical, or thermal processes. In the optical synthesis method, a laser is typically used to ablate a carbon target either in water or solvents. For instance, Goncalves et al. [78] reported the preparation of CDs by irradiating carbon targets using a pulsed UV laser for 60 s. The dimension of the resultant CDs is determined by the separation between the focusing lens and the carbon target, where long distance yields CDs with larger particle sizes and vice versa. However, the resultant CDs are not fluorescent and require some post functionalization to attain fluorescent property. On the other hand, Li et al. [79] prepared CDs by carrying out the laser ablation process in different solvents such as water, ethanol, and acetone. The group discovered CDs prepared in ethanol and acetone exhibit fluorescence while no fluorescence was seen from CDs prepared in water. Therefore, the group attributed the fluorescence emission to the surface moieties generated during the synthesis process.

The chemical synthesis route generally employs strong oxidative chemicals such as concentrated sulfuric or phosphoric acid to oxidize the carbon precursors to form fluorescent CDs. In one of the reports, human hair was used as the carbon precursor and dopant source to form fluorescent CDs [27]. The human hair was added to the concentrated sulfuric acid and sonicated before being stirred at 40, 100 and 140 °C for 24 h. It was found that smaller CDs can be obtained at higher temperatures. Green precursors have also been used to produce CDs by chemical oxidation. For instance, sucrose was chemically oxidized by concentrated phosphoric acid to produce CDs [80]. The as-synthesized CDs exhibit yellow emission at 560 nm under UV excitation and is stable from pH 4 – pH 11.4. In a separate report, Hu et al. [81] prepared CDs by dehydrating and oxidizing waste frying oil with concentrated sulfuric acid. The resultant CDs exhibit uniform dimensions, partially disordered graphite-like structure, and unique pH-sensitive photoluminescence.

There are several thermal synthesis routes for the preparation of CDs, namely hydrothermal, solvothermal, direct pyrolysis, and microwave-assisted pyrolysis. Hydrothermal synthesis that employs water as the solvent is one of the simplest and cost-effective methods to prepare CDs. Solvothermal, on the other hand, uses other solvents such as ethanol and dimethylformamide [82,83]. Li et al. [83] prepared CDs using a one-pot solvothermal technique with Taixi anthracite in DMF. The resultant CDs exhibited a strong photoluminescence quantum yield of 47% and a production yield of 25.6 wt%. Direct thermal treatment and microwave-assisted heat treatment has also been widely used to produce CDs with different optical properties. Typically, the carbon precursor undergoes several processes such as dehydration, polymerization, and carbonization prior to the formation of CDs [3]. Direct thermal treatment exposes the carbon precursor to a high temperature to induce the carbonization process. However, this pyrolysis process appears less favourable since it is lengthy, and the heating duration can range up to a few hours. Thus, microwave-assisted synthesis rises as a facile alternative since it is rapid, provides uniform heating and can be executed using a domestic microwave oven [84]. For instance, Chan et al. [85] prepared nitrogen and sulfur co-doped CDs by subjecting a mixture of citric acid and thiourea in water to 6 min of microwave-assisted heat treatment. The resultant CDs were found to be responsive towards ferric ion and were employed as a sensitive and selective ferric ion sensor.

3.3. Graphene

Typically, the preparation of graphene by mechanical exfoliation refers to a repeated peeling process using Scotch tape to produce thin graphene flakes. In a pioneering work by Novoselov et al. [41], a highly oriented pyrolytic graphite was exposed to dry etching using oxygen plasma to produce 5 μm deep mesas. They were then placed on a photoresist and heated up to adhere to the photoresist. Subsequently, the Scotch tape was used to exfoliate layers of graphene from the graphite sheet. Thin graphene flakes of single to few layers of graphene that adhered to the photoresist were released using acetone and transferred to a silicon substrate. Despite being a simple and effective method to produce monolayer or a few layers of graphene, this technique is limited by the low production yield.

Hernandez et al. [86] introduced a liquid exfoliation technique to obtain a single to a few layers of graphene sheets by dispersion and exfoliation in solvents. Ultrasonication enables the solvent such as N-methylpyrrolidone, N,N-dimethylacetamide, γ-butyrolactone, and 1,3-dimethyl-2-imidazolidinone that have similar surface energy to graphene, to intercalate the graphite layers. Subsequent centrifugation and decantation processes produce high-quality unoxidized graphene flakes that can be used as transparent electrodes and conductive polymers. The production yield of this method is approximately 1%, which can be further increased to 7–12% by sediment recycling. The exfoliation mechanism is governed by the fact that the energy required to exfoliate graphite into single-layer graphene is countered by the solvent-graphene interaction. Nevertheless, this liquid exfoliation method suffers from incapability to control the number of graphene layers, defects, as well as difficulty to remove the residual solvents, which can have an adverse effect when used as a sensing material on OFS.

Another chemical technique to obtain graphene is by the reduction of GO. GO can be obtained by oxidizing graphite using strong oxidizing chemicals, as reported by Brodie et al. [87], Staudenmaier et al. [88], and Hummers et al. [89], among others. The oxidation process adds various types of functional groups such as carboxyl and hydroxyl moieties to the graphitic surface. GO is easily exfoliated in water and can be easily reduced and converted back into graphene. The reduction process is carried out using reducing agents such as hydrazine, hydrides, and titanium under UV illumination [90].

The main setback of this technique is that the reduction process is unable to reduce the GO completely. This process also creates defects that are unremovable via a simple annealing process and is usually of a lower quality than pure graphene. The resultant product is also sometimes referred to as reduced graphene oxide (rGO). Other methods of preparing graphene include CVD [91,92], the intercalative expansion of graphite [93], heat treatment of SiC [94,95], and epitaxial growth technique [96,97].

3.4. Nanodiamonds

A popular method of preparing NDs is by detonating an explosive carbon precursor such as trinitrotoluene and hexogen (1,3,5-triazinane) [98–100]. The detonation process takes place in an enclosed chamber supplied with inert gas or water coolant, also known as the "dry" and "wet" synthesis, respectively [101]. After the detonation process, the carbon soot contains a mixture of NDs with a diameter of 4–5 nm, other carbon allotropes, and contaminants. The weight content of NDs in the carbon soot can be as high as 75%, while the NDs yield is about 4–10% of the weight of the explosive precursor [98,99]. In a study carried out by Danilenko [102], the pressures and temperatures were found to have a significant influence on the formation of the NDs. The temperature and pressure at the Jouguet point are insufficient to produce liquid bulk carbon but are capable of producing nanosized liquid carbon. The area of liquid carbon is moved to a lower temperature for nanocarbon while the area of NDs stability is marginally shifted to a higher pressure. As such, it is implied that the NDs are formed by homogeneous nucleation in supersaturated carbon vapor by condensation and crystallization of liquid carbon. A major drawback of this method is that the NDs tends to aggregate and are not dispersible in organic solvents or water. To make matters worse, the ND aggregates are often covered in a layer of graphitic material and further complicates the dispersion of NDs.

The NDs aggregates can be broken down by mechanical milling or ultrasonication. Krüger et al. [103] managed to reduce the size of NDs aggregate with diameters of 100–200 nm, 2–3 μm, and 20–30 μm to NDs with a dimension of 4–5 nm by stirred-media milling with microscale ceramic beads. The group speculated that the milling process is mainly based on the shearing action within the fast-turbulent flow.

Meanwhile, Ozawa et al. [104] introduced a bead-assisted high-power ultrasonication technique to break up NDs aggregates. The resultant nanosized NDs can be dispersed in various types of polar solvents such as water, dimethyl sulfoxide (DMSO), and ethanol. Nevertheless, these methods also introduce contaminants originating from the beads, blades, and vessels. Purification procedures such as reflux treatment in acid and centrifugation are required to remove the impurities.

The pulsed laser has also been reported to produce nanoscale NDs. Amans et al. [105] used a pulsed laser to ablate a graphite target in water. The resultant NDs have a dimension of 5–15 nm but is covered by a graphitic-like structure with a thickness between 3–4 nm. Findings from a separate study indicate the dimension of the NDs can be controlled by manipulating the laser light source [106]. With the same power density, a short pulse width laser will obtain single-crystal NDs with size between 3–4 nm, while long pulse width laser will obtain particles larger than 4 nm. In another report, Kumar et al. [107] developed a microplasma process to prepare NDs at near ambient environment. The NDs were homogeneously nucleated by dissociating ethanol vapor and quickly quenched with a reaction duration of less than 1 ms to achieve particle dimension in the nanoscale range. With the assistance of hydrogen gas, the non-diamond phase is removed while the diamond phase is retained and stabilized, resulting in a high-purity NDs. The resultant NDs have an average size of 3 nm, which is in accordance with theoretical calculations. Table 2 summarizes the synthesis approaches for each type of carbon allotrope.

Table 2. Comparison of synthesis approaches for carbon allotropes.

Types of Carbon Allotropes	Synthesis Strategy	Advantages	Disadvantages
Carbon nanotubes	Arc-discharge	• Can provide high-quality CNTs • Simple apparatus • High reproducibility	• High synthesis temperature • Contaminated with impurities
	Chemical vapor deposition	• Lower synthesis temperature • Uses hydrocarbon gas as the precursors • Different types of CNTs can be prepared by varying the precursors.	• Contaminated with impurities • Requires many optimization processes from the catalyst preparation to the CVD reaction
	Laser ablation	• Higher yield and greater purity than arc-discharge method	• High synthesis temperature • Contaminated with impurities
Carbon dots	Optical	• Simple experimental setup • Can tune the dimension of CDs by adjusting experiment parameters	• Low yield
	Chemical	• High yield • Large scale production • Inexpensive apparatus	• Requires hash chemicals • Environmentally unfriendly
	Thermal	• Rapid process • Can be prepared using simple apparatus such as domestic microwave oven • Can be carried out without hash chemical	• Uneven heating • Large size distribution
Graphene	Mechanical exfoliation	• Simple procedure • High quality graphene	• Low yield
	Liquid exfoliation	• Large scale production	• Difficult to control the number of layers and defects • Difficult to remove residual solvents
	Chemical treatment	• Large scale production	• Requires harsh chemical • Long duration
Nanodiamonds	Detonation	• Can be prepared from precursors such as old munitions	• Requires purification • High synthesis temperature
	Pulsed laser	• Can be carried out in water • Environmentally friendly procedure	• Resultant product contains impurities

4. Preparation Techniques of Carbon Allotrope-Based Optical Fiber Sensors

For chemical or biosensing applications, the coating conditions (i.e., temperature, pH, and duration), thickness, and uniformity are among the important factors in determining the sensor performance. For example, a non-uniform and thick coating is usually undesirable since it may lead to poor sensing performance. Furthermore, the response and recovery time of carbon allotrope-based OFS are equally affected by the coating thickness due to the adsorption dynamics between the target molecules and the nanocoating. To achieve a stable, repeatable, and high sensitivity carbon allotrope-based OFS, it is important to optimize the coating parameters by adopting suitable deposition techniques. In the following section, several established techniques for the deposition of carbon allotropes onto OFS will be discussed.

4.1. Langmuir-Blodgett (LB)

Irving Langmuir and Katharine Blodgett first introduced the Langmuir-Blodgett (LB) technique for the deposition of nanocoating onto a solid substrate. The deposition process is commonly performed at room temperature and usually involves amphiphilic molecules with hydrophobic tails and hydrophilic heads. Using this method, the optical fiber is usually prepared to have a hydrophilic surface and is placed in the sub-phase. Next, the receptors with water-insoluble amphiphilic molecules are prepared in volatile organic solvents and applied to the surface of the sub-phase. The molecules are oriented such that the hydrophilic part stays in the water while the hydrophobic part is facing upwards (Figure 3a), creating a floating monolayer of molecules in an arranged manner on the surface of sub-phase. Next, controlled compression is applied to the surface to form a condensed and stable monolayer film (Figure 3b) for the subsequent deposition onto the optical fiber that is vertically raised through the sub-phase. If multiple layers of coating are desired, the substrate is returned into the sub-phase to create head-to-head and tail-to-tail stack layer pattern, commonly known as *Y-type* (Figure 3d). Conversely, to obtain a monolayer coating, the deposition techniques known as *X-type* (monolayer transferred during downstroke only) (Figure 3c) or *Z-type* (monolayer transferred during upstroke only) (Figure 3e) can be carried out [108].

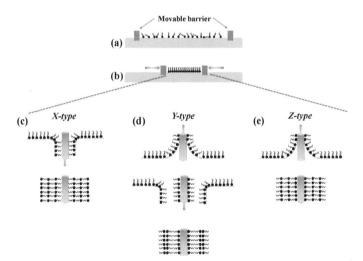

Figure 3. Langmuir-Blodgett film deposition scheme. (**a**) Spreading of molecules to the surface of sub-phase, (**b**) surface compression with constant pressure to yield a condensed and stable monolayer film, (**c**) *X-type* LB film deposition, (**d**) *Y-type* LB film deposition, and (**e**) *Z-type* LB film deposition.

The LB surface coating technique offers precise control over the deposition thickness, approximately 1–3 nm for each layer on the planar substrate. Deposition onto a single-mode optical fiber, for instance, is able to achieve 2.6 nm for each deposited molecular layer [109]. In order to achieve the desired film thickness, the amount of surface tension being applied and the material concentration are some of the important parameters that need to be considered [110]. Compare to other deposition techniques, fabrication of LB film on the optical fiber can be as simple as requiring only one chemical compound, given it an added value in terms of homogeneity. On the flip side, this could also mean that only a limited number of chemical compounds can be used with this technique. The fabricated LB film also suffers from poor thermal stability [111]. Moreover, this deposition process is tedious, slow, and requires skilled executor and sophisticated instrument to control the surface tension. For these reasons, the LB technique appears to be less popular unless nanometric precision of coating thickness is desirable.

4.2. Layer-by-Layer Electrostatic Self-Assembly (LbL-ESA)

LbL-ESA technique was first demonstrated by Decher and co-workers in preparing a multi-layered film on a solid substrate by alternate exposure to anionic and cationic polyelectrolyte with immediate adsorption of the oppositely charged ions [112]. This approach has attracted tremendous interest particularly in the field of surface material engineering due to its outstanding merits such as uniformity, stability, simplicity, and excellent controllability of the coating thickness at the nanometer scale. The driving factor to this deposition technique premises upon the electrostatic interaction between two materials of opposite charges. Typically, multi-layered nanocoating using LbL-ESA is usually performed at room temperature, and is independent of the size and shape of the substrate unlike the LB approach [113]. The fabrication of LbL-ESA begins with treating the substrate to obtain a negatively charged surface. The treatment varies according to the types of substrates, for instance, common silica-based substrates such as glass slide and silica optical fiber are usually done via piranha solution, a mixture of concentrated sulfuric acid (H_2SO_4) and 30% hydrogen peroxide (H_2O_2) at 3:1 v/v ratio. The strong dehydrating power of the H_2SO_4 and oxidizing power of H_2O_2 remove the organic residues from the surface, followed by generating a dense layer of hydroxyl groups (–OH), making it highly hydrophilic and favorable for subsequent electrostatic interaction with polycation electrolyte. Similarly, the immersion of plastic optical fiber with poly(methyl methacrylate) (PMMA) core material into 1M H_2SO_4 helps to develop carboxylic groups (–COOH) by reducing the methyl ester groups of PMMA [114]. After obtaining a negatively charged surface, the optical fiber is then immersed into a polycation solution for a sufficient amount of time to allow for molecules to adsorb. Next, the optical fiber will be washed thoroughly with deionized water and dried before immersing it into a polyanion electrolyte-containing solution. This will yield one bilayer LbL-ESA film, and the process can be repeated for several cycles to achieve the desired multi-layered thin-film structure.

In general, LbL-ESA process takes place in an aqueous solution but slowly diversify to nonpolar solvents due to the discovery of novel nanomaterials. Although the deposition mechanism remains unchanged, Lindgren et al. [115] recently reported a new insight into the significant role of the solvent towards the effectiveness of electrostatic assembly. The study explains different types of solvents may alter the electrostatic force between the interacting particles and surface, from attractive to repulsive or vice versa. The type of interaction is dependent on the permittivity of the particles and solvent. Briefly, a solvent with a large dielectric constant that is more polarizable than both interacting particles and surface promotes repulsive interaction, while a solvent with a small dielectric constant promotes attractive electrostatic force [116]. Therefore, it is rational for one to include the polarization effect of the solvent in designing the deposition system. Due to its vast applicability to most of the optical fiber platforms and the broad availability of molecules, LbL-ESA can be employed for entire surface area or end-face deposition on the optical fiber sensor probe to suit different sensing schemes. However, low molecular weight molecules alone are incapable of being assembled directly onto the optical fiber using the LbL-ESA method due to deficiency of charged groups in these molecules and are likely to face a substantial amount of loss in the rinsing step. To counter this problem, deposition with a single or combination of polyelectrolytes with long alkyl chain such as PAA, poly(allylamine hydrochloride) (PAH), polyethylenimine (PEI), etc. onto the substrate is preferred. For instance, Goncalves et al. [117] developed an Hg^{2+} sensor using a tapered tip silica fiber by immobilizing PEI as the polycation electrolyte followed by depositing CDs at the end tip of the sensor probe. Similarly, Alberto et al. [118] reported a GO-coated tilted Bragg grating prepared using the LbL-ESA method. Alberto et al. [118] first treated the optical fiber with sodium hydroxide to produce a negatively charged surface for subsequent deposition of poly(diallyldimethyammonium chloride) (PDDA) and poly(sodium 4-styrene-sulfonate) in an alternate manner. PDDA that served as the polycation electrolyte was coated to the external layer of the optical fiber before the deposition of the negatively charged GO. Besides polyelectrolyte-carbon allotropes multi-layered film, the LbL-ESA technique is also adopted for the development of metal oxide-polyelectrolyte-carbon allotrope films on optical fiber substrates. In a work done by Hernaez et al. [119], the deposition of PEI/GO multi-layered films onto a tin oxide-coated multimode fiber has significantly enhanced its sensitivity for ethanol sensing by 20% and 210% for one and four bilayers of PEI/GO, respectively. Henceforth, LbL-ESA method of fabricating carbon-allotrope coatings have become one of the most favorable deposition technique to develop a wide range of OFS.

4.3. Chemical Vapor Deposition (CVD)

The fundamental of plasma-assisted vapor deposition lies in the activation of a precursor in a glow discharge (i.e., plasma) environment. The growth of a thin-film using this method generally exhibits less contamination as compared to other wet techniques [120]. In a generic case, carbon nanomaterial thin film produced using the CVD process is formed from the chemical reaction of gaseous reactants in the close vicinity of a lightly heated substrate (~20–50 °C). Briefly, the initiator and the target material, usually in liquid form, are vaporized by either heating or reducing the air pressure, followed by passing it to a vacuum chamber where the substrate is placed. The initiator functions to accelerate the film growth rate and finally, the target material will be deposited on the cold substrate. The thickness and the refractive index of the CVD film can be easily adjusted by altering the pressure and temperature of the deposition process. In many generic cases, synthesis of carbon allotropes such as graphene thin film using the CVD approach usually involves reaction gases like methane and dilute hydrogen environment, on a copper foil as a catalyst substrate at over 1000 °C [6]. Transfer of the thin film can be done in several ways. To attach the graphene thin film onto the fiber structure, graphene is transferred to a low refractive index substrate, such as MgF_2, followed by adhering onto the optical fiber via van der Waals bond [121]. If wrapping the graphene thin film surrounding the fiber structure is desired, one way to achieve this is by spin-coating a layer of PMMA on the surface of the graphene forming a PMMA/graphene/copper hybrid. Next, the copper under graphene is removed using iron (III) chloride solvent followed by wrapping the PMMA/carbon nanomaterial thin film on the fiber. Finally, the PMMA is removed with acetone leaving only the graphene thin film on the fiber.

Nevertheless, the merits of CVD are obvious, particular its ability to fabricate dense and amorphous films, and more importantly, good uniformity [122]. Despite these advantages, CVD is not with no limitations. Temperature or UV sensitive materials are not suitable for these techniques, thus there is a limited number of materials that can be deposited on the optical fiber using this approach. Moreover, high accuracy and high-resolution instruments are required to regulate important parameters such as temperature, pressure, current, and others that may significantly affect the reproducibility of the nanocoating. For these reasons, the CVD setup is costly and may be inaccessible to some laboratories due to its high operating cost.

4.4. Optical Deposition

The optical deposition method utilizes the light guided by an optical fiber to draw the sensing material in close proximity to the optical fiber surface, and finally, depositing onto the external surface of the optical fiber. Commonly, the optical deposition technique involves dispersing the sensing material into solvents such as ethanol, isopropanol, dimethylformamide, and others, followed by immersing the fiber into the mixture. Kashiwagi et al. [123] reported the coating of CNTs onto the end facet and tapered region of the optical microfiber via the optical deposition method. Together with the optical forces from the injection of light into the sensing material-dispersed solution, the Brownian motion of the sensing material became highly oriented, and the swirl and convection tend to draw them toward the surface of optical fiber. Kashiwagi et al. [123] also deduced that the trapping of this sensing material on the microfiber surface is jointly resulted from the optical tweezer effect due to optical intensity diversion in the solution caused by the evanescent field of the optical microfiber. Since the size of the CNTs is much smaller than the wavelength of the light, the CNTs are treated as a point dipole. Two forces acting on this dipole, the scattering force that pushes the particle along the light propagation and the Lorentz force which moves the particle toward the region of higher optical intensity [124,125]. Centered on this principle, the evanescent field of the optical microfiber that had the optical intensity diversion may trap the CNTs by optical tweezer effect and consequently, immobilized them onto the surface of the desired area [126].

Generally, the film thickness produced using optical deposition technique is controllable by adjusting the injected light power and the deposition time. Besides, in-situ monitoring of the transmitted or reflected power is often performed to monitor the insertion loss caused by the deposited sensing material. Ideally, it is recommended that the insertion loss falls within 3 to 5 dB. Overall, the simple and economical process of the optical deposition technique is capable of achieving the area-selective deposition of sensing material.

4.5. Crosslinking

In the context of conjugation, a crosslinker is used to mediate the attachment of one molecule to another, usually through the covalent bond to create a complex comprising of both molecules linked together. Generally, the design of the conjugation process is dependent on the reactive groups present on the reactive crosslinking agents and the functional groups present on the target molecules. The conjugation process is unfeasible without the availability and chemical compatibility of both reactive and functional groups. Examples of these functional groups include amine, thiol, carboxylate, aldehyde, and hydroxyl. Meanwhile, some reactive groups that are often employed in the conjugation process include isothiocyanate, isocyanates, NHS ester, maleimide, and glutaraldehyde [127].

In many instances, the final conjugate complex is bound by a crosslinker that introduces foreign chemical components to the molecules being crosslinked. The first crosslinking agent introduced for the conjugation of macromolecules, known as the homobifunctional crosslinker consists of bireactive compounds of the same functionality at both ends of the spacer arm (Figure 4a) [128]. The use of a homobifunctional crosslinker in a one-step conjugation protocol, however, provides the least control over the final product of a conjugation reaction and may yield a broad range of poorly defined conjugates [129]. This is because when crosslinking two molecules, for instance, the homobifunctional crosslinker first reacts with either one of the molecules, forming an active intermediate. Ideally, this activated molecule may crosslink with the second molecule, however, it may also react intramolecularly with other functional groups on part of its own. To circumvent this, a two-step conjugation protocol may alleviate the problem. This can be done by removing the excess crosslinker and byproducts before introducing the second molecule to the activated molecule to allow the final conjugation reaction to take place. This solution, however, may raise another problem where the activated molecule experience degradation before the second phase of crosslinking commence due to hydrolysis phenomena. Moreover, chances of the problem associated with the one-step conjugation protocol may persist in the two-step conjugation protocol since the first molecule may crosslink with itself long before the introduction of the second molecule. Thereafter, another type of crosslinker known as heterobifunctional crosslinking agent that contains two different reactive groups at the end of the spacer arm (Figure 4b) is introduced for targeted coupling between two different functional targets on macromolecules. Heterobifunctional crosslinker exhibits the ability to yield direct crosslinking reaction to selected parts of target molecules and thus, warrant better control over the resultant product of the conjugation reaction.

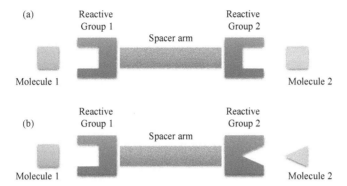

Figure 4. General design of crosslinking agent. (**a**) Homobifunctional crosslinker with identical reactive groups at the end of spacer arm and (**b**) heterofunctional crosslinker with two different reactive groups at the either end of spacer arm.

Another commonly used crosslinker would be the zero-length crosslinker, also known as the smallest crosslinking agent available to perform a conjugation process. A zero-length crosslinker allows one atom of a molecule to attach covalently to another atom of a second molecule with no additional intervening atoms or spacer in between the bond. This is advantageous, since in some cases, the presence of these intervening linkers or spacers in between the established bond may cause cross-reactivity with undesired reactive/functional groups. Carbodiimide such as 1-ethyl-3-(3-dimethylaminopropyl) carbodiimide hydrochloride (EDC), is most popularly adopted for conjugating substances containing carboxylate to molecules containing amine functional groups. While EDC alone can be used in a one-step conjugation protocol, the efficiency of the conjugation process can be enhanced through the use of N-hydroxysuccinimide (NHS) that increases the solubility and stability of the intermediate before conjugating with the targeted amine groups [130,131].

For silica-based optical fiber, the foundation layer to facilitate the surface functionalization process is solely dependent on the hydroxyl groups present on the external surface. However, these hydroxyl groups are rather weak and insufficient for conjugation with other functional groups via crosslinking. Therefore, many works reported an additional step of performing oxygen plasma or acid treatment such as using a piranha solution to form a dense and active layer of hydroxyl groups for subsequent crosslinking use. Some also reported the use of silane coupling agents to create amine or carboxylate terminal end groups to ease the conjugation reaction with carbon allotropes via zero-length crosslinker [132].

4.6. Drop-Casting

Drop casting is another simple yet economical approach of depositing sensing material onto the surface of the optical fiber. In a typical manner, the as-prepared carbon allotrope is dispersed in a volatile solvent. Meanwhile, the bare optical fiber is cleaned with an alcohol-based solvent followed by annealing in an oven before initiating the deposition process. Next, the carbon allotrope solution will be drop-casted at the desired deposition area on the fiber and left undisturbed at the ambient environment for the solvent to evaporate naturally. This process can be repeated depending on the desired amount of deposited material. Alternatively, different concentrations of carbon allotrope solution can be used to achieve the desired coating thickness. Post deposition process usually followed by annealing the optical fiber for a second time to enhance the coating adhesion. Carbon allotrope-based OFS fabricated using this deposition technique is vastly reported in the past few years [133–135].

5. Sensing Mechanisms of Carbon Allotrope-Based Optical Fiber Sensors

Many research groups have utilized the unique properties of carbon allotropes to develop OFS for diverse applications. This section will mainly discuss the different sensing mechanisms employed in carbon allotrope-based OFS.

5.1. Thermo-Optic

CNTs, NDs, and graphene are known to exhibit excellent thermal conductivity attributable to their C-C covalent bond and phonon scattering characteristics [8]. The high thermal conductivity of these nanomaterials makes them interesting candidates and can be exploited as a temperature-sensitive material for temperature sensing. For instance, Zhang et al. [1] developed an all-fiber temperature sensor based on rGO. The group described that as the temperature increases, the availability of thermally excited electrons-holes also increases. This changes the Fermi-Dirac distribution of electrons in the rGOs that sequentially decreases the dynamic conductivity. Theoretically, the real part of dynamic conductivity affects the amount of light absorbed by the intraband and interband transitions in the rGO. Therefore, a rise in temperature will reduce the amount of light being absorbed and hence, resulting in a reduction of transmission loss and an increase in transmitted optical power. As a result, temperature measurement can be attained using rGO film on an optical fiber. Despite the availability of different carbon allotrope-based platforms such as resistive-based sensors that exploit the thermal conductivity for sensory applications, there are still some major setbacks that restrict its translation for practical application. For instance, resistive-based sensors are costly, energy-intensive and often require elevated temperature to introduce metal oxide layers as part of the device configuration [136]. Contrarily, the configuration of temperature sensors using optical fiber is relatively simple, and the carbon allotrope-based sensing layers can be introduced via many facile strategies as described in Section 4. However, careful selection of light wavelength and source power are needed to minimize the potential of self-heating cause by the absorption of the injected light source that can affect the measurement accuracy.

5.2. Surface Plasmon Resonance

In a typical surface plasmon resonance (SPR)-based OFS, precious metals such as gold or silver that serve as surface plasmon materials are coated on the optical fiber to form SPR structure. Gold coating is beneficial as it introduces a larger resonance shift to the changes in refractive index at the sensing layer while silver coating which exhibits a smaller SPR curve width will result in a higher signal to noise ratio. However, the procedure to introduce these metallic coatings on an optical fiber is complicated and costly. For example, silver is easily oxidized when exposed to oxygen in ambient air, elevated temperature, or water vapor due to their poor chemical stability and resulted in the oxidation of silver to silver oxide [8,137]. This will affect the sensor's reproducibility and thus remains impractical for real-life sensing. To overcome this, a bilayer metallic coating configuration that comprises of a silver and gold (outer layer) coating may solve the problem for sensing applications [138]. However, careful optimization of the layer thickness is vital to achieve high signal to noise ratio as well as optimal sensitivity of the proposed sensor. Thus, carbon allotropes, especially graphene, rises as a potential alternative due to their superior properties and resistance towards oxidation. Furthermore, it has also been reported that graphene can enhance the SPR signal [139]. By depositing a layer of graphene on the optical fiber, a fiber-graphene interface that supports charge density oscillation upon light excitation can be obtained. It should be noted that the excitation light should have an identical polarization state as well as matching momentum and wave vector to the surface plasmon [140]. As a result, a resonance is generated at a particular wavelength, and a sharp wavelength dip, also identified as resonance wavelength can be measured in the output spectrum, and the presence of target molecules can be detected as a shift in the resonance wavelength. On top of this, the concentration of the target molecules can also be detected and quantified by correlating to the amount of shift in the resonance wavelength.

5.3. Fluorescence

CDs, NDs, and graphene are among the fluorescence carbon allotropes that are commonly integrated into an optical fiber-based fluorescence sensor. The emission of these nanomaterial ranges from the UV and visible spectrum under various excitation wavelengths. In general, a fluorescence sensor operates based on the perturbation in optical characteristics such as the fluorescence intensity in the presence of the target molecules. This process can take place based on several mechanisms, such as photoinduced electron transfer (PET) or fluorescence resonance electron transfer (FRET) [3]. PET can be described when a new complex is formed between an electron donor and an electron acceptor. Upon excitation, the new complex will return to the ground state without the emission of photons. On the other hand, FRET is an energy transfer process that occurs between a donor molecule and an acceptor molecule via dipole-dipole interactions [141]. The energy received will excite the donor molecule to the lowest unoccupied molecular orbital (LUMO). Subsequently, the energy will be transferred to the acceptor molecule while the donor molecule relaxes to the ground state. For this to happen, the absorption spectrum of the acceptor molecule requires an overlap with the emission spectrum of the donor molecule. Furthermore, since the energy transfer efficiency is inversely proportional to the sixth power of the distance between the donor and acceptor molecules, both donor and acceptor molecules need to be in close proximity to allow the occurrence of this process.

5.4. Molecular Adsorption

Carbon allotropes such as graphene possess a high density of hexagonal ring structure that can be functionalized on the optical fiber surface to absorb molecules such as gas molecules, heavy metal ions, and organic pollutants. The number of molecules being absorbed can be correlated to the changes in the refractive index and detected using an optical fiber. For example, evanescence wave-based optical fiber is well known for its sensitivity to the perturbation in the surrounding refractive index. In a typical manner, light launched into the core of the fiber first propagates as fundamental mode, HE_{11}. As the light reaches the sensing region, a substantial amount of light energy will be coupled into the next high-order mode, HE_{12}, as a result of the morphology and local refractive index change. Unlike HE_{11} mode, the HE_{12} mode is not confined within the core but exposed to the outer surface and become cladding guided. As the light travels down the fiber, a second coupling occurs between the HE_{11} and the HE_{12} modes and generates a phase difference between the two modes that will result in a modal interference spectrum governed by the $I = I_1 + I_2 + 2\sqrt{I_1 I_2} \cos \phi$ [142,143]. In the event where target molecules are absorbed on the carbon allotropes, a perturbation of localized refractive index on the sensing region will occur and can be detected as a spectrum wavelength shift in the output signal of the fiber. An example can be seen from a work reported by An et al. [144] using a D-shaped fiber coated with a thin gold film followed by a layer of graphene to measure the refractive index range from 1.38 to 1.39. The presented sensor showed a maximum sensitivity of 4391 nm/RIU at a resolution of 2.28×10^{-5} using the wavelength interrogation method. Likewise, Fu et al. also simulated a D-shaped fiber for refractive index sensing in the range of 1.33 to 1.39 [145]. The D-shaped fiber is designed such that silver nano-columns coated with graphene layers are deposited on the side polished area to enhance the sensitivity to the surrounding refractive index change. Maximum refractive index response sensitivity can reach up to 8860.93 nm/RIU when the diameter of the silver nano-column is fixed 90 nm and 23 layers of graphene, each layer with a thickness of 0.34, is coated on each silver nano-column. In another study, two different etched fiber Bragg grating (FBG) sensors coated with SWCNTs and GO, respectively, demonstrated high specificity to protein concanavalin A (Con A) via the mannose-functionalized poly(propyl ether imine) dendrimers attached to the coated sensors [146]. Even in the presence of interfering proteins such as bovine serum albumin and lectin peanut agglutinin, SWCNTs and GO coated etched FBG sensors showed great selectivity to Con A and are able to achieve LODs of 1 nM and 500 pM, and affinity constant of ~4×10^7 M^{-1} and ~3×10^8 M^{-1}, respectively. Overall, these studies have widely proven the possibility of using carbon allotrope-based OFS not just in chemical or environmental sensing but also in biological sensing applications.

6. Sensing Applications of Carbon Allotrope-Based Optical Fiber Sensors

6.1. Humidity

Humidity sensing is vastly employed for domestic and industrial applications. For instance, humidity sensors are often employed in smart buildings, food processing plants, and microelectronics industries. Relative humidity (RH) is defined as the amount of water vapor present in the air, expressing the ratio of the actual moisture in the air to the maximum amount of moisture that the air can retain at that temperature and is quantified in terms of percentage. Optical fiber-based humidity sensor rises as an alternative solution to overcome the drawbacks of conventional humidity sensors such as hygrometer and psychrometer that exhibit long response time and suffer from electromagnetic interference. Moreover, the versatility of optical fibers to different functional nanocoatings has led to the integration with carbon allotropes for the development of optical fiber-based humidity sensors. Shivananju et al. [147] reported an etched fiber Bragg grating (FBG) coated with CNTs at the etched region for humidity sensing. Due to the interaction between water molecules and the CNTs coating, the effective refractive index surrounding the core will change, resulting in a shift of the Bragg wavelength and a sensitivity of 31 pm/%RH within a linear detection range from 20–90 %RH. Mohamed et al. [148] prepared an optical microfiber coated with MWCNTs slurry using the drop-casting technique. The fabricated sensor demonstrated a linear detection range from 45 to 80 %RH and an improvement of 1.3 times (5.17 µW/%RH) when compared to a bare tapered fiber. Alternatively, doping of MWCNTs onto a PMMA microfiber to form a thin layer of nanocoating on the optical microfiber for RH sensing was reported by Isa et al. [149]. The MWCNTs/PMMA functional coating increases the index contrast between the microfiber core and the surrounding air cladding causing more water molecules to be adsorbed on the sensing surface. Consequently, more electrons are transferred from the water molecules to MWCNTs, diminishing the available holes in MWCNTs and thus, decreasing the output optical power of the MWCNTs/PMMA microfiber sensor. The sensor showed a good linear detection range between 45 to 80 %RH and sensitivity of 0.3341 dBm/%RH, demonstrating an approximately 4-fold sensitivity improvement over a undoped PMMA microfiber sensor. Similarly, Ma et al. [150] also developed a CNT/polyvinyl alcohol (PVA) coated at the end tip of the thin core fiber (TCF) for RH detection. The as-developed sensor demonstrated good reversibility and output stability after 12 consecutive exposures to different RH levels. Moreover, the sensor showed a good linear detection range from 70 to 86 %RH with a measured sensitivity of 0.4573 dB/%RH.

On the other hand, GO-based OFS have also attracted significant interest for their prominent use in RH sensing. Gao et al. [151] presented a hollow-core fiber coated with rGO in which the sensing mechanism is based upon the adsorption of water molecules on the rGO surface that serves as electron acceptors. Along with the adsorption of water molecules, the surface charge carrier density of rGO will increase, further inducing a change in the chemical potential and dynamic conductivity of the rGO. The changes in these parameters will then influence the effective refractive index of the rGO and hence, altering the output signal of the sensor. The sensitivity of the fabricated rGO-based hollow-core fiber sensor was found to increase with increasing sensor length, and a maximum sensitivity of 0.229 dB/%RH was achieved within the linear detection range from 60–90 %RH at the fiber length of 12.6 cm. Furthermore, the sensor showed good reversibility and was unaffected by the surrounding temperature fluctuation. When tested with human breath alone, a swift response time of 5.2 s and a recovery time of 8.1 s was recorded. Xing et al. [152] prepared rGO nanosheet-coated polystyrene (PS) microsphere via a thermodynamically-driven hetero-coagulation approach to create a three-dimensional graphene network (3-DGN) coating surrounding the taper waist region of an optical microfiber. It is interesting to note that PS microsphere alone is hydrophobic in nature. However, upon functionalized with rGO, the chemically active defect sites of rGO that exhibit hydrophilic functional groups such as carboxylic and carbonyl groups are likely to absorb water molecules present in the surrounding environment. Moreover, the constructed 3-DGN was able to achieve much higher sensitivity as compared to single rGO or GO nanocoating. The measured results obtained from the rGO/PS-coated optical microfiber exhibited a sensitivity of −0.224 dB/%RH and −4.118 dB/%RH for detection range 50.5–70.6 %RH and 79.5–85 %RH, correspondingly. Overall, Table 3 summarizes all carbon allotrope-based RH OFS and their respective sensing performance [134,147–162].

Table 3. Summary of carbon allotrope-based OFS for RH sensing.

Sensing Material	Optical Fiber	Linear Detection Range (%)	Sensitivity (% RH^{-1})	Ref.
CNT	Etched FBG	20–90	31 pm	[147]
CNTs doped PMMA	Optical microfiber	45–80	0.3341 dBm	[149]
MWCNTs slurry	Optical microfiber	45–80	5.17 µW	[148]
CNT/PVA	Thin core fiber	70–86	0.4573 dB	[150]
rGO	Hollow core fiber	60–90	0.229 dB	[151]
rGO	Side polished fiber	70–95	0.31 dB	[134]
rGO	Microfiber resonator	30–50	0.0537 nm	[153]
rGO/PS	Optical microfiber	50.5–70.6 79.5–85.0	0.224 dB 4.118 dB	[152]
GO	Side polished fiber	32–85 85–97.6	0.145 nm 0.915 nm	[154]
GO	Side polished twin core fiber	40–75	2.720 nm	[155]
GO	Polarization maintaining fiber	60–77	0.349 dB	[156]
GO/PEI	Multimode fiber	20–70 70–90	0.317 nm 0.311 nm	[157]
GO	Tilted FBG	10–80	0.129 dB	[158]
GO	Tilted FBG	30–80	0.027 dB 18.5 pm	[159]
GO	Single mode fiber	30–60	0.104 dB 0.0272 nm	[160]
GO/PVA	Optical microfiber	40–60	0.0606 dBm	[161]
GO/PVA	Waist enlarged taper SMF	25–80	0.193 dB	[162]

6.2. Temperature and Pressure

An rGO-based side polished fiber sensor using the refractive index change scheme for temperature sensing was developed by Zhang et al. [1]. Briefly, when the surrounding temperature increase, the concentration of thermally excited electrons-holes increases and the change in the Fermi-Dirac distribution of electrons in the rGO will reduce the real part of its dynamic conductivity. Since the real part of the dynamic conductivity correlates to the light absorption induced by intra and interband transitions in the rGO, reduction in the real part of the dynamic conductivity will consequently decrease the light absorption and reduce the transmission loss of the rGO-coated side polished fiber [163]. In other words, as the surrounding temperature increases, the transmitted optical power of the rGO-coated side polished fiber also increases. Therefore, a linear relationship of surrounding temperature as a function of output transmitted optical power was obtained for the rGO-coated side polished fiber within the range of −7.8 to 77 °C with a maximum sensitivity of 0.134 dB °C^{-1} was achieved. Other optical fiber structures such as etched FBG [164] and suspended core hollow fiber [165] coated with rGO have also been employed to demonstrate temperature sensing capability. On the other hand, Sun et al. [166] reported a graphene-coated optical microfiber temperature sensor constructed with a thin graphene film adhered onto the optical microfiber surface via the strong evanescent field and the electrostatic force. To reduce the insertion loss, the graphene thin film was transferred to MgF$_2$ with a lower refractive index before introducing them to the optical microfiber surface. As the surrounding temperature fluctuates, the graphene thin film and MgF$_2$ substrate alter the effective refractive index of the optical microfiber. Since the thermo-optic coefficient of MgF_2 (3.2×10^{-7} °C^{-1}) is much smaller than graphene (7.385×10^{-6} °C^{-1}), the effect of the temperature change on MgF_2 substrate is negligible. Good linearity and temperature sensitivity of 0.1018 dB °C^{-1} and 0.1052 dB °C^{-1} were measured when surrounding temperature increase and decrease, respectively, in steps of 5 °C between 30–80 °C. In a similar context, Wang et al. [167] proposed the inclusion of polydimethylsiloxane (PDMS) to produce a PDMS-graphene pliable composite film wrapping around the microfiber ring resonator. Similar to the MgF_2 substrate, PDMS exhibits low refractive index and high thermal stability, making it a good candidate to work with graphene to achieve better performance of the temperature sensor. Moreover, PDMS is a highly transparent film with a high degree of flexibility. As a result, these features enable very close contact between the graphene and the fiber that helps to improve the temperature response of the proposed sensor. The proposed sensor exhibits excellent temperature sensitivity of 0.541 dB °C^{-1} under an incremental temperature environment and 0.542 dB °C^{-1} under gradually decreasing temperature environment for the temperature range of 30–60 °C.

Some studies also reported the simple fabrication of graphene diaphragm integrated into an optical fiber sensor for both temperature and pressure sensing. For instance, Ameen et al. [168] proposed FBG-based temperature and water level sensors of different configurations, as shown in Figure 5. The study revealed that FBGI is responsive to water levels that are associated with the hydrostatic pressure, while FBGII is sensitive to the surrounding temperature. For both parameter measurements, increasing the number of graphene diaphragm layers, each with a thickness of 25 µm, tends to decrease the sensing performance owing to the reduction in graphene diaphragm elasticity. This attributes to the fact that for a thinner diaphragm, stronger diaphragm deflection is expected in response to the external pressure, while thicker diaphragm will bend less under the same amount of applied pressure leading to a smaller responsivity. Thus, a single-layer graphene diaphragm was found to deliver the best detection sensitivity of 13.31 pm °C^{-1} for temperature sensing within the range of 27 to 75 °C and 253.21 pm kPa^{-1} for pressure or water level sensing. However, this single-layer graphene diaphragm is only able to resist up to a maximum of 9.81 kPa that is equivalent to 100 cm of water level. To enhance the detection range sensor configuration as illustrated in Figure 5a(iii) has a higher tolerance up to 135 cm of water level without damaging the diaphragm but with a lower sensitivity of 99.18 pm kPa^{-1}. Alternatively, Dong et al. [7] presented a simpler design of a Fabry-Perot interferometer with an integrated FBG to measure pressure and temperature changes simultaneously. Graphene sheet was coated onto the end facet of the fiber ferrule via van der Waals reaction to form a reflecting surface of a sealed Fabry-Perot microcavity. The study revealed that as the surrounding temperature increases, the cavity length increases together with the red shifting of the resonant wavelength. However, for pressure increase, only the cavity length decreases while the resonant wavelength remains unchanged. Therefore, by using matrix inversion calculation, the variation in surrounding pressure and temperature can be identified simultaneously. The proposed sensor was reported to exhibit temperature and pressure sensitivity of 306.2 nm °C^{-1} and 501.4 nm kPa^{-1}, respectively.

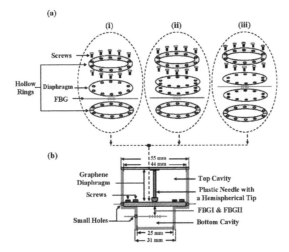

Figure 5. Experimental architecture: (**a**) diaphragm incorporated FBG positioned between two hollow rings, (i) a single sheet of graphene diaphragm on the top of FBG, (ii) two sheets of graphene on the top of FBG, (iii) two sheets of graphene between FBG called sandwich layer. (**b**) Sandwich layers of the sensor head structure [168]. Copyright © 2016, Elsevier B.V.

Cui et al. [169] recently reported an analytical model that can predict the critical diaphragm thickness wherein the responsivity of the pressure sensor is independent on the elasticity property when the diaphragm thickness is below the critical thickness value. In other words, further reduction of the diaphragm thickness will not help to improve the sensor's sensitivity if diaphragm thickness is much smaller than the diaphragm deflection. Redesigning the cavity shape other than cylindrical appears to be a possible solution for this, where the sensitivity of the sensor can be enhanced if the cavity shape design yields a larger cavity volume without altering the resonator length and diaphragm radius. The denouement of this study has raised the importance of placing more effort on optimizing cavity shape rather than opting for smaller single-layer diaphragm thickness that will complicate the sensor design. Tables 4 and 5 summarize some of the research works related to temperature and pressure detection using carbon allotrope-based OFS.

Table 4. Summary of carbon allotrope-based OFS for temperature sensing.

Sensing Material	Optical Fiber	Linear Detection Range (° C)	Sensitivity (° C^{-1})	Ref.
rGO	Side polished fiber	−7.8–77	0.134 dB	[1]
rGO	Etched FBG	−100–300	33 pm	[164]
rGO	Suspended core hollow fiber	30–80	179.4 pm	[165]
Graphene	Optical microfiber	30–80	0.1018 dB	[166]
Graphene/PDMS	Microfiber ring resonator	30–60	0.544 dB	[167]
Graphene/Ag	Hollow core fiber	22–47	9.44 nm	[170]
PU/Graphene	FBG	25–60	6 pm	[171]
Graphene diaphragm	Fabry-Perot interferometer	20–60	352 nm	[172]
Graphene diaphragm	Fabry-Perot interferometer	500–510 1000–1008	1.56 nm 1.87 nm	[173]
Graphene diaphragm	FBG	27–77	13.31 pm	[168]
Graphene diaphragm	FBG	20–100	306.2 nm	[7]

Table 5. Summary of carbon allotrope-based OFS for pressure sensing.

Sensing Material	Optical Fiber	Linear Detection Range (kPa)	Sensitivity (nm kPa^{-1})	Ref.
Graphene diaphragm	Fabry-Perot interferometer	0–3.5	1096	[174]
Graphene diaphragm	Fabry-Perot interferometer	0–2.5	80	[175]
Graphene diaphragm	Fabry-Perot interferometer	0–13	65.71	[169]
Graphene diaphragm	Fabry-Perot interferometer	0–5	39.4	[176]
Graphene diaphragm	FBG	0–2	501.4	[7]
Graphene diaphragm	FBG	0–9.81	0.25	[168]

6.3. Other Physical Parameter Sensing Applications

Besides the above-mentioned physical parameters, current, acoustic, wind speed, and magnetic are some other physical parameters that can adopt carbon allotrope-based OFS to execute the measurement. For example, Zheng et al. [177] prepared a graphene membrane coated fiber tip probe sensor for current sensing, as shown in Figure 6. Having a gold electrode and graphene membrane covering the end face of the etched fiber, electric current was coupled to the fiber sensor via two contact pads. The functional coating was heated up due to the applied current, and thus, the linear temperature change of the graphene membrane can be correlated to the square of current applied. Due to the negative thermal expansion coefficient of the graphene membrane, the graphene membrane will contract uniformly with increasing temperature, and further increasing the cavity length. These physical changes were reflected as a resonance wavelength shift in the reflectance output spectrum. The proposed sensor exhibited a current sensitivity of 2.2×10^5 nm/A^2 within the detection range from 0 to 2 mA and a short response time of 0.25 s. Even though higher sensitivity can be achieved by reducing the size of the graphene membrane to expedite the heating process, the operating range of the developed sensor remains limited to 2 mA to avoid damage to the fiber probe sensor.

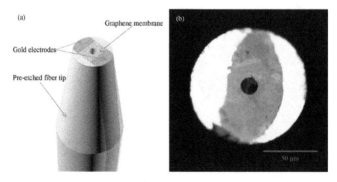

Figure 6. Schematic of (**a**) Pre-etched fiber tip coated with graphene membrane that covers the hole and two gold electrodes and (**b**) optical image of the tip's end facet [177]. Copyright © 2015, WILEY-VCH Verlag GmbH & Co. KGaA.

The practicability of graphene to be used as a deflectable diaphragm to sense pressure change has also inspired its feasible use for acoustic sensing. Ma et al. [176] designed a Fabry-Perot interferometer for acoustic sensing using a 125 μm multilayer graphene diaphragm with a thickness of 100 nm coated at the end face of the single-mode fiber. The sensor was placed in front of a speaker that acted as the acoustic pressure source, and the reflectance output spectrum was measured. The sensor demonstrated approximately 1100 nm/kPa of acoustic pressure sensitivity and noise-limited detectable pressure of approximately 60 μPA/Hz$^{1/2}$ at 10 kHz. The study further revealed that the sensitivity of the fiber sensor is dependent on the alignment angle of which the speaker was placed. Maximum sensitivity was achieved when the fiber sensor is aligned to the central axis of the speaker. Meanwhile, the sensitivity was found to decrease as the speaker is moved to a certain angle from the central point of the fiber sensor. Tan et al. also investigated the effect of graphene diaphragm's diameter by comparing the acoustic sensing performance of two Fabry-Perot interferometers each formed by immobilizing a 100 nm thick graphene diaphragm of 2.5 mm and 125 μm diameter respectively onto the end face of a single-mode fiber [178]. The work revealed that graphene diaphragm with a larger diameter exhibited 31 times improvement in sensitivity that agrees well with the linear deflection model.

Hot wire fiber anemometer is another promising sector demonstrating the potential use of OFS for measurement of wind speed in the coal mine, power transmission, and agricultural industries. In general, OFS based on hot wire anemometer correlates the measured cooling rate or temperature variation to the wind flow rate. Typically, the fiber sensor that incorporates nanomaterials with good heat conversion coefficients such as metal films and CNTs is usually heated up and subjected to wind flow. The amount of temperature drop measured by the sensor can then be used to approximate the wind flow rate. Based on the aforementioned detection scheme, Zhang et al. [179] proposed a simple hot wire fiber anemometer using SWCNTs coated tilted FBG. A pump laser of center wavelength 1550 nm acted as the heating and broadband light source to obtain a transmission spectrum of the tilted FBG were combined using a 3 dB coupler and launched into the core of the probe sensor. The coupled pumped light interacted with and absorbed by the SWCNTs film, which then raised the local temperature of the fiber sensor. When the heated sensor is subjected to a wind field, the cooling rate associated with the wind speed is detected by a wavelength shift in the resonance peak of the tilted FBG. Further, it was discovered that higher pump power of laser source, larger tilted angle of probe sensor, and thicker SWCNTs coating can enhance the wind speed sensitivity. However, considering the chirped effect of grating-inscribed fiber that may weaken the sensor response and harder dissipation of heat generated within the thick SWCNTs, pump power of 97.76 mW and SWCNTs coating of 1.6 μm thick was found to be optimal. The developed sensor was found to achieve a wind speed sensitivity of -0.3667 nm/(m/s) at the wind speed of 1 m/s.

Liu et al. [180] also worked on fabricating Au/SWCNT-based tilted FBG anemometer sensor with an LOD of 0.05 m/s on account to the SPR effect of the gold coating that provides sensitivity enhancement. A similar study has also been performed by Liu et al. [181] who reported a SWCNT-based long period grating anemometer sensor with a simpler sensing setup and a sensitivity of 102.5 pm/(m/s) at the wind speed of 1 m/s.

Although OFS is well-known for its immunity to electromagnetic interference, Ruan et al. [182] highlighted the advantages of magnetically-sensitive OFS in their work that provide a cheaper, robust, and ambient alternative for remote magnetic sensing use. This work reported a novel technique to fabricate an ND-doped tellurite glass fiber for magnetic field sensing to provide an isolated sensing platform since the magnetic sensitive materials are immobilized in the tellurite glass matrix, as shown in Figure 7a. The NDs used in this work were developed via a non-detonation approach with an average particle size of 40–50 nm and a substantial density of negatively-charged NV defects were introduced to the NDs by means of high energy irradiation with 2 MeV electrons followed by annealing under vacuum at 800 °C to mobilize the vacancies to the intrinsic nitrogen. The negatively charged NV centers in the ND exhibits unique electron-spin properties in which they possess non-zero electronic spin and favorable energy level structure that support spin polarization to occur due to manipulation with magnetic field that can be interrogated by visible light. Since a single NV center enables magnetic surface mapping with nanometric spatial resolution, this resulted in NDs with a high density of NV centers to offer extremely high magnetometric sensitivity [183]. The ability to sense the magnetic field using the ND-doped tellurite glass fiber is highly dependent on the fluorescence emission of the NV centers that can be observed due to the control of the ground state population [184]. Figure 7b illustrates the inhomogeneous distribution of optical image captured from the fiber end face that validated the emission was originated from the collection of spatially distributed incoherent sources of the NVs in the doped ND. When an external magnetic field was applied, a decrease in the fluorescence emission intensity at 637 nm was observed (Figure 7c). This is attributed to the mixing of 'bright' ($m_s = 0$) and 'dark' ($m_s = 1$) spin states that occur when spins and external magnetic field are misaligned [185]. Apart from the emission wavelength, the overall reduction in the fluorescence intensity with the increasing magnetic field was also observed at off-resonant fluorescence intensity, as seen in Figure 7d. The presented sensor demonstrated an inherently magnetic field sensitivity of 10 $\mu T/Hz^{1/2}$. Bai et al. [186] recently published a novel ND-doped lead silicate glass fiber that not only yielded an improved magnetic field sensitivity of 350 $nT/Hz^{1/2}$, but also introduce a new avenue of coupling fluorescence emission from ND to guided modes of conventional step-index optical fiber.

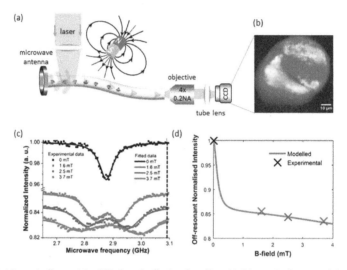

Figure 7. Magnetically sensitive ND-doped tellurite glass fiber. (**a**) Schematic diagram of the optically detected magnetic resonance using ND-doped tellurite glass fiber. (**b**) Image of the NV emission from the fiber end face captured using an sCMOS camera. (**c**) Output spectra of the sensor with varying magnetic field strength. (**d**) Theoretical and experimental results for the off-resonant fluorescence intensity versus magnetic field [182]. Copyright © 2018, Springer Nature.

6.4. Heavy Metal Ions

Heavy metal pollution in the environment arises from both natural phenomena and anthropogenic activities such as volcanic eruptions, weathering, soil erosion, mining, coal burning, and other industrial, and agricultural production activities. Although some of the heavy metals like iron, copper, and zinc are essential nutrients for various biochemical and physiological functions, exposures to a high level of these metal ions can result in lethal effects such as organ failure and slowing the progression of physical and neurological degenerative processes [4]. Furthermore, heavy metals are persistent, bioaccumulate, and have negative impacts on the ecological health that may lead to contamination of food chains. For these reasons, heavy metal sensing is critical, particularly for water and soil quality monitoring applications. To date, many research efforts are channeled to the development of highly sensitive and selective sensors, ranging from portable handheld sensors to paper-based heavy metal sensors [187–189]. OFS incorporating carbon allotropes as the sensing element to detect heavy metal ions is particularly interesting since it does not only retain the advantages of conventional OFS but also improves the detection sensitivity, owing to the large surface to volume ratio of these carbon allotropes. Alwahib et al. [190] proposed a nickel ion sensor using a gold-coated D-shaped optical fiber that incorporates a functional coating comprising of rGO/Fe_2O_3 composite. The rGO composite is rich in hydroxyl, OH^-, and carboxyl, $COOH^-$ groups, thus, creating a strong affinity to the positive charge Pb^{2+} ions due to the electrostatic interaction between them. It was further discovered that the sensor operates on a dual-mode interrogation scheme, intensity change, and resonant wavelength shift, depending on the target detection range. At lower Pb^{2+} concentration range, output intensity was found to be more responsive as compared to output resonant wavelength shift that was constrained by the low resolution. Thus, an LOD of 0.3 µg/L and 1 mg/L were obtained for a linear detection range of 0–1 mg/L and 1–15 mg/L, correspondingly. Yao et al. [191] also developed a "FRET on fiber" sensing platform to detect various analytes, including Cd^{2+}. The sensor was prepared by immobilizing partially rGO (prGO) onto the fiber, followed by Rhodamine 6G (R6G). In the absence of Cd^{2+}, no fluorescence scattering is observed from the R6G due to the quenching effect of prGO. However, when Cd^{2+} ions are present, R6G will be released due to the higher affinity of prGO to Cd^{2+}, hence the fluorescence intensity will be restored. A pulsed laser of center wavelength 532 nm is focused on the sensing region, and the output fluorescence signals are captured by optical lenses and measured using a spectrometer. In addition to the fluorescence signal detection, binding of Cd^{2+} onto the prGO also induces refractive index change surrounding the prGO sensing region that is reflected as an output wavelength shift at wavelengths between 1510–1590 nm. This suggests that the developed sensor can be used as a fluorometric or an interferometric sensor. Moreover, the established sensing platform exhibits an ultra-high resolution of wavelength detection of 1 pm, achieving an LOD of 1.2 nM for Cd^{2+} detection.

Apart from graphene, a pioneering work conducted by Goncalves et al. [117] focused on fabricating CDs via laser ablation method followed by functionalization with NH_2-polyethylene glycol (PEG_{200}) and N-acetyl-L-cysteine. The end product was then deposited onto the fiber end tip using the LbL coating technique for Hg^{2+} detection in water. It was deduced that the responsivity of the sensor towards the presence of Hg^{2+} is primarily contributed by the special coordination chemistry between the thiol groups belong to the N-acetyl-L-cysteine and Hg^{2+}. This occurrence is detected when the fluorescence emission of CDs measured by the optical fiber decrease gradually due to the fluorescence quenching by Hg^{2+}. To achieve optimal sensing performance, the proposed sensor was prepared in different configurations. It was discovered that the sensor with 6 layers of CDs/PEG_{200}/N-acetyl-L-cysteine performed the best with an LOD of 0.01 µM and a linear detection range of 0.01 to 2.69 µM. Moreover, the fabricated sensor also showed high physical and chemical stability where no signal degradation or residual fluorescence was observed despite multiple cycles of drying and hydration. In another interesting work, Yap et al. [132] developed an evanescent wave-based optical microfiber sensor that uses CDs as the chelator for detecting Fe^{3+} in aqueous and biological samples. The as-synthesized nitrogen- and sulfur-co-doped CDs were prepared via a microwave-assisted pyrolysis approach and is the first instance which reports the use of CDs as a chelator to perform metal ion sensing at the solid-liquid interface without relying on its fluorescence properties. The abundance of hydroxyl, amine, and carboxyl groups present on the surface of the CDs yield specific coordination to Fe^{3+} that promotes an excellent detection limit of 0.77 µg/L and good selectivity to Fe^{3+} despite the presence of other interfering ions in aqueous samples. Quantification of Fe^{3+} concentration was obtained by correlating the amount of output wavelength shift that arises from the effective refractive index change of the sensing region due to the chelation of Fe^{3+} by the CDs. The authors also demonstrated the feasibility of integrating the CD-functionalized optical microfiber sensor with a portable optical interrogation system to facilitate on-site application use. When tested against Fe^{3+} spiked samples such as tap water, biological buffers, and animal serums, the test results were in good agreement with ICPMS results, further confirming the reliability and practicability of the proposed sensor for Fe^{3+} detection. Other studies that involved the integration of carbon-allotropes with OFS can be found at [78,192,193].

6.5. Alcohol

Alcohol such as ethanol is colorless, flammable, and is commonly employed in extensive industrial applications ranging from household products, biomedical supplies, beverage and chemical industries. Ethanol sensors that are simple, practical, and reproducible have become important for these manufacturing industries to ensure consistent production and a safe working environment. Even though exposures to ethanol will not cause any deleterious effects, it will still result in minor health impacts such as headache, drowsiness, eye irritation, and difficulty in breathing. In view of this, Girei et al. [194] presented a tapered multimode fiber coated with GO for ethanol detection in water. The graphene was prepared using the electrochemical exfoliation of graphite rod using sodium dodecyl benzene sulfonate, followed by a simplified Hummers method to synthesized GO. The end product was then drop-casted onto the taper waist of the fiber and annealed to strengthen the coating. To quantitate the measured ethanol concentration, the absorbance response within the visible wavelength range of the sensor was found to increase linearly with increasing ethanol concentration from 5 to 40%. It is believed that the ethanol detection occurs on account of the electron transfer mechanism between the graphene layer and oxygen molecules that consequently induces a change in the coating surface index. Girei et al. [194] also prepared two different fiber sensors, one coated with graphene while another with GO for comparison purposes. Tapered fiber sensor coated with GO demonstrated an approximately two-fold increase in sensitivity with a sensitivity of 1.33 a.u./% as compared to graphene-coated tapered OFS. Nevertheless, both sensors showed good reproducibility after three consecutive cycles of exposure to 5% of ethanol solvent, indicating the high reversibility and stability of the coating. The response and recovery time for graphene-coated fibers were 15 and 18 s, respectively. Meanwhile, for GO-coated fiber, the recorded time was 25 and 12 s correspondingly.

On the other hand, Aziz et al. [195] developed an Ag/rGO coated tapered multimode fiber to perform ethanol sensing via the LSPR operation principle. Similar to an earlier report by Girei et al. [194], when the fiber sensor was introduced to increasing ethanol concentration from 1 to 100%, the output absorption spectrum in the range of 350 to 800 nm showed an increase in intensity. The authors also prepared an rGO-only coated tapered multimode fiber and conducted the same test for ethanol sensing. It was later found that the Ag/rGO coated fiber exhibited higher sensitivity than rGO-only coated fiber. This is attributed to the presence of optical and catalytically active Ag nanoparticles that are sensitive to the local refractive index change induced by the binding of ethanol molecules, and thus, enhances the output signal. Aside from attaining an LOD of 1%, the Ag/rGO coated fiber sensor also recorded a response time of 11 s and a rapid recovery time of only 6 s. Hernaez et al. [119] also studied lossy mode resonance (LMR)-based SnO$_2$/PEI/GO coated multimode fiber for ethanol sensing (Figure 8). The LMRs yield an absorption band/resonance peak at a specific wavelength in the transmission spectrum and is modulated by the variation in the surrounding refractive index stimulated by the ethanol molecules being absorbed onto the surface coating.

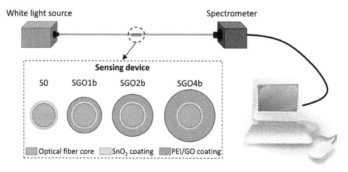

Figure 8. Ethanol sensing experimental setup using SnO$_2$/PEI/GO coated fiber [119]. Copyright © 2018, MDPI.

For the sake of comparison, fiber sensors of one, two, and four bilayers of PEI/GO were prepared to study the effect of coating thickness towards the sensing performance. With reference to a bare SnO_2 fiber, the study revealed that a significant sensitivity enhancement was observed in the sensor with 4 bilayers coating as the amount of resonant wavelength shift of the LMR peak was the highest (176.47%) among other sensors for ethanol concentration ranging from 0 to 100%. The amount of sensitivity enhancement is even higher (210%) when the range of ethanol concentration of interest is reduced to only 0–40%. Furthermore, the dynamic response of the sensor when exposed to 4 consecutive cycles of 40% ethanol showed the sensor exhibits good reusability and had a swift response and recovery time of 1 s and 2.75 s, respectively.

CNTs have also been reported for aqueous ethanol sensing when Shabaneh et al. [196] fabricated a tapered multimode fiber tip coated with CNTs using the drop cast approach. The CNTs coating was uniformly deposited on the end facet of the fiber probe sensor, and the thickness was approximately 370 nm. The reflectance signal of the fiber sensor was analyzed and found to decline with increasing concentration of ethanol solution. The fact that the CNTs are sensitive to ethanol can be explained since the pristine CNT used in this work was treated with nitric acid that serves as an oxidizing agent to introduce covalent attachment of carboxylic groups, −COOH, on the CNT surface. In the presence of ethanol molecules, hydrogen bonds are formed between the −COOH groups of the CNTs and the −OH groups belong to the ethanol molecules. As a result, this dipole-dipole interaction will cause a change in the surface index of the CNT coating and was reflected as a decrease in the reflectance spectrum. The developed sensors demonstrated good reproducibility and achieved ethanol sensitivity of 0.1441 a.u./%. The response and recovery time of the sensor was measured as 50 and 53 s, respectively. In summary, Table 6 collates information on all carbon nanomaterial-based optical fiber ethanol sensors and their respective sensing performance [197].

Table 6. Summary of carbon allotrope-based OFS for ethanol sensing.

Sensing Material	Optical Fiber	Detection Range (%)	LOD* (%)	Sensitivity	Ref.
Graphene	Tapered MMF	0–40	-	0.829 a.u. / %	[194]
GO	Plastic cladding silica fiber	10–80	5	-	[198]
GO	Tapered MMF	5–40	5	-	[199]
GO	FBG	0–80	10	-	[200]
GO	Tapered MMF	0–40	-	1.330 a.u. / %	[194]
SnO2/GO/PEI	MMF	0–100	20	-	[119]
Ag/rGO	MMF	0–100	1	-	[195]
CNTs	Tapered tip fiber	5–80	-	0.144 a.u. / %	[196]
CNTs	POF	20–100	0.2	-	[201]
CNTs	MMF	5–80	0.02	-	[133]

* Limit of Detection (LOD) in percentage concentration.

6.6. Ammonia and Volatile Organic Compounds

Ammonia is widely employed in the production of explosives, fertilizers, and as an industrial refrigerant even though it is known to be a highly toxic colorless gas, and inhalation of only a small amount of ammonia vapor may create potential hazards to human health. On the other hand, volatile organic compounds (VOCs) are categorized as carbon compounds with a low boiling point, high vapor pressure, toxic, and may result in serious environmental problems such as the formation of photochemical ozone smog, greenhouse effects, stratospheric ozone depletion, and others. The contribution of VOCs rising from anthropogenic emission has been increasing significantly due to the development of the industry. Common sources of VOCs emissions are often linked to the exploitation, transport, and usage of fossil fuels. Furthermore, non-point source VOCs caused by leakage, evaporation during production, and storage activities are equally significant and harder to control. In recent years, great efforts have been made to establish efficient solutions to mitigate ammonia and VOCs pollution as well as reliable sensors to monitor the level of ammonia and VOCs concentration in the air. The crossover between the attractive properties of carbon allotropes and the benefits offered by optical fibers has led to an attractive platform for ammonia and VOCs sensing. This can be seen from a study conducted by Kavinkumar and Manivannan, who synthesized Ag- decorated GO sheets by reducing the AgNO$_3$ with vitamin C in the presence of GO [202]. The abundance of epoxide and hydroxyl functional groups present on the GO enabled the in-situ formation of positively charged Ag nanoparticles to be adsorbed onto the negatively charged GO surface via the strong electrostatic interaction. Subsequently, when ammonia molecules (NH$_3$) interact with the Ag-decorated GO sheets, electron transfer reaction occurs between the two entities, making Ag-decorated GO sheets naturally responsive to the NH$_3$ vapor. Premised on this sensing mechanism, the as-developed Ag-decorated GO sheets-coated multimode fiber sensor was able to achieve a sensitivity of 0.17 counts/ppm for NH$_3$ concentration ranging from 0 to 500 ppm.

The same year, Kavinkumar and Manivannan also prepared a GO-MWCNTs-coated multimode fiber sensor for VOCs detection at room temperature [203]. The unique combination of GO and MWCNTs provides an improved gas sensing performance due to the large surface to volume ratio and high amount of oxygen functional groups of GO composites that served as the binding sites for the target gas molecules. The variation of output intensity of the fiber sensor is largely affected by the physical adsorption of gas molecules on the surface of the GO-MWCNTs coated on the outer surface of the fiber sensor. The adsorbed gas molecules act as an electron donor to the functional coating and subsequently lead to an increase in charge density of the functional coating that concomitantly changes the refractive index of the modified fiber. The proposed sensor is capable of detecting NH$_3$, ethanol, methanol vapor with sensitivities of 0.41, 0.36, and 0.23 counts/ppm, respectively.

Embedding CNTs within host-matrices of foreign material is a way to improve the gas sensing performance such as enhanced sensitivity and selectivity and extension of detection capabilities to variables of gas molecules. Consales et al. [204] designed and coated a cadmium arachidate(CdA)/SWCNTs composite onto the distal end of a single-mode fiber using the LB deposition technique and capable of detecting xylene, toluene, ethanol, and isopropanol. Even though the incorporation of CdA successfully yields detection sensitivity enhancement over the SWCNTs-coated fiber, the CdA matrix also affects the adsorption dynamics of the gas molecules onto the functional coating and this lengthens the response time of the sensor. Recently, Zhang et al. [205] prepared graphene-doped tin oxide nanocomposites for methane sensing using a side polished single-mode fiber as the sensing platform. Typically, SnO$_2$ is an n-type semiconductor material that contains electrons as its majority carrier. When methane molecules are in contact with SnO$_2$, the adsorbed molecules will diffuse freely across the thin film and loses its motion energy. Electrons from the methane molecules are transferred into SnO$_2$, turning into positive ions that lead to the increase of charge carriers in SnO$_2$ and consequently increase the conductivity and effective refractive index of the thin film. Nevertheless, the resistivity of pristine SnO$_2$ thin film alone is very high such that the amount of carrier concentration is hard to control since it is determined by the number of oxygen vacancies. For this reason, doping of graphene into the SnO$_2$ thin film appears to be a way to increase high carrier concentration for SnO$_2$ as well as improving the conductivity of the thin film that gives rise to an enhanced methane molecules sensitivity. The transmittance intensity of the output signal was found to increase with increasing methane concentration (0–55%) to which the fiber was exposed to, and a sensitivity of 200 a.u./% was obtained. Table 7 summarizes all recent advances in the fabrication of carbon nanomaterial-based OFS for VOC detection.

Table 7. Summary of carbon allotrope-based OFS for VOCs sensing.

Target Molecule	Sensing Material	Optical Fiber	Detection Range	LOD	Sensitivity*	Ref.
NH₃	GO-MWCNTs	MMF	0–500 ppm	–	0.41 cts ppm⁻¹	[202]
	GO	MMF	0–500 ppm	–	–0.32 cts ppm⁻¹	[206]
	rGO	MMF	0–500 ppm	–	0.08 cts ppm⁻¹	[202]
	Ag/GO	MMF	0–500 ppm	–	0.17 cts ppm⁻¹	[202]
	MWCNTs	MMF	0–500 ppm	–	0.31 cts ppm⁻¹	[207]
	SWCNTs	MMF	50–500 ppm	–	0.22 cts ppm⁻¹	[208]
	Graphene	Optical microfiber	0–300 ppm	–	0.015 nm/ppm	[209]
	Graphene/PANI	Side polished fiber	0–1%	–	132.8 a.u./%	[210]
	Graphene	D-shaped fiber	0–1000 ppm	0.04 ppm	–	[211]
	Graphene	Microfiber Bragg grating	0–100 ppm	0.2 ppm	–	[212]
	Fe₃O₄ - graphene	SMF	1.5–150 ppm	7 ppb	–	[213]
	Graphene	Microfiber hybrid waveguide	0–360 ppm	0.3 ppm	–	[214]
CH₄	Graphene/CNTs	MMF	10–100 ppm	–	0.3 nm/ppm	[215]
	PAA-NTs/PAH	PCF-LPG	0–3.5%	0.18%	1.078 nm/%	[216]
	SnO₂/Graphene	Side polished fiber	0–55%	–	200 a.u./%	[205]
Ethanol	GO-MWCNTs	MMF	0–500 ppm	–	0.36 cts ppm⁻¹	[203]
	GO	MMF	0–500 ppm	–	-0.26 cts ppm⁻¹	[206]
	rGO	MMF	0–500 ppm	–	0.065 cts ppm⁻¹	[217]
	MWCNTs	MMF	0–500 ppm	–	0.52 cts ppm⁻¹	[218]
	SWCNTs	MMF	50–500 ppm	–	0.20 cts ppm⁻¹	[208]
	CdA/SWCNTs	Fabry-Perot interferometer	–	–	0.5 10⁻³ / ppm	[204]
Methanol	GO-MWCNTs	MMF	0–500 ppm	–	0.23 cts ppm⁻¹	[203]
	GO	MMF	0–500 ppm	–	–0.20 cts ppm⁻¹	[206]
	rGO	MMF	0–500 ppm	–	0.038 cts ppm⁻¹	[217]
	MWCNTs	MMF	0 – 500 ppm	–	0.14 cts ppm⁻¹	[218]
	SWCNTs	MMF	50–500 ppm	–	0.01 cts ppm⁻¹	[208]
Toluene	rGO	Side polished fiber	50–200 ppm	79 ppm	–	[219]
	CdA/SWCNTs	Fabry-Perot interferometer	–	–	1.3 10⁻³ / ppm	[204]
Xylene	Graphene	Microfiber Bragg grating	0–100 ppm	0.5 ppm	–	[212]
	CdA/SWCNTs	Fabry-Perot interferometer	–	–	3.2 10⁻³ / ppm	[204]

* cts ppm⁻¹ = counts ppm⁻¹, Limit of Detection (LOD).

6.7. Other Biomolecule Sensing

By exploiting the evanescent wave of an optical fiber, carbon allotropes can serve as a functional coating capable of detecting bio-analytes and alter the output light properties for interrogation purposes. For instance, Qiu et al. [220] synthesized graphene film using the CVD technique and transferred directly onto the tapered PMMA core section of the plastic optical fiber. The graphene layer serves as a good molecule enricher or an absorbable film supported by the plastic fiber that can firmly absorb glucose molecules onto its surface and alters the output light intensity. A gradual decrease of output light intensity was observed when tested against 1–40% of glucose concentration. Similarly, Jiang et al. [221] also demonstrated a U-bent plastic optical fiber sensor for glucose sensing. Briefly, PVA/graphene/AgNPs thin film was coated onto the U-bent plastic optical fiber via the dip-coating method. The study investigated the effect of bent inner diameter and found that the refractive index sensitivity of the fiber sensor decreases with an increasing inner diameter, thus, U-bent plastic optical fiber with 5 mm inner bent diameter was chosen for subsequent glucose detection study. The sensitivity of the as-developed LSPR-based fiber sensor against glucose solutions of concentration ranging from 1.25 to 20% resulted in glucose sensitivity of approximately 0.9375 nm/%. Sharma and Gupta presented a graphene-based chalcogenide fiber optic sensor for hemoglobin detection in human blood samples [222]. The evanescent wave of the optical fiber interacts with the graphene monolayer that serves as a bio-enricher material in the proposed sensor. In the presence of hemoglobin of various concentrations, a significant portion of light was absorbed and output light intensity was reduced. At

1000 nm wavelength, an LOD of 18 μg/dL and hemoglobin sensitivity of 6.71×10^{-4} per g/dL were obtained. On top of that, the detection of uric acid [223], dopamine [224], human IgG [2] have been reported using graphene and CNTs-based OFS using the evanescent wave detecting scheme.

7. Future Outlook

The studies on carbon allotrope-based OFS have provided us with a deeper insight into the current trend of carbon allotrope-based sensing schemes and applications. Undoubtedly, carbon is an element with the widest and most diverse types of structure. Many carbon allotropes with different morphologies can be formed by the various arrangement of sp^2 and sp^3 carbon atoms, resulting in many different nanostructures with unique properties. Even though carbon allotropes have immense commercial potential in the sensing industries, there are still some challenges that need to be addressed before the commercialization of carbon allotrope-based sensors is possible. Firstly, extensive investigation into the preparation of the carbon allotropes is required to achieve uniform and reproducible nanostructures. For example, CDs can be prepared via various carbon precursors and can be doped with single or multiple dopant atoms to modulate the optical properties [225]. However, intensive studies are required to warrant a deeper understanding of the formation mechanism as well as the effects of the precursor types and synthesis methods on the resultant CDs to gain better control of the optimization work for sensing performance enhancement. Next, the stability of carbon allotropes is also a matter of concern. rGO, for instance, can form via chemically or thermally reduced route from GO. For this reason, the operating temperature of the GO-based OFS is often limited to prevent the undesired reduction process of the GO structure. Thus, even though it is advantageous for OFS to operate at high temperatures to allow quicker molecules adsorption and desorption process, most of the reported GO-based OFS were tested at ambient temperature. Besides, the sensitivity and selectivity of carbon allotrope-based sensors can also be further enhanced by proper surface treatment or ion doping. In a work by Ham et al. [226], it was reported that oxygen plasma treatment on CNTs improved the immuno-sensing detection limit by almost 1000 times when compared to the standard ELISA assay. On the other hand, non-treated CNTs exhibit no detectable signal. These results show that the oxygen-containing moieties introduced during the oxygen plasma process have a favorable effect on the sensing performance of the CNT-based sensors.

The integration of these carbon allotropes with optical fiber has given rise to many different OFS that are capable of detecting a wide range of molecules/parameters such as toxic gas, biomolecules, heavy metal ions, humidity, pressure, and others. As compared to other currently available technologies for sensing industries, carbon allotrope-based OFS sensing platform is advantageous in the aspect of resistance to corrosion, immune to electromagnetic interference, distributed and multiplexing sensing capabilities, and large dynamic operating range. However, the development of the OFS sensing platform is slightly more sophisticated than other sensing technologies such as electrochemical-based sensors that are excellent in terms of generating an output signal that can be used directly for hardware circuit, or microwave-based sensors such as microwave resonators that can also be easily linked to wireless sensing network platform due to the compatible operating frequencies. Table 8 summarizes the advantages and limitations of optical fiber sensing technology as compared to other conventional methods such as electrochemical, acoustic, microwave, and mass spectroscopy.

In summary, although OFS have humongous potential for sensing applications, there are still some obvious limitations that need to be addressed. Firstly, cross-sensitivity to factors such as temperature and pressure can significantly affect the sensing performance of the OFS. Thus, careful design of the sensory system is required to minimize interferences. One may implement an array of OFS, each functionalized with different sensing materials that exhibit different affinity levels to a group of target molecules in order to obtain multivariable responses which allow statistical method such as linear discriminant analysis to be done to improve sensor's selectivity [227]. A multiplexing system also enables the translation of the existing optical fiber sensing platform for distributed sensing applications. This can be done by integrating commercial optical frequency domain reflectometry system to spatially discriminate the measurand at different locations along a fiber length.

Secondly, traditional optical fiber measurement requires an optical spectrometer that is costly and bulky because it comprises many optical parts such as mirrors, beam splitters, and other assemblies to deliver broad wavelength coverage. Not only that these characters make the entire sensing system difficult for on-site measurement, but they may also appear to be redundant for some applications. In fact, a narrower wavelength span requirement is likely to ease the realization of a compact optical interrogator for the development of a portable OFS. Certainly, intensity interrogation will be much easier to implement given the fact that a simple and cost-effective photodetector is sufficient to deliver the output signal in electrical form. If wavelength interrogation is unavoidable, several novel interrogator designs involving the use of post-inscription of fiber gratings in the optical sensor probe can be used [228,229], but at the cost of complicated fabrication procedures and the possibility of affecting the sensor performance.

Table 8. Advantages and limitations of different sensing methods for environmental and biological sensing applications.

Sensing Methods	Advantages	Limitations
• Optical fiber [5]	• Small size • Long-term cost savings • Multiplexing capabilities • Distributed sensing capabilities • Multifunctional sensing capabilities • Large dynamic operating range • Self-referencing • Fast response time • Resistant to electromagnetic interference • Corrosion resistant • Can be used in harsh environments • Versatile sensing platform • Miniaturized and portable sensor • Feasible on-site or in-situ monitoring	• Requires custom design of the interrogation system • Cross sensitivities to physical parameters
• Electrochemical • (e.g., potential, current, conductivity, resistance [230])	• Chemical signal converts to electrical signal that can be used immediately for electrical hardware • Fast response time • Miniaturized and portable sensor • Potential for mass production • Feasible on-site or in-situ monitoring	• Complicated sensor configuration • Cross sensitivities to physical parameters
• Acoustic • (e.g., surface acoustic wave [231,232])	• Low power consumption • Small size • Able to work without batteries	• Cross sensitivities to physical parameters • Requires costly electronic detection system
• Microwave • (e.g., microwave resonator [233,234])	• Capable of non-contact sensing • Miniaturization is easy due to high operating frequencies • Easily linked with commercial mobile communication system / wireless sensing network • Inherently stable because the resonant frequency is related to the physical dimension	• Cross sensitivities to physical parameters • Susceptible to electromagnetic interference • Requires high degree of specialization • Cost of sensor configuration increase with frequency • Long wavelengths cause the achievable resolution is limited
• Mass spectroscopy [235]	• Ultra-high sensitivity and selectivity • High precision and accuracy	• Expensive setup • Bulky • Requires high degree of specialization • Tedious sample preparation

Sensors **2020**, *20*, 2046

Alternatively, one may incorporate fiber Bragg grating fibers into the design of the optical sensing system, serving as line markers that measure the light intensity at the pre-determined wavelengths of the output spectrum to estimate the amount of wavelength shift [187].

In short, future research works requires the synergistic effort from the material science groups and optical fiber technologists to integrate the benefits of carbon allotropes with optical fiber as well as to incorporate multiplexing capabilities, distributed sensing and Internet of Thing (IoT) to produce the next generation of carbon allotrope-based OFS.

Author Contributions: Conceptualization, S.H.K.Y. and K.K.C.; Writing, S.H.K.Y. and K.K.C.; Review and editing, S.H.K.Y., K.K.C., S.C.T., and K.-T.Y.; Supervision, S.C.T., and K.-T.Y. All authors have read and agreed to the published version of the manuscript.

Funding: This research received no external funding.

Conflicts of Interest: The authors declare no conflict of interest.

References

1. Zhang, J.; Liao, G.; Jin, S.; Cao, D.; Wei, Q.; Lu, H.; Yu, J.; Cai, X.; Tan, S.; Xiao, Y.; et al. All-fiber-optic temperature sensor based on reduced graphene oxide. *Laser Phys. Lett.* **2014**, *11*, 035901. [CrossRef]
2. Wang, Q.; Wang, B.-T. Surface plasmon resonance biosensor based on graphene oxide/silver coated polymer cladding silica fiber. *Sens. Actuators B Chem.* **2018**, *275*, 332–338. [CrossRef]
3. Chan, K.K.; Yap, S.H.K.; Yong, K.-T. Biogreen Synthesis of Carbon Dots for Biotechnology and Nanomedicine Applications. *Nano Micro Lett.* **2018**, *10*, 72. [CrossRef]
4. Chien, Y.-H.; Chan, K.K.; Yap, S.H.K.; Yong, K.-T. NIR-responsive nanomaterials and their applications; upconversion nanoparticles and carbon dots: A perspective. *J. Chem. Technol. Biotechnol.* **2018**, *93*, 1519–1528. [CrossRef]
5. Grattan, K.T.V. *Optical Fiber Sensor Technology Chemical and Environmental Sensing*; Springer: Dordrecht, The Netherlands, 1999; pp. 15–112.
6. Wu, Y.; Yao, B.; Yu, C.; Rao, Y. Optical Graphene Gas Sensors Based on Microfibers: A Review. *Sensors* **2018**, *18*, 941. [CrossRef] [PubMed]
7. Dong, N.; Wang, S.; Jiang, L.; Jiang, Y.; Wang, P.; Zhang, L. Pressure and Temperature Sensor Based on Graphene Diaphragm and Fiber Bragg Gratings. *IEEE Photonics Technol. Lett.* **2018**, *30*, 431–434. [CrossRef]
8. Zhao, Y.; Li, X.-g.; Zhou, X.; Zhang, Y.-n. Review on the graphene based optical fiber chemical and biological sensors. *Sens. Actuators B Chem.* **2016**, *231*, 324–340. [CrossRef]
9. Consales, M.; Cutolo, A.; Penza, M.; Aversa, P.; Giordano, M.; Cusano, A. Fiber optic chemical nanosensors based on engineered single-walled carbon nanotubes. *J. Sens.* **2008**, *2008*, 936074. [CrossRef]
10. Iijima, S. Helical microtubules of graphitic carbon. *Nature* **1991**, *354*, 56–58. [CrossRef]
11. Iijima, S.; Ichihashi, T. Single-shell carbon nanotubes of 1-nm diameter. *Nature* **1993**, *363*, 603–605. [CrossRef]
12. Niyogi, S.; Hamon, M.A.; Hu, H.; Zhao, B.; Bhowmik, P.; Sen, R.; Itkis, M.E.; Haddon, R.C. Chemistry of Single-Walled Carbon Nanotubes. *Acc. Chem. Res.* **2002**, *35*, 1105–1113. [CrossRef] [PubMed]
13. Bahr, J.Y.; Yang, J.; Kosynkin, D.V.; Bronikowski, M.J.; Smalley, R.E.; Tour, J.M. Functionalization of carbon nanotubes by electrochemical reduction of aryl diazonium salts: A bucky paper electrode. *J. Am. Chem. Soc* **2001**, *123*, 6536–6542. [CrossRef] [PubMed]
14. Agnihotri, S.; Mota, J.P.; Rostam-Abadi, M.; Rood, M.J. Structural characterization of single-walled carbon nanotube bundles by experiment and molecular simulation. *Langmuir* **2005**, *21*, 896–904. [CrossRef] [PubMed]
15. Gao, S.; Zhuang, R.C.; Zhang, J.; Liu, J.W.; Mäder, E. Glass fibers with carbon nanotube networks as multifunctional sensors. *Adv. Funct. Mater.* **2010**, *20*, 1885–1893. [CrossRef]
16. Lee, R.S.; Kim, H.J.; Fischer, J.; Thess, A.; Smalley, R.E. Conductivity enhancement in single-walled carbon nanotube bundles doped with K and Br. *Nature* **1997**, *388*, 255–257. [CrossRef]
17. Claye, A.S.; Fischer, J.E.; Huffman, C.B.; Rinzler, A.G.; Smalley, R.E. Solid-state electrochemistry of the Li single wall carbon nanotube system. *J. Electrochem. Soc.* **2000**, *147*, 2845–2852. [CrossRef]
18. Hirsch, A. Functionalization of Single-Walled Carbon Nanotubes. *Angew. Chem. Int. Ed.* **2002**, *41*, 1853–1859. [CrossRef]

19. Kamaras, K.; Itkis, M.E.; Hu, H.; Zhao, B.; Haddon, R.C. Covalent Bond Formation to a Carbon Nanotube Metal. *Science* **2003**, *301*, 1501. [CrossRef]
20. Xu, X.; Ray, R.; Gu, Y.; Ploehn, H.J.; Gearheart, L.; Raker, K.; Scrivens, W.A. Electrophoretic analysis and purification of fluorescent single-walled carbon nanotube fragments. *J. Am. Chem. Soc.* **2004**, *126*, 12736–12737. [CrossRef]
21. Ponomarenko, L.A.; Schedin, F.; Katsnelson, M.I.; Yang, R.; Hill, E.W.; Novoselov, K.S.; Geim, A.K. Chaotic Dirac billiard in graphene quantum dots. *Science* **2008**, *320*, 356–358. [CrossRef]
22. Jiang, J.; He, Y.; Li, S.; Cui, H. Amino acids as the source for producing carbon nanodots: Microwave assisted one-step synthesis, intrinsic photoluminescence property and intense chemiluminescence enhancement. *Chem. Commun.* **2012**, *48*, 9634–9636. [CrossRef] [PubMed]
23. Hsu, P.-C.; Chang, H.-T. Synthesis of high-quality carbon nanodots from hydrophilic compounds: Role of functional groups. *Chem. Commun.* **2012**, *48*, 3984–3986. [CrossRef] [PubMed]
24. Salinas-Castillo, A.; Ariza-Avidad, M.; Pritz, C.; Camprubí-Robles, M.; Fernández, B.; Ruedas-Rama, M.J.; Megia-Fernández, A.; Lapresta-Fernández, A.; Santoyo-Gonzalez, F.; Schrott-Fischer, A. Carbon dots for copper detection with down and upconversion fluorescent properties as excitation sources. *Chem. Commun.* **2013**, *49*, 1103–1105. [CrossRef]
25. Hu, S.-L.; Niu, K.-Y.; Sun, J.; Yang, J.; Zhao, N.-Q.; Du, X.-W. One-step synthesis of fluorescent carbon nanoparticles by laser irradiation. *J. Mater. Chem.* **2009**, *19*, 484–488. [CrossRef]
26. Dong, Y.; Wang, R.; Li, H.; Shao, J.; Chi, Y.; Lin, X.; Chen, G. Polyamine-functionalized carbon quantum dots for chemical sensing. *Carbon* **2012**, *50*, 2810–2815. [CrossRef]
27. Sun, D.; Ban, R.; Zhang, P.-H.; Wu, G.-H.; Zhang, J.-R.; Zhu, J.-J. Hair fiber as a precursor for synthesizing of sulfur-and nitrogen-co-doped carbon dots with tunable luminescence properties. *Carbon* **2013**, *64*, 424–434. [CrossRef]
28. Xu, Q.; Pu, P.; Zhao, J.; Dong, C.; Gao, C.; Chen, Y.; Chen, J.; Liu, Y.; Zhou, H. Preparation of highly photoluminescent sulfur-doped carbon dots for Fe (III) detection. *J. Mater. Chem. A* **2015**, *3*, 542–546. [CrossRef]
29. Shan, X.; Chai, L.; Ma, J.; Qian, Z.; Chen, J.; Feng, H. B-doped carbon quantum dots as a sensitive fluorescence probe for hydrogen peroxide and glucose detection. *Analyst* **2014**, *139*, 2322–2325. [CrossRef]
30. Jiang, G.; Jiang, T.; Li, X.; Wei, Z.; Du, X.; Wang, X. Boronic acid functionalized N-doped carbon quantum dots as fluorescent probe for selective and sensitive glucose determination. *Mater. Res. Express* **2014**, *1*, 025708. [CrossRef]
31. Pan, L.; Sun, S.; Zhang, A.; Jiang, K.; Zhang, L.; Dong, C.; Huang, Q.; Wu, A.; Lin, H. Truly fluorescent excitation-dependent carbon dots and their applications in multicolor cellular imaging and multidimensional sensing. *Adv. Mater.* **2015**, *27*, 7782–7787. [CrossRef]
32. Li, X.; Zhang, S.; Kulinich, S.A.; Liu, Y.; Zeng, H. Engineering surface states of carbon dots to achieve controllable luminescence for solid-luminescent composites and sensitive Be 2+ detection. *Sci. Rep.* **2014**, *4*, 4976. [CrossRef]
33. Chien, Y.H.; Chan, K.K.; Anderson, T.; Kong, K.V.; Ng, B.K.; Yong, K.T. Advanced Near-Infrared Light-Responsive Nanomaterials as Therapeutic Platforms for Cancer Therapy. *Adv. Ther.* **2019**, *2*, 1800090. [CrossRef]
34. Sun, Y.-P.; Zhou, B.; Lin, Y.; Wang, W.; Fernando, K.S.; Pathak, P.; Meziani, M.J.; Harruff, B.A.; Wang, X.; Wang, H. Quantum-sized carbon dots for bright and colorful photoluminescence. *J. Am. Chem. Soc.* **2006**, *128*, 7756–7757. [CrossRef] [PubMed]
35. Wang, H.; Sun, C.; Chen, X.; Zhang, Y.; Colvin, V.L.; Rice, Q.; Seo, J.; Feng, S.; Wang, S.; William, W.Y. Excitation wavelength independent visible color emission of carbon dots. *Nanoscale* **2017**, *9*, 1909–1915. [CrossRef]
36. Wang, C.; Xu, Z.; Cheng, H.; Lin, H.; Humphrey, M.G.; Zhang, C. A hydrothermal route to water-stable luminescent carbon dots as nanosensors for pH and temperature. *Carbon* **2015**, *82*, 87–95. [CrossRef]
37. Yu, P.; Wen, X.; Toh, Y.-R.; Tang, J. Temperature-dependent fluorescence in carbon dots. *J. Phys. Chem. C* **2012**, *116*, 25552–25557. [CrossRef]
38. Meng, X.; Chang, Q.; Xue, C.; Yang, J.; Hu, S. Full-colour carbon dots: From energy-efficient synthesis to concentration-dependent photoluminescence properties. *Chem. Commun.* **2017**, *53*, 3074–3077. [CrossRef]
39. Jia, X.; Li, J.; Wang, E. One-pot green synthesis of optically pH-sensitive carbon dots with upconversion luminescence. *Nanoscale* **2012**, *4*, 5572–5575. [CrossRef]

40. Panwar, N.; Soehartono, A.M.; Chan, K.K.; Zeng, S.; Xu, G.; Qu, J.; Coquet, P.; Yong, K.-T.; Chen, X. Nanocarbons for biology and medicine: Sensing, imaging, and drug delivery. *Chem. Rev.* **2019**, *119*, 9559–9656. [CrossRef]

41. Novoselov, K.S.; Geim, A.K.; Morozov, S.V.; Jiang, D.; Zhang, Y.; Dubonos, S.V.; Grigorieva, I.V.; Firsov, A.A. Electric field effect in atomically thin carbon films. *Science* **2004**, *306*, 666–669. [CrossRef]

42. Han, T.-H.; Lee, Y.; Choi, M.-R.; Woo, S.-H.; Bae, S.-H.; Hong, B.H.; Ahn, J.-H.; Lee, T.-W. Extremely efficient flexible organic light-emitting diodes with modified graphene anode. *Nat. Photonics* **2012**, *6*, 105–110. [CrossRef]

43. Pumera, M. Graphene-based nanomaterials for energy storage. *Energy Environ. Sci.* **2011**, *4*, 668–674. [CrossRef]

44. Bonaccorso, F.; Colombo, L.; Yu, G.; Stoller, M.; Tozzini, V.; Ferrari, A.C.; Ruoff, R.S.; Pellegrini, V. Graphene, related two-dimensional crystals, and hybrid systems for energy conversion and storage. *Science* **2015**, *347*, 1246501. [CrossRef] [PubMed]

45. Shen, H.; Zhang, L.; Liu, M.; Zhang, Z. Biomedical applications of graphene. *Theranostics* **2012**, *2*, 283. [CrossRef] [PubMed]

46. Yang, C.; Chan, K.K.; Xu, G.; Yin, M.; Lin, G.; Wang, X.; Lin, W.-J.; Birowosuto, M.D.; Zeng, S.; Ogi, T. Biodegradable Polymer-Coated Multifunctional Graphene Quantum Dots for Light-Triggered Synergetic Therapy of Pancreatic Cancer. *ACS Appl. Mater. Interfaces* **2018**, *11*, 2768–2781. [CrossRef] [PubMed]

47. Hong, G.; Diao, S.; Antaris, A.L.; Dai, H. Carbon Nanomaterials for Biological Imaging and Nanomedicinal Therapy. *Chem. Rev.* **2015**, *115*, 10816–10906. [CrossRef] [PubMed]

48. Liu, J.; Liu, Z.; Barrow, C.J.; Yang, W. Molecularly engineered graphene surfaces for sensing applications: A review. *Anal. Chim. Acta* **2015**, *859*, 1–19. [CrossRef]

49. Huang, Y.; Dong, X.; Shi, Y.; Li, C.M.; Li, L.-J.; Chen, P. Nanoelectronic biosensors based on CVD grown graphene. *Nanoscale* **2010**, *2*, 1485–1488. [CrossRef]

50. Kwak, Y.H.; Choi, D.S.; Kim, Y.N.; Kim, H.; Yoon, D.H.; Ahn, S.-S.; Yang, J.-W.; Yang, W.S.; Seo, S. Flexible glucose sensor using CVD-grown graphene-based field effect transistor. *Biosens. Bioelectron.* **2012**, *37*, 82–87. [CrossRef]

51. Ling, X.; Xie, L.; Fang, Y.; Xu, H.; Zhang, H.; Kong, J.; Dresselhaus, M.S.; Zhang, J.; Liu, Z. Can graphene be used as a substrate for Raman enhancement? *Nano Lett.* **2009**, *10*, 553–561. [CrossRef]

52. Dresselhaus, M.S.; Jorio, A.; Hofmann, M.; Dresselhaus, G.; Saito, R. Perspectives on carbon nanotubes and graphene Raman spectroscopy. *Nano Lett.* **2010**, *10*, 751–758. [CrossRef] [PubMed]

53. Qiu, C.; Zhou, H.; Yang, H.; Chen, M.; Guo, Y.; Sun, L. Investigation of n-layer graphenes as substrates for Raman enhancement of crystal violet. *J. Phys. Chem. C* **2011**, *115*, 10019–10025. [CrossRef]

54. Ju, L.; Geng, B.; Horng, J.; Girit, C.; Martin, M.; Hao, Z.; Bechtel, H.A.; Liang, X.; Zettl, A.; Shen, Y.R. Graphene plasmonics for tunable terahertz metamaterials. *Nat. Nanotechnol.* **2011**, *6*, 630. [CrossRef]

55. Fei, Z.; Andreev, G.O.; Bao, W.; Zhang, L.M.; McLeod, A.S.; Wang, C.; Stewart, M.K.; Zhao, Z.; Dominguez, G.; Thiemens, M. Infrared nanoscopy of Dirac plasmons at the graphene–SiO2 interface. *Nano Lett.* **2011**, *11*, 4701–4705. [CrossRef]

56. Wallace, P.R. The band theory of graphite. *Phys. Rev.* **1947**, *71*, 622. [CrossRef]

57. Neto, A.C.; Guinea, F.; Peres, N.M.; Novoselov, K.S.; Geim, A.K. The electronic properties of graphene. *Rev. Mod. Phys.* **2009**, *81*, 109–162. [CrossRef]

58. Marini, A.; Silveiro, I.n.; García de Abajo, F.J. Molecular sensing with tunable graphene plasmons. *ACS Photonics* **2015**, *2*, 876–882. [CrossRef]

59. Shang, J.; Ma, L.; Li, J.; Ai, W.; Yu, T.; Gurzadyan, G.G. The origin of fluorescence from graphene oxide. *Sci. Rep.* **2012**, *2*, 792. [CrossRef]

60. Hernaez, M.; Zamarreño, C.; Melendi-Espina, S.; Bird, L.; Mayes, A.; Arregui, F. Optical fibre sensors using graphene-based materials: A review. *Sensors* **2017**, *17*, 155. [CrossRef]

61. Ruan, Y.; Ding, L.; Duan, J.; Ebendorff-Heidepriem, H.; Monro, T.M. Integration of conductive reduced graphene oxide into microstructured optical fibres for optoelectronics applications. *Sci. Rep.* **2016**, *6*, 21682. [CrossRef]

62. Zhang, Y.; Rhee, K.Y.; Hui, D.; Park, S.-J. A critical review of nanodiamond based nanocomposites: Synthesis, properties and applications. *Compos. Part B: Eng.* **2018**, *143*, 19–27. [CrossRef]

63. Shenderova, O.; Zhirnov, V.; Brenner, D. Carbon nanostructures. *Crit. Rev. Solid State Mater. Sci.* **2002**, *27*, 227–356. [CrossRef]

64. Greiner, N.R.; Phillips, D.; Johnson, J.; Volk, F. Diamonds in detonation soot. *Nature* **1988**, *333*, 440–442. [CrossRef]

65. Mochalin, V.N.; Neitzel, I.; Etzold, B.J.; Peterson, A.; Palmese, G.; Gogotsi, Y. Covalent incorporation of aminated nanodiamond into an epoxy polymer network. *ACS Nano* **2011**, *5*, 7494–7502. [CrossRef] [PubMed]

66. Liang, Y.; Ozawa, M.; Krueger, A. A general procedure to functionalize agglomerating nanoparticles demonstrated on nanodiamond. *ACS Nano* **2009**, *3*, 2288–2296. [CrossRef] [PubMed]

67. Pentecost, A.; Gour, S.; Mochalin, V.; Knoke, I.; Gogotsi, Y. Deaggregation of nanodiamond powders using salt-and sugar-assisted milling. *ACS Appl. Mater. Interfaces* **2010**, *2*, 3289–3294. [CrossRef]

68. Morita, Y.; Takimoto, T.; Yamanaka, H.; Kumekawa, K.; Morino, S.; Aonuma, S.; Kimura, T.; Komatsu, N. A Facile and Scalable Process for Size-Controllable Separation of Nanodiamond Particles as Small as 4 nm. *Small* **2008**, *4*, 2154–2157. [CrossRef]

69. Plakhotnik, T. Diamonds for quantum nano sensing. *Curr. Opin. Solid State Mater. Sci.* **2017**, *21*, 25–34. [CrossRef]

70. Plakhotnik, T.; Aman, H.; Chang, H.-C. All-optical single-nanoparticle ratiometric thermometry with a noise floor of 0.3 K Hz− 1/2. *Nanotechnology* **2015**, *26*, 245501. [CrossRef]

71. Plakhotnik, T.; Doherty, M.W.; Cole, J.H.; Chapman, R.; Manson, N.B. All-optical thermometry and thermal properties of the optically detected spin resonances of the NV–center in nanodiamond. *Nano Lett.* **2014**, *14*, 4989–4996. [CrossRef]

72. Aman, H.; Plakhotnik, T. Accuracy in the measurement of magnetic fields using nitrogen-vacancy centers in nanodiamonds. *Josa B* **2016**, *33*, B19–B27. [CrossRef]

73. Ando, Y.; Iijima, S. Preparation of carbon nanotubes by arc-discharge evaporation. *Jpn. J. Appl. Phys. Part 2 Lett.* **1993**, *32*, L107–L109. [CrossRef]

74. Wang, Y.; Yeow, J.T. A review of carbon nanotubes-based gas sensors. *J. Sens.* **2009**, *2009*. [CrossRef]

75. Mubarak, N.; Abdullah, E.; Jayakumar, N.; Sahu, J. An overview on methods for the production of carbon nanotubes. *J. Ind. Eng. Chem.* **2014**, *20*, 1186–1197. [CrossRef]

76. Ebbesen, T.W. Production and purification of carbon nanotubes. *ChemInform* **1997**, *28*, 139–162. [CrossRef]

77. Li, J.; Lu, Y.; Ye, Q.; Cinke, M.; Han, J.; Meyyappan, M. Carbon nanotube sensors for gas and organic vapor detection. *Nano Lett.* **2003**, *3*, 929–933. [CrossRef]

78. Gonçalves, H.; Jorge, P.A.; Fernandes, J.; da Silva, J.C.E. Hg (II) sensing based on functionalized carbon dots obtained by direct laser ablation. *Sens. Actuators B Chem.* **2010**, *145*, 702–707. [CrossRef]

79. Li, X.; Wang, H.; Shimizu, Y.; Pyatenko, A.; Kawaguchi, K.; Koshizaki, N. Preparation of carbon quantum dots with tunable photoluminescence by rapid laser passivation in ordinary organic solvents. *Chem. Commun.* **2010**, *47*, 932–934. [CrossRef]

80. Chang, M.M.F.; Ginjom, I.R.; Ngu-Schwemlein, M.; Ng, S.M. Synthesis of yellow fluorescent carbon dots and their application to the determination of chromium (III) with selectivity improved by pH tuning. *Microchim. Acta* **2016**, *183*, 1899–1907. [CrossRef]

81. Hu, Y.; Yang, J.; Tian, J.; Jia, L.; Yu, J.-S. Waste frying oil as a precursor for one-step synthesis of sulfur-doped carbon dots with pH-sensitive photoluminescence. *Carbon* **2014**, *77*, 775–782. [CrossRef]

82. Dutta, P.; Ghosh, T.; Kumar, H.; Jain, T.; Singh, Y. Hydrothermal and solvothermal synthesis of carbon dots from chitosan-ethanol system. *Asian Chitin J* **2015**, *11*, 1–4.

83. Li, M.; Yu, C.; Hu, C.; Yang, W.; Zhao, C.; Wang, S.; Zhang, M.; Zhao, J.; Wang, X.; Qiu, J. Solvothermal conversion of coal into nitrogen-doped carbon dots with singlet oxygen generation and high quantum yield. *Chem. Eng. J.* **2017**, *320*, 570–575. [CrossRef]

84. Chan, K.K.; Yap, S.H.K.; Yong, K. Solid State Carbon Dots-Based Sensor Using Optical Microfiber for Ferric Ion Detection. In Proceedings of the 2019 IEEE International Conference on Sensors and Nanotechnology, Penang, Malaysia, 24–25 July 2019; pp. 1–4.

85. Chan, K.K.; Yang, C.; Chien, Y.-H.; Panwar, N.; Yong, K.-T. A Facile Synthesis of Label-Free Carbon Dots with Unique Selectivity-Tunable Characteristics for Ferric Ion Detection and Cellular Imaging Application. *New J. Chem. Commun.* **2019**, *43*, 4734–4744. [CrossRef]

86. Hernandez, Y.; Nicolosi, V.; Lotya, M.; Blighe, F.M.; Sun, Z.; De, S.; McGovern, I.; Holland, B.; Byrne, M.; Gun'Ko, Y.K. High-yield production of graphene by liquid-phase exfoliation of graphite. *Nat. Nanotechnol.* **2008**, *3*, 563–568. [CrossRef] [PubMed]

87. Brodie, B. Note sur un nouveau procédé pour la purification et la désagrégation du graphite. *Ann. Chim. Phys.* **1855**, *45*, 351–353.

88. Staudenmaier, L. Verfahren zur darstellung der graphitsäure. *Ber. Der Dtsch. Chem. Ges.* **1898**, *31*, 1481–1487. [CrossRef]

89. Hummers, W.S., Jr.; Offeman, R.E. Preparation of graphitic oxide. *J. Am. Chem. Soc.* **1958**, *80*, 1339. [CrossRef]

90. Jang, J.W.; Cho, S.; Moon, G.h.; Ihm, K.; Kim, J.Y.; Youn, D.H.; Lee, S.; Lee, Y.H.; Choi, W.; Lee, K.H. Photocatalytic Synthesis of Pure and Water-Dispersible Graphene Monosheets. *Chem. A Eur. J.* **2012**, *18*, 2762–2767. [CrossRef]

91. Sutter, P.W.; Flege, J.-I.; Sutter, E.A. Epitaxial graphene on ruthenium. *Nat. Mater.* **2008**, *7*, 406–411. [CrossRef]

92. Dedkov, Y.S.; Fonin, M.; Rüdiger, U.; Laubschat, C. Rashba effect in the graphene/Ni (111) system. *Phys. Rev. Lett.* **2008**, *100*, 107602. [CrossRef]

93. Li, X.; Wang, X.; Zhang, L.; Lee, S.; Dai, H. Chemically derived, ultrasmooth graphene nanoribbon semiconductors. *Science* **2008**, *319*, 1229–1232. [CrossRef] [PubMed]

94. Huang, H.; Chen, W.; Chen, S.; Wee, A.T.S. Bottom-up growth of epitaxial graphene on 6H-SiC (0001). *ACS Nano* **2008**, *2*, 2513–2518. [CrossRef] [PubMed]

95. Berger, C.; Song, Z.; Li, X.; Wu, X.; Brown, N.; Naud, C.; Mayou, D.; Li, T.; Hass, J.; Marchenkov, A.N. Electronic confinement and coherence in patterned epitaxial graphene. *Science* **2006**, *312*, 1191–1196. [CrossRef] [PubMed]

96. First, P.N.; de Heer, W.A.; Seyller, T.; Berger, C.; Stroscio, J.A.; Moon, J.-S. Epitaxial graphenes on silicon carbide. *Mrs Bull.* **2010**, *35*, 296–305. [CrossRef]

97. Zangwill, A.; Vvedensky, D.D. Novel growth mechanism of epitaxial graphene on metals. *Nano Lett.* **2011**, *11*, 2092–2095. [CrossRef]

98. Dolmatov, V.Y. Detonation synthesis ultradispersed diamonds: Properties and applications. *Russ. Chem. Rev.* **2001**, *70*, 607–626. [CrossRef]

99. Shenderova, O.A.; Gruen, D.M. *Ultrananocrystalline Diamond: Synthesis, Properties and Applications*; William Andrew: Norwich, NY, USA, 2012; pp. 79–114.

100. Danilenko, V.V. On the history of the discovery of nanodiamond synthesis. *Phys. Solid State* **2004**, *46*, 595–599. [CrossRef]

101. Mochalin, V.N.; Shenderova, O.; Ho, D.; Gogotsi, Y. The properties and applications of nanodiamonds. *Nat. Nanotechnol.* **2012**, *7*, 11–23. [CrossRef]

102. Gruen, D.M.; Shenderova, O.A.; Vul, A.Y. Synthesis, Properties and Applications of Ultrananocrystalline Diamond. In Proceedings of the NATO ARW on Synthesis, Properties and Applications of Ultrananocrystalline Diamond, St. Petersburg, Russia, 7–10 June 2004; Springer Science & Business Media: Amsterdam, The Netherlands, 2006.

103. Krüger, A.; Kataoka, F.; Ozawa, M.a.a.; Fujino, T.; Suzuki, Y.; Aleksenskii, A.E.; Vul, A.Y.; Ōsawa, E. Unusually tight aggregation in detonation nanodiamond: Identification and disintegration. *Carbon* **2005**, *43*, 1722–1730. [CrossRef]

104. Ozawa, M.; Inaguma, M.; Takahashi, M.; Kataoka, F.; Krueger, A.; Ōsawa, E. Preparation and behavior of brownish, clear nanodiamond colloids. *Adv. Mater.* **2007**, *19*, 1201–1206. [CrossRef]

105. Amans, D.; Chenus, A.-C.; Ledoux, G.; Dujardin, C.; Reynaud, C.; Sublemontier, O.; Masenelli-Varlot, K.; Guillois, O. Nanodiamond synthesis by pulsed laser ablation in liquids. *Diam. Relat. Mater.* **2009**, *18*, 177–180. [CrossRef]

106. Bai, P.; Hu, S.; Zhang, T.; Sun, J.; Cao, S. Effect of laser pulse parameters on the size and fluorescence of nanodiamonds formed upon pulsed-laser irradiation. *Mater. Res. Bull.* **2010**, *45*, 826–829. [CrossRef]

107. Kumar, A.; Lin, P.A.; Xue, A.; Hao, B.; Yap, Y.K.; Sankaran, R.M. Formation of nanodiamonds at near-ambient conditions via microplasma dissociation of ethanol vapour. *Nat. Commun.* **2013**, *4*, 2618. [CrossRef] [PubMed]

108. Petty, M.C. *Langmuir-Blodgett films an introduction*; Cambridge University Press: Cambridge, UK, 1996; pp. 39–41.
109. Rees, N.D.; James, S.W.; Tatam, R.P.; Ashwell, G.J. Optical fiber long-period gratings with Langmuir–Blodgett thin-film overlays. *Opt. Lett.* **2002**, *27*, 686–688. [CrossRef] [PubMed]
110. Peterson, I.R. Langmuir-Blodgett films. *J. Phys. D Appl. Phys.* **1990**, *23*, 379–395. [CrossRef]
111. Smietana, M.; Bock, W.J.; Szmidt, J.; Pickrell, G.R. Nanocoating Enhanced Optical Fiber Sensors. In *Advances in Materials Science for Environmental and Nuclear Technology*; The American Ceramic Society: Pittsburgh, PA, USA, 2010; pp. 275–286.
112. Decher, G.; Hong, J.D.; Schmitt, J. Buildup of ultrathin multilayer films by a self-assembly process: III. Consecutively alternating adsorption of anionic and cationic polyelectrolytes on charged surfaces. *Thin Solid Films* **1992**, *210–211*, 831–835. [CrossRef]
113. Paul, P.K.; Hansda, C.; Hussain, S.A. Layer-by-Layer Electrostatic Self-assembly Method: A Facile Approach of Preparing Nanoscale Molecular Thin Films. *Invertis J. Sci. Technol.* **2014**, *7*, 104–113.
114. Gowri, A.; Sai, V.V.R. Development of LSPR based U-bent plastic optical fiber sensors. *Sens. Actuators B Chem.* **2016**, *230*, 536–543. [CrossRef]
115. Lindgren, E.B.; Derbenev, I.N.; Khachatourian, A.; Chan, H.-K.; Stace, A.J.; Besley, E. Electrostatic Self-Assembly: Understanding the Significance of the Solvent. *J. Chem. Theory Comput.* **2018**, *14*, 905–915. [CrossRef]
116. Verveniotis, E.; Kromka, A.; Ledinský, M.; Čermák, J.; Rezek, B. Guided assembly of nanoparticles on electrostatically charged nanocrystalline diamond thin films. *Nanoscale Res. Lett.* **2011**, *6*, 144. [CrossRef]
117. Gonçalves, H.M.R.; Duarte, A.J.; Davis, F.; Higson, S.P.J.; Esteves da Silva, J.C.G. Layer-by-layer immobilization of carbon dots fluorescent nanomaterials on single optical fiber. *Anal. Chim. Acta* **2012**, *735*, 90–95. [CrossRef] [PubMed]
118. Alberto, N.; Vigário, C.; Duarte, D.; Almeida, N.A.F.; Gonçalves, G.; Pinto, J.L.; Marques, P.A.A.P.; Nogueira, R.; Neto, V. Characterization of Graphene Oxide Coatings onto Optical Fibers for Sensing Applications. *Mater. Today Proc.* **2015**, *2*, 171–177. [CrossRef]
119. Hernaez, M.; Mayes, A.G.; Melendi-Espina, S. Graphene Oxide in Lossy Mode Resonance-Based Optical Fiber Sensors for Ethanol Detection. *Sensors* **2018**, *18*, 58. [CrossRef] [PubMed]
120. Martin, P.M. *Handbook of Deposition Technologies for Films and Coatings: Science, Applications and Technology*, 3rd ed.; Elsevier: Amsterdam, The Netherlands; Boston, MA, USA, 2010; pp. 314–363.
121. Yao, B.; Wu, Y.; Wang, Z.; Cheng, Y.; Rao, Y.; Gong, Y.; Chen, Y.; Li, Y. Demonstration of complex refractive index of graphene waveguide by microfiber-based Mach–Zehnder interferometer. *Opt. Express* **2013**, *21*, 29818–29826. [CrossRef]
122. Jeschkowski, U.; Niederwald, H.; Möhl, W.; Beier, W.; Disam, J.; Gohlke, D.; Lübbers, K. Coating Technologies. In *Thin Films on Glass*; Bach, H., Krause, D., Eds.; Springer: Berlin/Heidelberg, Germany, 2003; pp. 51–98.
123. Kashiwagi, K.; Yamashita, S.; Set, S.Y. In-situ monitoring of optical deposition of carbon nanotubes onto fiber end. *Opt. Express* **2009**, *17*, 5711–5715. [CrossRef]
124. Malagnino, N.; Pesce, G.; Sasso, A.; Arimondo, E. Measurements of trapping efficiency and stiffness in optical tweezers. *Opt. Commun.* **2002**, *214*, 15–24. [CrossRef]
125. Jonáš, A.; Zemánek, P. Light at work: The use of optical forces for particle manipulation, sorting, and analysis. *Electrophoresis* **2008**, *29*, 4813–4851. [CrossRef]
126. Kashiwagi, K.; Yamashita, S. Deposition of carbon nanotubes around microfiber via evanescent light. *Opt. Express* **2009**, *17*, 18364–18370. [CrossRef]
127. Hermanson, G.T. Chapter 3—The Reactions of Bioconjugation. In *Bioconjugate Techniques*, 3rd ed.; Hermanson, G.T., Ed.; Academic Press: Boston, MA, USA, 2013; pp. 229–258.
128. Hartman, F.C.; Wold, F. Bifunctional Reagents. Cross-Linking of Pancreatic Ribonuclease with a Diimido Ester1. *J. Am. Chem. Soc.* **1966**, *88*, 3890–3891. [CrossRef]
129. Avrameas, S. Coupling of enzymes to proteins with glutaraldehyde: Use of the conjugates for the detection of antigens and antibodies. *Immunochemistry* **1969**, *6*, 43–52. [CrossRef]

130. Hermanson, G.T. Chapter 4 - Zero-Length Crosslinkers. In *Bioconjugate Techniques*, 3rd ed.; Hermanson, G.T., Ed.; Academic Press: Boston, MA, USA, 2013; pp. 259–273.

131. Yap, S.H.K.; Chan, K.K.; Chien, Y.; Yong, K. Factors Influencing Metal Binding Efficiency at Solid/Liquid Interface: An Investigation for the Prediction of Heavy Metal Ion Sensing Performance. In Proceedings of the 2019 IEEE International Conference on Sensors and Nanotechnology, Penang, Malaysia, 24–25 July 2019. [CrossRef]

132. Yap, S.H.K.; Chan, K.K.; Zhang, G.; Tjin, S.C.; Yong, K.-T. Carbon Dot-functionalized Interferometric Optical Fiber Sensor for Detection of Ferric Ions in Biological Samples. *ACS Appl. Mater. Interfaces* **2019**, *11*, 28546–28553. [CrossRef] [PubMed]

133. Shabaneh, A.A.; Girei, S.H.; Arasu, P.T.; Rashid, S.A.; Yunusa, Z.; Mahdi, M.A.; Paiman, S.; Ahmad, M.Z.; Yaacob, M.H. Reflectance Response of Optical Fiber Coated With Carbon Nanotubes for Aqueous Ethanol Sensing. *IEEE Photonics J.* **2014**, *6*, 1–10. [CrossRef]

134. Xiao, Y.; Zhang, J.; Cai, X.; Tan, S.; Yu, J.; Lu, H.; Luo, Y.; Liao, G.; Li, S.; Tang, J.; et al. Reduced graphene oxide for fiber-optic humidity sensing. *Opt. Express* **2014**, *22*, 31555–31567. [CrossRef] [PubMed]

135. Azzuhri, S.R.A.; Amiri, I.S.; Zulkhairi, A.S.; Salim, M.A.M.; Razak, M.Z.A.; Khyasudeen, M.F.; Ahmad, H.; Zakaria, R.; Yupapin, P. Application of graphene oxide based Microfiber-Knot resonator for relative humidity sensing. *Results Phys.* **2018**, *9*, 1572–1577. [CrossRef]

136. Kuo, C.Y.; Chan, C.L.; Gau, C.; Liu, C.; Shiau, S.H.; Ting, J. Nano Temperature Sensor Using Selective Lateral Growth of Carbon Nanotube Between Electrodes. *IEEE Trans. Nanotechnol.* **2007**, *6*, 63–69. [CrossRef]

137. Choi, S.H.; Kim, Y.L.; Byun, K.M. Graphene-on-silver substrates for sensitive surface plasmon resonance imaging biosensors. *Opt. Express* **2011**, *19*, 458–466. [CrossRef]

138. Sharma, A.K.; Gupta, B.D. On the sensitivity and signal to noise ratio of a step-index fiber optic surface plasmon resonance sensor with bimetallic layers. *Opt. Commun.* **2005**, *245*, 159–169. [CrossRef]

139. Zeng, S.; Hu, S.; Xia, J.; Anderson, T.; Dinh, X.-Q.; Meng, X.-M.; Coquet, P.; Yong, K.-T. Graphene–MoS$_2$ hybrid nanostructures enhanced surface plasmon resonance biosensors. *Sens. Actuators B Chem.* **2015**, *207*, 801–810. [CrossRef]

140. Prabowo, B.; Purwidyantri, A.; Liu, K.-C. Surface plasmon resonance optical sensor: A review on light source technology. *Biosensors* **2018**, *8*, 80. [CrossRef]

141. Ikeda, S.; Okamoto, A. Hybridization-Sensitive On–Off DNA Probe: Application of the Exciton Coupling Effect to Effective Fluorescence Quenching. *Chem. Asian J.* **2008**, *3*, 958–968. [CrossRef]

142. Ji, W.B.; Tan, Y.C.; Lin, B.; Tjin, S.C.; Chow, K.K. Nonadiabatically Tapered Microfiber Sensor With Ultrashort Waist. *IEEE Photonics Technol. Lett.* **2014**, *26*, 2303–2306. [CrossRef]

143. Ji, W.B.; Liu, H.H.; Tjin, S.C.; Chow, K.K.; Lim, A. Ultrahigh Sensitivity Refractive Index Sensor Based on Optical Microfiber. *IEEE Photonics Technol. Lett.* **2012**, *24*, 1872–1874. [CrossRef]

144. An, G.; Li, S.; Cheng, T.; Yan, X.; Zhang, X.; Zhou, X.; Yuan, Z. Ultra-stable D-shaped Optical Fiber Refractive Index Sensor with Graphene-Gold Deposited Platform. *Plasmonics* **2019**, *14*, 155–163. [CrossRef]

145. Fu, H.; Zhang, M.; Ding, J.; Wu, J.; Zhu, Y.; Li, H.; Wang, Q.; Yang, C. A high sensitivity D-type surface plasmon resonance optical fiber refractive index sensor with graphene coated silver nano-columns. *Opt. Fiber Technol.* **2019**, *48*, 34–39. [CrossRef]

146. Sridevi, S.; Vasu, K.S.; Jayaraman, N.; Asokan, S.; Sood, A.K. Optical bio-sensing devices based on etched fiber Bragg gratings coated with carbon nanotubes and graphene oxide along with a specific dendrimer. *Sens. Actuators B Chem.* **2014**, *195*, 150–155. [CrossRef]

147. Shivananju, B.N.; Yamdagni, S.; Fazuldeen, R.; Kumar, A.K.S.; Nithin, S.P.; Varma, M.M.; Asokan, S. Highly Sensitive Carbon Nanotubes Coated Etched Fiber Bragg Grating Sensor for Humidity Sensing. *IEEE Sens. J.* **2014**, *14*, 2615–2619. [CrossRef]

148. Mohamed, H.; Irawati, N.; Ahmad, F.; Ibrahim, M.H.; Ambran, S.; Rahman, M.A.A.; Harun, S.W. Optical humidity sensor based on tapered fiber with multi-walled carbon nanotubes slurry. *Indones. J. Electr. Eng. Comput. Sci.* **2017**, *6*, 97–103. [CrossRef]

149. Isa, N.M.; Irawati, N.; Harun, S.W.; Ahmad, F.; Rahman, H.A.; Yusoff, M.H.M. Multi-walled carbon nanotubes doped Poly(Methyl MethAcrylate) microfiber for relative humidity sensing. *Sens. Actuators A Phys.* **2018**, *272*, 274–280. [CrossRef]

150. Ma, Q.F.; Tou, Z.Q.; Ni, K.; Lim, Y.Y.; Lin, Y.F.; Wang, Y.R.; Zhou, M.H.; Shi, F.F.; Niu, L.; Dong, X.Y.; et al. Carbon-nanotube/Polyvinyl alcohol coated thin core fiber sensor for humidity measurement. *Sens. Actuators B Chem.* **2018**, *257*, 800–806. [CrossRef]

151. Gao, R.; Lu, D.-f.; Cheng, J.; Jiang, Y.; Jiang, L.; Qi, Z.-m. Humidity sensor based on power leakage at resonance wavelengths of a hollow core fiber coated with reduced graphene oxide. *Sens. Actuators B Chem.* **2016**, *222*, 618–624. [CrossRef]

152. Xing, Z.; Zheng, Y.; Yan, Z.; Feng, Y.; Xiao, Y.; Yu, J.; Guan, H.; Luo, Y.; Wang, Z.; Zhong, Y.; et al. High-sensitivity humidity sensing of microfiber coated with three-dimensional graphene network. *Sens. Actuators B Chem.* **2019**, *281*, 953–959. [CrossRef]

153. Ahmad, H.; Rahman, M.T.; Sakeh, S.N.A.; Razak, M.Z.A.; Zulkifli, M.Z. Humidity sensor based on microfiber resonator with reduced graphene oxide. *Optik* **2016**, *127*, 3158–3161. [CrossRef]

154. Huang, Y.; Zhu, W.; Li, Z.; Chen, G.; Chen, L.; Zhou, J.; Lin, H.; Guan, J.; Fang, W.; Liu, X.; et al. High-performance fibre-optic humidity sensor based on a side-polished fibre wavelength selectively coupled with graphene oxide film. *Sens. Actuators B Chem.* **2018**, *255*, 57–69. [CrossRef]

155. Chu, R.; Guan, C.; Bo, Y.; Shi, J.; Zhu, Z.; Li, P.; Yang, J.; Yuan, L. All-optical graphene-oxide humidity sensor based on a side-polished symmetrical twin-core fiber Michelson interferometer. *Sens. Actuators B Chem.* **2019**, *284*, 623–627. [CrossRef]

156. Wang, Y.; Shen, C.; Lou, W.; Shentu, F. Polarization-dependent humidity sensor based on an in-fiber Mach-Zehnder interferometer coated with graphene oxide. *Sens. Actuators B Chem.* **2016**, *234*, 503–509. [CrossRef]

157. Hernaez, M.; Acevedo, B.; Mayes, A.G.; Melendi-Espina, S. High-performance optical fiber humidity sensor based on lossy mode resonance using a nanostructured polyethylenimine and graphene oxide coating. *Sens. Actuators B Chem.* **2019**, *286*, 408–414. [CrossRef]

158. Wang, Y.; Shen, C.; Lou, W.; Shentu, F.; Zhong, C.; Dong, X.; Tong, L. Fiber optic relative humidity sensor based on the tilted fiber Bragg grating coated with graphene oxide. *Appl. Phys. Lett.* **2016**, *109*, 031107. [CrossRef]

159. Jiang, B.; Bi, Z.; Hao, Z.; Yuan, Q.; Feng, D.; Zhou, K.; Zhang, L.; Gan, X.; Zhao, J. Graphene oxide-deposited tilted fiber grating for ultrafast humidity sensing and human breath monitoring. *Sens. Actuators B Chem.* **2019**, *293*, 336–341. [CrossRef]

160. Liu, S.; Meng, H.; Deng, S.; Wei, Z.; Wang, F.; Tan, C. Fiber Humidity Sensor Based on a Graphene-Coated Core-Offset Mach–Zehnder Interferometer. *IEEE Sens. Lett.* **2018**, *2*, 1–4. [CrossRef]

161. Mohamed, H.; Hussin, N.; Ahmad, F.; Ambran, S.; Harun, S.W. Optical based relative humidity sensor using tapered optical fiber coated with graphene oxide. In Proceedings of the AIP Conference, Melville, NY, USA, 15–17 August 2016; p. 050006. [CrossRef]

162. Weimin, L.; Youqing, W.; Changyu, S.; Fengying, S. An optical fiber humidity sensor based on Mach-Zehnder interferometer coated with a composite film of graphene oxide and polyvinyl alcohol. In Proceedings of the 15th International Conference on Optical Communications and Networks (ICOCN), Hangzhou, China, 24–27 September 2016.

163. Falkovsky, L.A. Optical properties of graphene. In Proceedings of the the International Conference on Theoretical Physics 'DUBNA-NANO2008', Moscow, Russia, 7–11 July 2008; IOP Publishing: Bristol, UK; p. 012004. [CrossRef]

164. Vasu, K.S.; Asokan, S.; Sood, A.K. Enhanced strain and temperature sensing by reduced graphene oxide coated etched fiber Bragg gratings. *Opt. Lett.* **2016**, *41*, 2604–2607. [CrossRef]

165. Chu, R.; Guan, C.; Bo, Y.; Shi, J.; Zhu, Z.; Li, P.; Yang, J.; Yuan, L. Temperature Sensor in Suspended Core Hollow Fiber Covered With Reduced Graphene Oxide. *IEEE Photonics Technol. Lett.* **2019**, *31*, 553–556. [CrossRef]

166. Sun, Q.; Sun, X.; Jia, W.; Xu, Z.; Luo, H.; Liu, D.; Zhang, L. Graphene-Assisted Microfiber for Optical-Power-Based Temperature Sensor. *IEEE Photonics Technol. Lett.* **2016**, *28*, 383–386. [CrossRef]

167. Wang, M.; Li, D.; Wang, R.; Zhu, J.; Ren, Z. PDMS-assisted graphene microfiber ring resonator for temperature sensor. *Opt. Quantum Electron.* **2018**, *50*, 132. [CrossRef]

168. Ameen, O.F.; Younus, M.H.; Aziz, M.S.; Azmi, A.I.; Raja Ibrahim, R.K.; Ghoshal, S.K. Graphene diaphragm integrated FBG sensors for simultaneous measurement of water level and temperature. *Sens. Actuators A Phys.* **2016**, *252*, 225–232. [CrossRef]

169. Cui, Q.; Thakur, P.; Rablau, C.; Avrutsky, I.; Cheng, M.M. Miniature Optical Fiber Pressure Sensor With Exfoliated Graphene Diaphragm. *IEEE Sens. J.* **2019**, *19*, 5621–5631. [CrossRef]

170. Xu, W.; Yao, J.; Yang, X.; Shi, J.; Zhao, J.; Zhang, C. Analysis of Hollow Fiber Temperature Sensor Filled with Graphene-Ag Composite Nanowire and Liquid. *Sensors* **2016**, *16*, 1656. [CrossRef]

171. Jasmi, F.; Azeman, N.H.; Bakar, A.A.A.; Zan, M.S.D.; Badri, K.H.; Su'ait, M.S. Ionic Conductive Polyurethane-Graphene Nanocomposite for Performance Enhancement of Optical Fiber Bragg Grating Temperature Sensor. *IEEE Access* **2018**, *6*, 47355–47363. [CrossRef]

172. Li, C.; Liu, Q.; Peng, X.; Fan, S. Analyzing the temperature sensitivity of Fabry-Perot sensor using multilayer graphene diaphragm. *Opt. Express* **2015**, *23*, 27494–27502. [CrossRef]

173. Li, L.; Feng, Z.; Qiao, X.; Yang, H.; Wang, R.; Su, D.; Wang, Y.; Bao, W.; Li, J.; Shao, Z.; et al. Ultrahigh Sensitive Temperature Sensor Based on Fabry–Pérot Interference Assisted by a Graphene Diaphragm. *IEEE Sens. J.* **2015**, *15*, 505–509. [CrossRef]

174. Li, C.; Xiao, J.; Guo, T.; Fan, S.; Jin, W. Interference characteristics in a Fabry–Perot cavity with graphene membrane for optical fiber pressure sensors. *Microsyst. Technol.* **2015**, *21*, 2297–2306. [CrossRef]

175. Li, C.; Xiao, J.; Guo, T.; Fan, S.; Jin, W. Effects of graphene membrane parameters on diaphragm-type optical fibre pressure sensing characteristics. *Mater. Res. Innov.* **2015**, *19*, S5-17–S5-23. [CrossRef]

176. Ma, J.; Jin, W.; Ho, H.L.; Dai, J.Y. High-sensitivity fiber-tip pressure sensor with graphene diaphragm. *Opt. Lett.* **2012**, *37*, 2493–2495. [CrossRef] [PubMed]

177. Zheng, B.-C.; Yan, S.-C.; Chen, J.-H.; Cui, G.-X.; Xu, F.; Lu, Y.-Q. Miniature optical fiber current sensor based on a graphene membrane. *Laser Photonics Rev.* **2015**, *9*, 517–522. [CrossRef]

178. Wang, D.; Fan, S.; Jin, W. Graphene diaphragm analysis for pressure or acoustic sensor applications. *Microsyst. Technol.* **2015**, *21*, 117–122. [CrossRef]

179. Zhang, Y.; Wang, F.; Liu, Z.; Duan, Z.; Cui, W.; Han, J.; Gu, Y.; Wu, Z.; Jing, Z.; Sun, C.; et al. Fiber-optic anemometer based on single-walled carbon nanotube coated tilted fiber Bragg grating. *Opt. Express* **2017**, *25*, 24521–24530. [CrossRef] [PubMed]

180. Liu, Y.; Liang, B.; Zhang, X.; Hu, N.; Li, K.; Chiavaioli, F.; Gui, X.; Guo, T. Plasmonic Fiber-Optic Photothermal Anemometers With Carbon Nanotube Coatings. *J. Lightwave Technol.* **2019**, *37*, 3373–3380. [CrossRef]

181. Liu, Z.; Wang, F.; Zhang, Y.; Jing, Z.; Peng, W. Low-Power-Consumption Fiber-Optic Anemometer Based on Long-Period Grating With SWCNT Coating. *IEEE Sens. J.* **2019**, *19*, 2592–2597. [CrossRef]

182. Ruan, Y.; Simpson, D.A.; Jeske, J.; Ebendorff-Heidepriem, H.; Lau, D.W.M.; Ji, H.; Johnson, B.C.; Ohshima, T.; Afshar, V.S.; Hollenberg, L.; et al. Magnetically sensitive nanodiamond-doped tellurite glass fibers. *Sci. Rep.* **2018**, *8*, 1268. [CrossRef]

183. Wojciechowski, A.M.; Nakonieczna, P.; Mrózek, M.; Sycz, K.; Kruk, A.; Ficek, M.; Głowacki, M.; Bogdanowicz, R.; Gawlik, W. Optical Magnetometry Based on Nanodiamonds with Nitrogen-Vacancy Color Centers. *Materials* **2019**, *12*, 2951. [CrossRef]

184. Rondin, L.; Tetienne, J.P.; Hingant, T.; Roch, J.F.; Maletinsky, P.; Jacques, V. Magnetometry with nitrogen-vacancy defects in diamond. *Rep. Prog. Phys.* **2014**, *77*, 056503. [CrossRef]

185. Ruan, Y.; Simpson, D.A.; Jeske, J.; Ebendorff-Heidepriem, H.; Lau, D.W.; Ji, H.; Johnson, B.C.; Ohshima, T.; Hollenberg, L.; Greentree, A.D. Remote nanodiamond magnetometry. *arXiv* **2016**, arXiv:1602.06611.

186. Bai, D.; Capelli, M.; Huynh, H.; Ebendorff-Heidepriem, H.; Foster, S.; Greentree, A.D.; Gibson, B.C. Hybrid Diamond-Glass Optical Fibres for Magnetic Sensing. In Proceedings of the 26th International Conference on Optical Fiber Sensors, Lausanne, Switzerland, 24–28 September 2018. [CrossRef]

187. Yap, S.H.K.; Chien, Y.-H.; Tan, R.; bin Shaik Alauddin, A.R.; Ji, W.B.; Tjin, S.C.; Yong, K.-T. An Advanced Hand-Held Microfiber-Based Sensor for Ultrasensitive Lead Ion Detection. *ACS Sens.* **2018**, *3*, 2506–2512. [CrossRef] [PubMed]

188. Wang, C.; Sun, Y.; Jin, J.; Xiong, Z.; Li, D.; Yao, J.; Liu, Y. Highly selective, rapid-functioning and sensitive fluorescent test paper based on graphene quantum dots for on-line detection of metal ions. *Anal. Methods* **2018**, *10*, 1163–1171. [CrossRef]

189. Ji, W.B.; Yap, S.H.K.; Panwar, N.; Zhang, L.L.; Lin, B.; Yong, K.T.; Tjin, S.C.; Ng, W.J.; Majid, M.B.A. Detection of low-concentration heavy metal ions using optical microfiber sensor. *Sens. Actuators B Chem.* **2016**, *237*, 142–149. [CrossRef]

190. Abdulkhaleq Alwahib, A.; Fawzi Alhasan, S.; Yaacob, M.H.; Lim, H.N.; Adzir Mahdi, M. Surface plasmon resonance sensor based on D-shaped optical fiber using fiberbench rotating wave plate for sensing pb ions. *Optik* **2020**, *202*, 163724. [CrossRef]

191. Yao, B.C.; Wu, Y.; Yu, C.B.; He, J.R.; Rao, Y.J.; Gong, Y.; Fu, F.; Chen, Y.F.; Li, Y.R. Partially reduced graphene oxide based FRET on fiber-optic interferometer for biochemical detection. *Sci. Rep.* **2016**, *6*, 23706. [CrossRef] [PubMed]

192. Lin, Y.; Dong, X.; Yang, J.; Maa, H.; Zu, P.; So, P.L.; Chan, C.C. Detection of Ni^{2+} with optical fiber Mach-Zehnder interferometer coated with chitosan/MWCNT/PAA. In Proceedings of the 16th International Conference on Optical Communications and Networks (ICOCN), Wuzhen, China, 7–10 August 2017. [CrossRef]

193. Gonçalves, H.M.R.; Duarte, A.J.; Esteves da Silva, J.C.G. Optical fiber sensor for Hg(II) based on carbon dots. *Biosens. Bioelectron.* **2010**, *26*, 1302–1306. [CrossRef]

194. Girei, S.H.; Shabaneh, A.A.; Ngee-Lim, H.; Hamidon, M.N.; Mahdi, M.A.; Yaacob, M.H. Tapered optical fiber coated with graphene based nanomaterials for measurement of ethanol concentrations in water. *Opt. Rev.* **2015**, *22*, 385–392. [CrossRef]

195. Aziz, A.; Lim, H.N.; Girei, S.H.; Yaacob, M.H.; Mahdi, M.A.; Huang, N.M.; Pandikumar, A. Silver/graphene nanocomposite-modified optical fiber sensor platform for ethanol detection in water medium. *Sens. Actuators B Chem.* **2015**, *206*, 119–125. [CrossRef]

196. Shabaneh, A.; Girei, S.; Arasu, P.; Mahdi, M.; Rashid, S.; Paiman, S.; Yaacob, M. Dynamic Response of Tapered Optical Multimode Fiber Coated with Carbon Nanotubes for Ethanol Sensing Application. *Sensors* **2015**, *15*, 10452–10464. [CrossRef]

197. Khalaf, A.L.; Shabaneh, A.A.A.; Yaacob, M.H. Chapter 10 - Carbon Nanotubes and Graphene Oxide Applications in Optochemical Sensors. In *Synthesis, Technology and Applications of Carbon Nanomaterials*; Rashid, S.A., Raja Othman, R.N.I., Hussein, M.Z., Eds.; Elsevier: Selangor, Malaysia, 2019; pp. 223–246.

198. Gao, S.S.; Qiu, H.W.; Zhang, C.; Jiang, S.Z.; Li, Z.; Liu, X.Y.; Yue, W.W.; Yang, C.; Huo, Y.Y.; Feng, D.J.; et al. Absorbance response of a graphene oxide coated U-bent optical fiber sensor for aqueous ethanol detection. *RSC Adv.* **2016**, *6*, 15808–15815. [CrossRef]

199. Girei, S.H.; Shabaneh, A.A.; Lim, H.M.; Huang, N.H.; Mahdi, M.A.; Yaacob, M.H. Absorbance response of graphene oxide coated on tapered multimode optical fiber towards liquid ethanol. *J. Eur. Opt. Soc. Rapid Publ.* **2015**, *10*, 15019. [CrossRef]

200. Arasu, P.T.; Noor, A.S.; Shabaneh, A.A.; Yaacob, M.H.; Lim, H.N.; Mahdi, M.A. Fiber Bragg grating assisted surface plasmon resonance sensor with graphene oxide sensing layer. *Opt. Commun.* **2016**, *380*, 260–266. [CrossRef]

201. Khalaf, A.L.; Arasu, P.T.; Lim, H.N.; Paiman, S.; Yusof, N.A.; Mahdi, M.A.; Yaacob, M.H. Modified plastic optical fiber with CNT and graphene oxide nanostructured coatings for ethanol liquid sensing. *Opt. Express* **2017**, *25*, 5509–5520. [CrossRef] [PubMed]

202. Kavinkumar, T.; Manivannan, S. Uniform decoration of silver nanoparticle on exfoliated graphene oxide sheets and its ammonia gas detection. *Ceram. Int.* **2016**, *42*, 1769–1776. [CrossRef]

203. Kavinkumar, T.; Manivannan, S. Synthesis, Characterization and Gas Sensing Properties of Graphene Oxide-Multiwalled Carbon Nanotube Composite. *J. Mater. Sci. Technol.* **2016**, *32*, 626–632. [CrossRef]

204. Consales, M.C.A.; Penza, M.; Aversa, P.; Veneri, P.D.; Giordano, M.; Cusano, A. SWCNT nano-composite optical sensors for VOC and gas trace detection. *Sens. Actuators B Chem.* **2009**, *138*, 351–361. [CrossRef]

205. Zhang, J.Y.D.; Ding, E.J.; Xu, S.C.; Li, Z.H.; Wang, X.X.; Song, F. Sensitization of an optical fiber methane sensor with graphene. *Opt. Fiber Technol.* **2017**, *37*, 26–29. [CrossRef]

206. Kavinkumar, T.; Sastikumar, D.; Manivannan, S. Effect of functional groups on dielectric, optical gas sensing properties of graphene oxide and reduced graphene oxide at room temperature. *RSC Adv.* **2015**, *5*, 10816–10825. [CrossRef]

207. Manivannan, S.; Shobin, L.; Saranya, A.; Renganathan, B.; Sastikumar, D.; Park, K.C. Carbon nanotubes coated fiber optic ammonia gas sensor. In Proceedings of the Integrated Optics: Devices, Materials, and Technologies XV, San Francisco, CA, USA, 24–26 January 2011; p. 79410M. [CrossRef]

208. Manivannan, S.; Saranya, A.M.; Renganathan, B.; Sastikumar, D.; Gobi, G.; Park, K.C. Single-walled carbon nanotubes wrapped poly-methyl methacrylate fiber optic sensor for ammonia, ethanol and methanol vapors at room temperature. *Sens. Actuators B Chem.* **2012**, *171–172*, 634–638. [CrossRef]

209. Zhao, Y.; Zhang, S.-Y.; Wen, G.-F.; Han, Z.-X. Graphene-based optical fiber ammonia gas sensor. *Instrum. Sci. Technol.* **2018**, *46*, 12–27. [CrossRef]

210. Khalaf, A.L.; Mohamad, F.S.; Rahman, N.A.; Lim, H.N.; Paiman, S.; Yusof, N.A.; Mahdi, M.A.; Yaacob, M.H. Room temperature ammonia sensor using side-polished optical fiber coated with graphene/polyaniline nanocomposite. *Opt. Mater. Express* **2017**, *7*, 1858–1870. [CrossRef]

211. Wu, Y.; Yao, B.C.; Zhang, A.Q.; Cao, X.L.; Wang, Z.G.; Rao, Y.J.; Gong, Y.; Zhang, W.; Chen, Y.F.; Chiang, K.S. Graphene-based D-shaped fiber multicore mode interferometer for chemical gas sensing. *Opt. Lett.* **2014**, *39*, 6030–6033. [CrossRef]

212. Wu, Y.; Yao, B.; Zhang, A.; Rao, Y.; Wang, Z.; Cheng, Y.; Gong, Y.; Zhang, W.; Chen, Y.; Chiang, K.S. Graphene-coated microfiber Bragg grating for high-sensitivity gas sensing. *Opt. Lett.* **2014**, *39*, 1235–1237. [CrossRef] [PubMed]

213. Pawar, D.; Rao, B.V.B.; Kale, S.N. Fe$_3$O$_4$-decorated graphene assembled porous carbon nanocomposite for ammonia sensing: Study using an optical fiber Fabry–Perot interferometer. *Analyst* **2018**, *143*, 1890–1898. [CrossRef]

214. Yao, B.W.Y.; Cheng, Y.; Zhang, A.; Gong, Y.; Rao, Y.-J.; Wang, Z.; Chen, Y. All-optical Mach–Zehnder interferometric NH3 gas sensor based on graphene/microfiber hybrid waveguide. *Sens. Actuators B Chem.* **2014**, *194*, 142–148. [CrossRef]

215. Mishra, S.K.; Tripathi, S.N.; Choudhary, V.; Gupta, B.D. Surface Plasmon Resonance-Based Fiber Optic Methane Gas Sensor Utilizing Graphene-Carbon Nanotubes-Poly(Methyl Methacrylate) Hybrid Nanocomposite. *Plasmonics* **2015**, *10*, 1147–1157. [CrossRef]

216. Yang, J.; Che, X.; Shen, R.; Wang, C.; Li, X.; Chen, W. High-sensitivity photonic crystal fiber long-period grating methane sensor with cryptophane-A-6Me absorbed on a PAA-CNTs/PAH nanofilm. *Opt. Express* **2017**, *25*, 20258–20267. [CrossRef]

217. Kavinkumar, T.; Sastikumar, D.; Manivannan, S. Reduced graphene oxide coated optical fiber for methanol and ethanol vapor detection at room temperature. In .Proceedings of the SPIE/COS Photonics, Beijing, China, 9–11 October 2014; p. 92700U.

218. Shobin, L.R.; Renganathan, B.; Sastikumar, D.; Park, K.C.; Manivannan, S. Pure and Iso-Butyl Methyl Ketone Treated Multi-Walled Carbon Nanotubes for Ethanol and Methanol Vapor Sensing. *IEEE Sens. J.* **2014**, *14*, 1238–1243. [CrossRef]

219. Xiao, Y.; Yu, J.; Shun, L.; Tan, S.; Cai, X.; Luo, Y.; Zhang, J.; Dong, H.; Lu, H.; Guan, H.; et al. Reduced graphene oxide for fiber-optic toluene gas sensing. *Opt. Express* **2016**, *24*, 28290–28302. [CrossRef]

220. Qiu, H.W.X.; Xu, S.C.; Jiang, S.Z.; Li, Z.; Chen, P.X.; Gao, S.S.; Zhang, C.; Feng, D.J. A novel graphene-based tapered optical fiber sensor for glucose detection. *Appl. Surf. Sci.* **2015**, *329*, 390–395. [CrossRef]

221. Jiang, S.; Li, Z.; Zhang, C.; Gao, S.; Li, Z.; Qiu, H.; Li, C.; Yang, C.; Liu, M.; Liu, Y. A novel U-bent plastic optical fibre local surface plasmon resonance sensor based on a graphene and silver nanoparticle hybrid structure. *J. Phys. D Appl. Phys.* **2017**, *50*, 165105. [CrossRef]

222. Sharma, A.K.; Gupta, J. Graphene based chalcogenide fiber-optic evanescent wave sensor for detection of hemoglobin in human blood. *Opt. Fiber Technol.* **2018**, *41*, 125–130. [CrossRef]

223. Batumalay, M. Tapered plastic optical fiber coated with single wall carbon nanotubes polyethylene oxide composite for measurement of uric acid concentration. *Sens. Rev.* **2014**, *34*, 75–79. [CrossRef]

224. Hu, W.H.Y.; Chen, C.; Liu, Y.; Guo, T.; Guan, B.-O. Highly sensitive detection of dopamine using a graphene functionalized plasmonic fiber-optic sensor with aptamer conformational amplification. *Sens. Actuators B Chem.* **2018**, *264*, 440–447. [CrossRef]

225. Liu, M.L.; Chen, B.B.; Li, C.M.; Huang, C.Z. Carbon dots: Synthesis, formation mechanism, fluorescence origin and sensing applications. *Green Chem.* **2019**, *21*, 449–471. [CrossRef]

226. Ham, S.W.; Hong, H.P.; Kim, J.H.; Min, S.J.; Min, N.K. Effect of Oxygen Plasma Treatment on Carbon Nanotube-Based Sensors. *J. Nanosci. Nanotechnol.* **2014**, *14*, 8476–8481. [CrossRef]

227. Pandit, S.; Banerjee, T.; Srivastava, I.; Nie, S.; Pan, D. Machine Learning-Assisted Array-Based Biomolecular Sensing Using Surface-Functionalized Carbon Dots. *ACS Sens.* **2019**, *4*, 2730–2737. [CrossRef]

228. Feder, K.S.; Westbrook, P.S.; Ging, J.; Reyes, P.I.; Carver, G.E. In-fiber spectrometer using tilted fiber gratings. *IEEE Photonics Technol. Lett.* **2003**, *15*, 933–935. [CrossRef]

229. Wagener, J.L.; Strasser, T.A.; Pedrazzani, J.R.; DeMarco, J.; DiGiovanni, D. Fiber grating optical spectrum analyzer tap. In Proceedings of the 11th International Conference on Integrated Optics and Optical Fibre Communications, and 23rd European Conference on Optical Communications, Edinburgh, UK, 22–25 September 1997; pp. 65–68. [CrossRef]

230. Yogeswaran, U.; Chen, S.-M. A Review on the Electrochemical Sensors and Biosensors Composed of Nanowires as Sensing Material. *Sensors* **2008**, *8*, 290–313. [CrossRef]

231. Bo, L.; Xiao, C.; Hualin, C.; Mohammad, M.A.; Xiangguang, T.; Luqi, T.; Yi, Y.; Tianling, R. Surface acoustic wave devices for sensor applications. *J. Semicond.* **2016**, *37*, 021001. [CrossRef]

232. Go, D.B.; Atashbar, M.Z.; Ramshani, Z.; Chang, H.-C. Surface acoustic wave devices for chemical sensing and microfluidics: A review and perspective. *Anal Methods* **2017**, *9*, 4112–4134. [CrossRef] [PubMed]

233. Park, J.-K.; Kang, T.-G.; Kim, B.-H.; Lee, H.-J.; Choi, H.H.; Yook, J.-G. Real-time Humidity Sensor Based on Microwave Resonator Coupled with PEDOT:PSS Conducting Polymer Film. *Sci. Rep.* **2018**, *8*, 439. [CrossRef] [PubMed]

234. Nyfors, E. Industrial Microwave Sensors—A Review. *Subsurf. Sens. Technol. Appl.* **2000**, *1*, 23–43. [CrossRef]

235. Bruno, T.; Svoronos, P. *CRC Handbook of Basic Tables for Chemical Analysis*; CRC Press: Boca Raton, FL, USA, 2010; p. 887.

Review

Relative Humidity Sensors Based on Microfiber Knot Resonators—A Review

Young-Geun Han [1,2]

1   Gimhae-Harvard Bioimaging Center, Gimhae Industry Promotion and Biomedical Foundation, Gimhae 50969, Korea; yghan@hanyang.ac.kr
2   Department of Physics, Hanyang University, Seoul 04763, Korea

Received: 5 November 2019; Accepted: 25 November 2019; Published: 27 November 2019

Abstract: Recent research and development progress of relative humidity sensors using microfiber knot resonators (MKRs) are reviewed by considering the physical parameters of the MKR and coating materials sensitive to improve the relative humidity sensitivity. The fabrication method of the MKR based on silica or polymer is briefly described. The many advantages of the MKR such as strong evanescent field, a high Q-factor, compact size, and high sensitivity can provide a great diversity of sensing applications. The relative humidity sensitivity of the MKR is enhanced by concerning the physical parameters of the MKR, including the waist or knot diameter, sensitive materials, and Vernier effect. Many techniques for depositing the sensitive materials on the MKR surface are discussed. The adsorption effects of water vapor molecules on variations in the resonant wavelength and the transmission output of the MKR are described regarding the materials sensitive to relative humidity. The sensing performance of the MKR-based relative humidity sensors is discussed, including sensitivity, resolution, and response time.

Keywords: fiber-optic sensor; microfiber; microfiber knot resonator; fiber-optic sensor; relative humidity sensor; gas sensor; sensitive materials

---

1. Introduction

Over the past few decades fiber-optic sensors have been intensively deployed in mechanical, chemical, and biological measurement because of their many advantages, including electromagnetic immunity, high accuracy, high sensitivity and flexibility, compactness, and low fabrication cost [1]. In spite of the feasible applications to fiber-optic sensors, controversy continues in terms of commercialization, for example, their small portion of the total sensor market. There is no doubt, however, about the great potential of fiber-optic sensor for future technologies. Recently, microfibers with strong evanescent field to enhance the sensitivity of fiber-optic sensors to external perturbations have been attracting growing interest [2–8]. The diameter of microfiber is usually in a range from hundreds nanometers to several micrometers [2–8]. High contrast of refractive index between the microfiber and the environment, diameter uniformity, and sidewall smoothness of the microfiber are attributed to the reduced optical loss, high field confinement, and high sensitivity regarding the strong fractional evanescent field [1–8].

The microfiber knot resonator (MKR) is readily fabricated by making a tie with a microfiber and has specific advantages, including high stability and easy fabrication, reliable response, and high Q-factors regarding effective mode coupling in the intertwisted overlap of the MKR [1,4–7]. Since light propagating and circulating in the MKR may induce a phase shift multiple of 2π, the periodic optical resonance in the MKR is generated essentially [1]. Since the MKR has good capability to improve the light-matter interaction, it is possible to realize all optical tunable devices based on the MKRs with the nonlinear material overlays [9–13]. In addition, the MKR-based optical sensing techniques have been intensively researched for the measurement of various physical parameters, including temperature, strain, pressure, and refractive index [14–22]. Table 1 summarizes various applications of the MKR-based sensing probes and their performance, including the physical parameters of the MKR, sensitivity, resolution, response time, and sensitive materials to enhance the sensitivity of the MKR to external perturbations. A temperature sensor with high sensitivity using the polydimethylsiloxane (PDMS)-packaged MKR was reported [15,16]. High thermal coefficients of the PDMS, including thermo-optic and thermal expansion factors further increased the wavelength shift of the MKR with variations in temperature [15,16]. The mechanical transverse load sensor using the PDMS-encapsulated MKR was proposed [17]. The Young's modulus of the PDMS can be used to change the knot diameter or length of the MKR with variations in the lateral load resulting in the wavelength shift of the MKR and the realization of MKR-based transverse load sensor [17]. Rather than the waist diameter, the knot diameter has a critical influence on the load sensitivity [17]. The MKR embedded in the steel blade was proposed as the bending sensor because the circular path-length and the effective index of the MKR are easily modified by changing the bending curvature [18]. A magnetic field sensor can be fabricated by inserting the MKR into a glass cell with a magnetic fluid [19]. The external magnetic field alters the transmission characteristics of the MKR depending on its knot diameter [19]. The knot diameter of the MKR regarding the structural bending-induced evanescent field is a significant factor for improving the magnetic field sensitivity of the MKR [19]. With the assistance of a NaCl solution, the MKR without any coating material was exploited to measure the increment of salinity [20]. The hydrogen gas sensor can be fabricated by depositing palladium on the MKR [21,22]. The formation of palladium hydride resulting from the adsorption of hydrogen molecules on the palladium may induce strain regarding the expansion of the palladium in the MKR and the red shift of the wavelength was observed [21,22]. Decreasing the waist diameter improves the hydrogen sensitivity of the MKR [22]. The MKR was also applied to measure electric current around the copper wire [23]. The resonant wavelength of the MKR has red-shift because of the thermal phase-shift induced by the electric current [23].

Table 1. Performance comparison of the MKR-based sensors.

Authors/ Publication Year	Sensor Type	Waist Diameter of Microfiber	Knot Diameter of MKR	Sensitive Material	Sensitivity Wavelength	Resolution/ Accuracy	Response Time
J. Li et al./ 2017 [15]	Temperature	Not Available	4.5 mm	Poly-Dimethylsiloxane (PDMS)	1.408 nm/°C for 24–38 °C 0.973 nm/°C for 40–54 °C	0.014 °C	Not Available
R. Fan et al./ 2019 [16]	Temperature	8.6 μm	4.5 mm	w/o Poly-Dimethylsiloxane (PDMS)	183 pm/°C w/o PDMS	Not Available	Not Available
				Poly-Dimethylsiloxane (PDMS)	1.67 nm/°C w PDMS		15 s
J. Li et al./ 2018 [17]	Load	6.92 μm	6.0 mm	w/o Poly-Dimethylsiloxane (PDMS)	6 pm/N w/o PDMS	Not Available	Not Available
		7.2 μm	4.0 mm	Poly-Dimethylsiloxane (PDMS)	90 pm/N w PDMS	0.38 N	Not Available
S. Dass et al./ 2018 [18]	Bending	16 μm	885 μm	None	3.04 nm/m^{-1}	3.29 × 10^{-3} m^{-1}	Not Available
Y. Ly et al./ 2018 [19]	Magnetic Field	4.0 μm	155 μm	Water-Based Magnetic Fluid	277 pm/mT	0.07 mT (Accuracy)	Not Available
Y. Liao et al./ 2015 [20]	Salinity	2.5 μm	855 μm	None	21.18 pm/‰	Not Available	Not Available
X. Wu et al./ 2015 [22]	Hydrogen	2.98 μm	7.25 mm	Palladium (Pd)	Not Available	Not Available	Not Available
K. S. Lim et al./ 2011 [23]	Current	2.0 μm	185 μm	None	51.3 pm/A^2	Not Available	3 s

Among a variety of sensing applications of the MKR, this manuscript specifically aims to review the recent progress of the MKR-based relative humidity sensors regarding their operating principle, structures, and sensitive materials to improve the relative humidity sensitivity. Humidity is an important sensing property in semiconductor and automotive industries, agriculture, chemical and medical areas [8,24–27]. Relative humidity sensors in medical field are necessary in various respiratory and sterilizer systems, biological products, etc. [28]. The microfiber fabrication method of a silica- or polymer-based MKR, such as a flame blushing or direct drawing technique, respectively, is briefly introduced. The MKR with a high Q-factor, high stability, simple structure, and small size is very suitable for precise measurement of humidity. The ambient index sensitivity is essential to achieve the MKR-based relative humidity sensor. The supplementary techniques to stabilize and enhance the performance and the external index sensitivity of the MKR are described, including Vernier effect or the waist diameter, which is related with the effective group index difference between optical modes. The silica- or polymer-based MKR responds to variation of relative humidity, but with somewhat low sensitivity. The sensitive materials to additionally increase the relative humidity sensitivity of the MKR are discussed, including the deposition method and procedure of sensitive materials, experimental results, and optical phenomena. Various MKR-based relative humidity sensors are compared regarding the physical parameters of the MKR, coating materials, and sensor specifications like sensitivity, resolution, and response time.

2. Fabrication, Operating Principle, and Supplementary Method for Improving the Ambient Index Sensitivity of the MKR

The key component of the MKR is a microfiber fabricated by using a micro-tapering technique with various heating sources, including a flame, a laser-based heating tube, and electric strip heater [2–6,29,30]. Two critical parameters in the micro-tapering technique are the temperature of the heating process and elongation of the pulling process. A computer-controlled heater using a flame generates high temperature to soften and melt a single-mode fiber (SMF) which is elongated simultaneously by two motorized stages as shown in Figure 1a. The flow rate of gas like oxygen or hydrogen and the pulling speeds of two motorized stages must be precisely controlled to produce the adiabatic or the non-adiabatic tapered structure of the microfiber. The use of a stereo optical microscope in situ enables observation of variations in the waist diameter. For the polymer optical fiber, a direct drawing method is usually employed because of its simple process as shown in Figure 1b [31]. After melting the polymer on the heating plate, the end of a silica or iron rod is immersed within the molten polymer on the heating plate, resulting in conglutination of polymer. The rod with polymer is then pulled to extend the polymer and produces polymer the micro/nanofiber, as seen in Figure 1b [31].

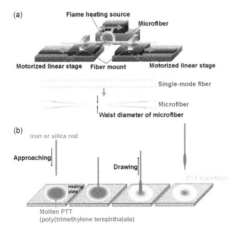

Figure 1. Fabrication of a microfiber [29,30,32]. (**a**) Flame blushing technique for fabrication of a silica-based microfiber [29,30]. Copyright 2016, IEEE; (**b**) direct drawing process for fabrication of a polymer-based micro/nanofiber (polytrimethylene terephthalate, PTT) [31].

After fabricating a silica or polymer-based microfiber, its two freestanding ends are assembled to form a comparatively large knot and tightened into the desired knot diameter by progressively pulling the free ends of the microfiber, as shown in Figure 2a, so that an MKR with a certain knot diameter can be achieved as shown in Figure 2b [1,29,30]. Mode coupling and interference occur in the microfiber knot region (coupling region) where the microfiber is twisted. Controlling the waist diameter regarding the effective group index difference between two modes (HE_{11} and HE_{12}) is capable of improving the sensitivity of the MKR to external perturbation as shown in Figure 3a–c [29,31]. The waist diameter regarding the V-parameter predominantly determines the number of guided modes regarding the cut-off wavelength and the effective indices of modes as shown in Figure 3b [29]. Figure 3a,b show that few modes are excited in the non-adiabatic down-tapered region when the waist diameter of the microfiber is larger than ~4 μm [29]. Then the few-mode MKR (FM-MKR) can be achieved by making a tie with the few-mode microfiber, which has two optical phenomena regarding optical modal interference in the few-mode microfiber and optical coupling in the FM-MKR. In the FM-MKR, the envelope shape in the transmission spectrum is generated by the modal interference between the HE_{11} and HE_{12} and the comb-like spectrum is created by optical coupling within the FM-MKR [29]. The sensitivity of the few-mode microfiber modal interferometer to ambient index regarding the waist diameter of the microfiber can be written as [29,32]:

$$S = \frac{\partial \lambda}{\partial n_{amb}} = \frac{\lambda}{\Delta n_g^{(m)}} \frac{\partial (n_{HE11} - n_{HE1m})}{\partial n_{amb}} \tag{1}$$

$$\Delta n_g^{(m)} = \Delta n_{eff}^{(m)} - \lambda \frac{\partial \Delta n_{eff}^{(m)}}{\partial \lambda} \tag{2}$$

where n_{HE11} and n_{HE1m} are the effective indices of HE_{11} and HE_{1m} modes, respectively. m is the mode order and $\Delta n_g^{(m)}$ is the difference in effective refractive group indices between the HE_{11} and HE_{1m} modes. The effective indices of the two modes (n_{HE11} and n_{HE1m}) and the value of $\Delta n_g^{(m)}$ can be changed by the waist diameter of the few-mode microfiber [29,32]. Equation (1) shows that the ambient index sensitivity can be improved by reducing $\Delta n_g^{(m)}$ regarding the waist diameter of the few-mode microfiber. Figure 3c exhibits the theoretical result on the ambient index sensitivity of the few-mode microfiber modal interferometer with respect to the waist diameter. Since $\Delta n_g^{(m)}$ becomes zero at 4-μm waist diameter, the ambient index sensitivity (S) is significantly enhanced, as shown in Figure 3c [29].

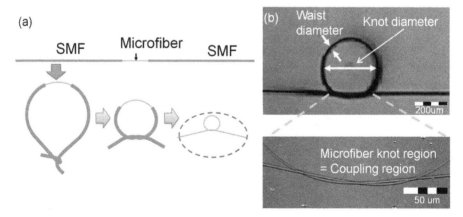

Figure 2. Fabrication of the MKR [29,30]; (**a**) knot procedure of the MKR; (**b**) microscopic image of the MKR. Copyright 2016, IEEE.

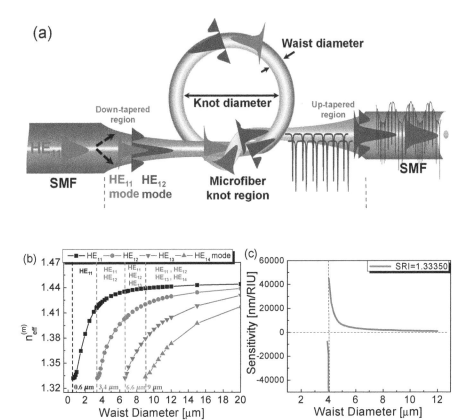

Figure 3. FM-MKR with the improved ambient index sensitivity [29]. (**a**) Operation principle of FM-MKR; (**b**) effective refractive indices ($n_{\text{eff}}^{(m)}$) with variations in the waist diameter depending on the radial mode number; (**c**) ambient index sensitivities with variations in the waist diameter (RIU: Relative index unit, SRI: Surrounding refractive index). Copyright 2018, IEEE.

By cascading two MKRs with slightly different free spectral ranges (FSRs), the Vernier effect based on optical spectrum interrogation can be induced to improve the ambient index sensitivity of the MKR [33]. After fabricating three pieces of microfibers, two MKRs were fulfilled by making ties with two microfibers as shown in Figure 4a ((I)–(III)). The other microfiber, in sequence, is carefully knotted with two MKRs in series, as shown in Figure 4a ((IV)–(VI)), to form the cascaded MKRs [33]. The output spectrum of the concatenated MKRs is created by the product of the individual transmission spectrum of the MKR, as seen in Figure 4b. Therefore, peak wavelengths of the cascaded MKRs are observed in the transmission spectrum such that two interference peaks are overlapped fractionally. The wavelength shift of the cascaded MKRs with variations in external perturbation can be increased by a magnification factor ($M = FSR_1/|FSR_1 - FSR_2|$) [33].

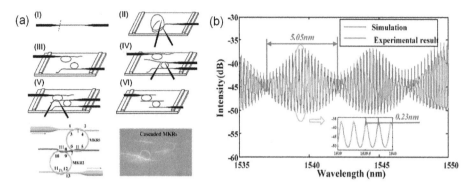

Figure 4. Cascaded MKRs with Vernier effect [33]. (**a**) Fabrication procedure, configuration, and photograph of the cascaded MKRs; (**b**) transmission spectrum. Theoretical analysis, including all mathematical models and parameters was described in [33] in detail. [Reprinted/Adapted] with permission from [33] © The Optical Society.

3. Relative Humidity Sensors Based on MKRs with Sensitive Materials

A silica- or polymer-based MKR without any humidity-sensitive material was applied to measure relative humidity in the surrounding environment [34]. A chamber with a hygrometer at a constant temperature is usually deployed in the experiment for relative humidity measurement. In the chamber, relative humidity is controlled by the amount of wet or dried air. Since the refractive index of silica or polymer is affected by relative humidity, the wavelength of the MKR is shifted. The density of the silica or polymer is changed by absorbing water vapor molecules so that refractive index is varied regarding the expansion or contract of the medium, as seen in Figure 5a [34]. Swelling the medium after adsorbing water molecules mitigates the density and the medium refractive index. Shrinking the medium after absorbing water vapor molecules increases the density and refractive index of medium because water vapor molecules fill the interstitial gaps of the medium [34]. For the silica-based MKR, increasing relative humidity shifts the wavelength into longer wavelengths [34]. This may be explained by the porous matrix of the silicon trapping water molecules on its interior surface [34]. Since water vapor molecules infiltrate into the porous matrix of the silicon, the average density of silicon regarding its refractive index is increased [34]. Consequently, a red-shift of the wavelength of the silica-based MKR was observed as shown in Figure 5b [34]. In a range of relative humidity from 20% to 60%, the linear behavior of the resonant wavelength shift regarding relative humidity was observed as shown in the inset of Figure 5b because of the variation of average density of the silicon resulting in the increase of its refractive index [34]. At a high relative humidity of more than 60%, the wavelength shift increased rapidly, which is attributed to the formation of clusters and the aggregation of water molecules on the silica-based MKR as shown in Figure 5b [34]. For the polymer-based MKR, as seen in Figure 5c, the peak wavelength also shifts to longer wavelengths and its relative humidity sensitivity is higher than that of the silica-based MKR because of the high hydrophilic feature of the polymer regarding the large molecule and the gap between molecules in the polymer [34]. At a high relative humidity of more than 90%, the amount of resonant wavelength shift increased because of the enhanced molecular state adsorption of water on the polymer surface [34].

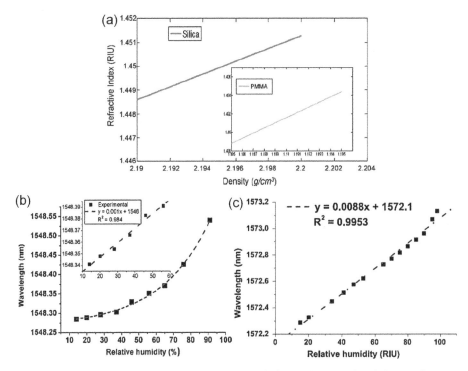

Figure 5. Silica or polymer MKR relative humidity sensor [34]. (**a**) Refractive index of silica or polymer (polymethyl methacrylate, PMMA) with variations in density; wavelength shift of the silica (**b**) or the polymer MKR (**c**) as a function of relative humidity. Copyright 2011, Elsevier.

The MKR-based relative humidity sensor at 2-μm waveband was proposed [35]. Since the strong water absorption exists at a wavelength of 1950 nm, the detection efficiency of relative humidity using the MKR can be improved without additional coating materials [35]. A tunable laser (OETLS-300) in a wavelength range from 1950 nm to 2050 nm was utilized as an input light source. An InGaAs photodetector was synchronized with a tunable laser by using a computer to recognize the wavelength shift of the MKR-based sensing probe [35]. The output spectrum of the 2-μm waveband MKR was changed by increasing relative humidity as depicted in Figure 6a [35]. Adsorption of water vapor molecules on the 2-μm MKR increases its refractive index, resulting in the red shift of wavelength, as shown in Figure 6b. Light absorption at 1950 nm reduces the extinction ratio of the MKR with increases in relative humidity. In Figure 6c, the fluctuation of the extinction ratio was approximately 0.19 dB for 30 consecutive measurements at a relative humidity of 57%. The rising and falling times were ~0.8 and ~1.55 s, respectively, as shown in Figure 6d [35].

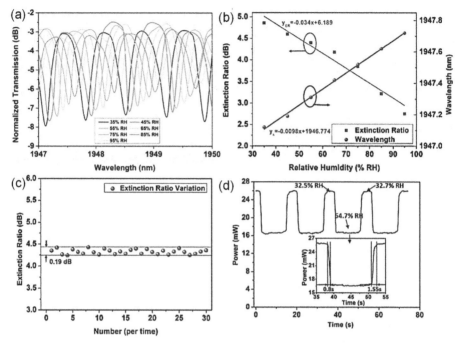

Figure 6. 2-μm waveband MKR relative humidity sensor [35]. (**a**) Normalized transmission spectra with variations in relative humidity; (**b**) variations of extinction ratio and peak wavelength as functions of relative humidity; (**c**) fluctuation of extinction ratio for 30 consecutive measurements at a relative humidity level of 57%; (**d**) temporal response. Copyright 2019, IEEE.

A supplementary coating layer as a receptor or transducer is exploited to apply the MKR to a relative humidity sensor [36]. The MKR with the strong evanescent field has an inherently high sensitivity to surrounding index change, so that implementing a humidity-sensitive material is effective for transforming the ambient index sensitivity to the relative humidity one. Nafion with high hydrophilicity, low refractive index, and high adhesivity to silica was exploited to realize the MKR-based relative humidity sensor [36]. A 3 μL of a 5 wt. % Nafion solution (in low alcohols and 10% water) was manually dropped on the MKR and then dried for 24 h at room temperature as shown in Figure 7a,b [36]. The Nafion-deposited MKR was placed in a chamber with a hygrometer. The relative humidity within the chamber was controlled by a gas blender. Wet air was provided by bubbling it into bubbling flasks containing distilled water. Dry air was obtained by passing an air flow through desiccant columns of silica gel and molecular sieve 5A (Sigma–Aldrich) [36]. Nafion as a perfluorosulfonate ion exchange polymer was developed by DuPont and has been extensively utilized in fuel cells [37] and electrochemical/optical sensors [36,38–41]. Nafion thin film has interesting properties, such as increasing relative humidity swells its structural dimension and reduces its refractive index [36,38–41]. The effect of relative humidity on the mechanical stress of the Nafion thin film predominantly induces a red shift of wavelength, as depicted in Figure 7c,d [36]. Two distinguished sensitive humidity regions are apparently observed in Figure 7d. At lower-mid relative humidity (30~45%), two relative humidity sensitivities (0.11 ± 0.02 nm/% for humidity increase and 0.08 ± 0.01 nm/% for humidity decrease) were exhibited. At higher-mid relative humidity (45~75% RH), two different sensitivities, such as 0.29 ± 0.01 nm/% for humidifying increase and 0.26 ± 0.01 nm/% for dehumidifying, are observed. A sensing hysteresis of ~1.9 nm was measured during the measurement. This is probably explained by the fact that the swelling tension of the Nafion thin film regarding the absorption of relative humidity degrades the sensitivity of relative humidity and induces hysteresis and two different sensitivities depending on the relative humidity ranges [36].

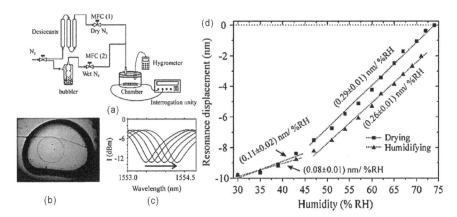

Figure 7. Nafion-deposited MKR relative humidity sensor [36]. (**a**) Experimental scheme for measurement of relative humidity using the MKR with Nafion; (**b**) photograph of the MKR deposited by Nafion; (**c**) transmission spectra with variations in relative humidity; (**d**) wavelength shift as a function of relative humidity. The system humidity calibration curve is shown in the inset. [Reprinted/Adapted] with permission from [36] © The Optical Society.

Two-dimensional materials, including graphene and reduced graphene oxide, were utilized as sensitive materials to ameliorate the relative humidity sensitivity of the MKR [42]. Graphene or reduced graphene oxide is suitable for gas- or biomolecule adsorption with large surface area, strong affinity, and high adsorption [42–47]. After dissolving the graphene oxide in deionized water, the graphene oxide solution was sonicated in an ultrasonic bath for 30 min [43]. The graphene oxide, in sequence, was dropped on the MKR fixed on the MgF_2 substrate and dried at a temperature of 40 °C as shown in Figure 8 [43]. For the MKR without the graphene oxide, variations of refractive index of the air regarding relative humidity alter the extinction ratio, output power, and the resonant wavelength, which are less distinct. Additional sensitive materials further improve the relative humidity sensitivity of the MKR. Since the graphene oxide layer has the high level of hygroscopicity, water molecules are absorbed by the graphene oxide matrix [42]. Therefore, the graphene oxide coating significantly develops the relative humidity sensitivity of the MKR [42]. The resonant wavelength of the MKR without or with the graphene oxide coating shifts into longer wavelengths as relative humidity increases as shown in Figure 9a,b. The MKR with the graphene oxide overlay has higher relative humidity sensitivity than the ordinary MKR [42]. The output power in the interference spectrum of the MKR without a graphene is barely reduced, as seen in Figure 9c. For the MKR with the graphene overlay, however, the output power is rapidly dropped from −30.7 dBm at a relative humidity of 0% to −31.3 dBm at a relative humidity of 60% as shown in Figure 9c [42]. As the relative humidity value is further increased to be 90%, the output power is slowly mitigated and dropped to −31.4 dBm because of the absorption saturation of water molecules in the graphene oxide layer as shown in Figure 9c [42]. The use of the graphene oxide overlay on the MKR successfully improves the sensitivity and linearity to relative humidity change.

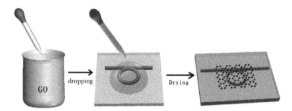

Figure 8. Coating process of graphene oxide on the MKR [43]. Reprinted/adapted with permission from The Optical Society ©.

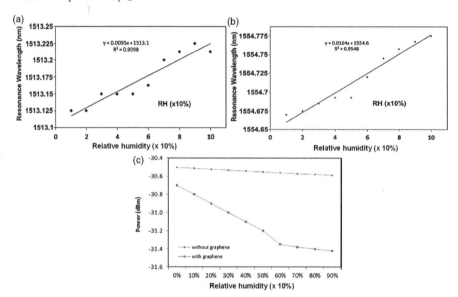

Figure 9. Graphene oxide-based MKR relative humidity sensor [42]. Wavelength shifts of MKRs without (**a**) and with (**b**) graphene oxide as a function of relative humidity; (**c**) output power variations of MKRs without and with graphene oxide as a function of relative humidity. Copyright 2018, Elsevier.

A polyvinyl alcohol (PVA) polymer was employed as an absorber of relative humidity to realize a MKR-based relative humidity sensor [29,30,48–54]. FM-MKR with two modes (HE_{11} and HE_{12}) was fabricated by using a flame-blushing method [29]. The non-adiabatic tapered shape in the microfiber is able to excite asymmetric modes (HE_{1m}) from a fundamental symmetric mode (HE_{11}) regarding the waist diameter [29]. The few-mode MKR on a low index MgF$_2$ disk was spin-coated by 9%-Cytop (n_{Cytop} = 1.34) at 700 rpm for 60 s and annealed in sequence for 120 min at 80 °C. Then the 8% PVA overlay with a thickness of 1.3 μm was spin-coated at 1200 rpm for 30 s and annealed again for 40 min at 60 °C and continuously for 180 min at 80 °C as shown in Figure 10a [29,30]. Figure 10b shows that the transmission of the FM-MKR was optically characterized by superimposing the spectrum of the modal interferometer with that of the MKR [29]. In the few-mode MKR, the envelope shape in the transmission spectrum is generated by the modal interference between the HE_{11} and HE_{12} regarding the slow-varying envelop in the transmission. Optical coupling within the few-mode MKR creates the comb-like spectrum regarding the fast-varying transmission as shown in Figure 10b. Since PVA is a water-soluble compound, absorbing water molecules swells the PVA overlay, which reduces its density. Consequently, the refractive index of the PVA is diminished by increasing relative humidity in the surrounding environment, which, in turn, changes the output spectrum of the MKR. Figure 10c shows that ascending the concentration of relative humidity shifts the resonant wavelength into shorter wavelengths [29]. Optimizing the waist diameter regarding the group index difference between the two modes improves the relative humidity sensitivity of the FM-MKR sensor [29]. Figure 10d exhibits the spatial frequency spectrum after converting the optical spectrum of the MKR with the fast Fourier transform. In Figure 10e, it is noticeable that all spatial frequencies shift to lower ones by increasing relative humidity because of the effective index reduction of the two modes.

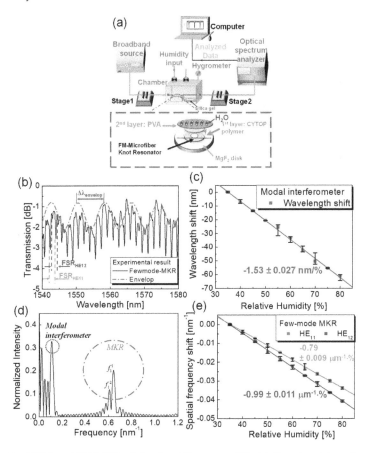

Figure 10. PVA-based FM-MKR relative humidity sensor [29]. (**a**) Experimental configuration; (**b**) transmission spectrum; (**c**) resonant wavelength shifts with variations in relative humidity; (**d**) spatial frequency spectrum; (**e**) spatial frequency shifts with variations in relative humidity. Copyright 2018, IEEE.

Titanium dioxide (TiO_2) nanoparticles were utilized to attain the MKR-based relative humidity sensor [55–59]. The TiO_2 nanoparticle-based overlay has fast adsorption capability for water vapor molecules because of its porous structure at room temperature [56–59]. The anatase phase of TiO_2 nanoparticles has better water adsorption ability than the two other phases, rutile and brookite [55]. The commercially available anatase TiO_2 powder was immersed with deionized water in a glass bottle [55–59]. The solution with TiO_2 nanoparticles was homogenized by the ultrasonication process and coated immediately on the surface of the MKR [55]. TiO_2 nanoparticles can be readily deposited on the MKR because of the evaporation of DI water, light injection, and the optical tweezing effects regarding the strong evanescent field of the MKR [55]. The output power of the MKR was measured to confirm the attachment of TiO_2 nanoparticles on the MKR. The transmission power of the MKR decreases monotonically as TiO_2 nanoparticles begin to be deposited on the surface of the MKR after immersing the MKR in the TiO_2 solution [55]. The transmission power is abruptly changed after taking the TiO_2 nanoparticle-coated MKR out of the TiO_2 solution and stabilized within a few seconds [55]. Increasing relative humidity shifts the wavelength of the TiO_2^- coated MKR into longer wavelengths because of the refractive index increment of the TiO_2 regarding the water filling process. The extinction ratio is reduced by increasing relative humidity. The deposition of TiO_2 nanoparticles on the MKR is effectively capable of improving the relative humidity sensitivity. The fast response and slow recovery time of the MKR with the TiO_2 coating are probably caused by the fast diffusion and slow desorption of water molecules regarding TiO_2 nanoparticles [55].

4. Conclusions

As presented, it is manifest that the MKR has great potential in various sensing applications to environmental, biomedical, and chemical monitoring. A great diversity of the MKR-based sensing probes with different sensitive materials has been researched. MKR is structurally based on microfiber. The fabrication methods of the silica- or polymer-based microfibers, including a flame blushing or direct drawing technique, respectively, were first presented. The MKR was simply realized by making a knot with a microfiber. The strong evanescent field and high Q-factor of the MKR enable the provision of various sensing capabilities to monitor physical, mechanical, and chemical parameters. The great adaptability of various coating materials and nanoparticles to the MKR facilitates the development of its sensitivity to external perturbation change and the extension of the MKR-based sensing capabilities. The applications and performance of the MKR-based sensors are summarized in Table 1. A variety of research and development of the MKR-based relative humidity sensors was overviewed regarding the physical properties of the MKR to improve the relative humidity sensitivity, sensitive materials to relative humidity, and the sensing performance like sensitivity, resolution, and response time. Adsorption of the water vapor molecules on the silica- or polymer-based MKR or the supplement coating materials may change the refractive index and the mechanical stress (swelling or shrinking) of materials and consequently modify the transmission characteristics of the MKR regarding the resonant wavelength and the output power. The performance of the MKR-based relative humidity sensors, including sensitivity, resolution, and response time is summarized in Table 2. Although the coating materials can effectively improve the relative humidity sensitivity of the MKR, they degrade the response time of the MKR-based sensor.

Table 2. Performance comparison of MKR-based relative humidity sensors with different sensitive materials and the physical parameters of MKRs.

Authors/ Publication Year	MKR Material	Waist Diameter of Microfiber	Knot Diameter of MKR	Humidity-Sensitive Material	Sensitivity		Resolution (*RH: Relative Humidity)	Response Time
					Wavelength	Transmission Power		
Y. Wu et al./ 2011 [34]	Silica MKR	1.2 μm	500 μm	Not Available	~12 pm/10% RH	Not Available	0.017% RH	Not Available
	Polymer MKR	2.1 μm	500 μm	Not Available	~88 pm/10% RH	Not Available	0.0023% RH	~0.5 s
K. Xu et al./ 2019 [35]	Silica MKR at 2-mm Waveband	1 μm	395 μm	Not Available	~10 pm/% RH	0.034 dB/% RH	Not Available	Rising: ~0.8 s Falling: ~1.55 s
M. Gouveia/ 2014 [36]	Silica MKR	3 μm	250 μm	Nafion	0.11 nm/% RH for low RH 0.29 nm/% RH for High RH	Not Available	Not Available	Not Available
S. Azzuhria et al./ 2018 [42]	Silica MKR	Not Available	3.02 μm	Graphene Oxide	0.0104 nm/% RH (0.0095 nm/% w/o Graphene Oxide)	Not Available	0.096 % RH (0.1% RH w/o graphene oxide)	Not Available
A. D. D. Le et al./ 2018 [29]	Silica MKR	4 μm	380 μm	Polyvinyl Alcohol (PVA)	-1.53 nm/% RH	Not Available	Not Available	Not Available
M. Faruki et al./ 2016 [55]	Silica MKR	2.1 μm	2.5 mm	Titanium Dioxide (TiO₂)	2.5 pm/% RH (1.3 pm/%RH w/o TiO₂)	0.0836 dB/% RH (0.0626 dB/%RH w/o TiO₂)	Not Available	Response: ~25 s Recovery: ~30 s

The small size of the MKR is advantageous for the integration and compatibility with fiber-optic sensors. We emphasize the need for additional research on the MKR-based relative humidity sensor focusing on practical and robust package techniques, stability and reproducibility, discrimination of multiple physical parameters, sensing signal interrogation, massive production, and automation. The micro/nano fabrication technology needs to be developed to stabilize all procedures involved in the MKR fabrication and the deposition of the sensitive materials. New materials sensitive to relative humidity change should be investigated to improve the response time of the MKR and the long-term stability. New configurations of the MKR-based humidity sensor incorporating new sensitive materials, new types of optical fibers (e.g., multicore fibers, few-mode fiber), and other fiber-optic sensors will be exported.

Funding: This research was supported by the Joint Research Project for Outstanding Research Institutions by the Gimhae Industry Promotion and Biomedical Foundation (GIBF).

Conflicts of Interest: The author declares no conflict of interest.

References

1. Jiang, X.; Tong, L.; Vienne, G.; Guo, X.; Tsao, A.; Yang, Q.; Yang, D. Demonstration of optical microfiber knot resonators. *Appl. Phys. Lett.* **2006**, *88*, 223501. [CrossRef]
2. Birks, T.A.; Li, Y.W. The shape of fiber tapers. *J. Light. Technol.* **1992**, *10*, 432–438. [CrossRef]
3. Donlagic, D. In-line higher order mode filters based on long highly uniform fiber tapers. *J. Light. Technol.* **2006**, *24*, 3532–3539. [CrossRef]
4. Brambilla, G. Optical fibre nanowires and microwires: A review. *J. Opt.* **2010**, *12*, 043001. [CrossRef]
5. Sumetsky, M.; Dulashko, Y.; Hale, A. Fabrication and study of bent and coiled free silica nanowires: Self-coupling microloop optical interferometer. *Opt. Express* **2004**, *12*, 3521–3531. [CrossRef] [PubMed]
6. Shi, L.; Chen, X.; Liu, H.; Chen, Y.; Ye, Z.; Liao, W.; Xia, Y. Fabrication of submicron-diameter silica fibers using electric strip heater. *Opt. Express* **2006**, *14*, 5055–5060. [CrossRef]
7. Guo, X.; Ying, Y.; Tong, L. Photonic nanowires: From subwavelength waveguides to optical sensors. *Acc. Chem. Res.* **2014**, *47*, 656–666. [CrossRef]
8. Li, J.H.; Chen, J.H.; Xu, F. Sensitive and wearable optical microfiber sensor for human health monitoring. *Adv. Mater. Technol.* **2018**, *3*, 1800296. [CrossRef]
9. Lu, H.; Tao, J.; Chen, L.; Li, Y.; Liu, L.; Dong, H.; Dong, J.; Qiu, W.; Zhu, W.; Yu, J.; et al. All-optical tuning of micro-resonator overlaid with MoTe$_2$ nanosheets. *J. Light. Technol.* **2019**, *37*, 3637–3646. [CrossRef]
10. Meng, Y.; Deng, L.; Liu, Z.; Xiao, H.; Guo, X.; Liao, M.; Guo, A.; Ying, T.; Tian, Y. All-optical tunable microfiber knot resonator with graphene-assisted sandwich structure. *Opt. Express* **2017**, *25*, 18451–18461. [CrossRef]
11. Wang, Y.; Gan, Z.; Zhao, C.; Fang, L.; Mao, D.; Zu, Y.; Zhang, F.; Zi, T.; Ren, L.; Zhao, J. All-optical control of microfiber resonator by graphene's photothermal effect. *Appl. Phys. Lett.* **2016**, *108*, 171905. [CrossRef]
12. Yi, Y.; Jiang, Y.; Lewis, E.; Brambilla, G.; Wang, P. Optical interleaver based on nested multiple knot microfiber resonators. *Opt. Lett.* **2019**, *44*, 1864–1867. [CrossRef]
13. Tong, L.; Gattass, R.R.; Ashcom, J.B.; He, S.; Lou, J.; Shen, M.; Maxwell, I.; Mazur, E. Subwavelength-diameter silica wires for low-loss optical wave guiding. *Nature* **2003**, *426*, 816–819. [CrossRef] [PubMed]
14. Lou, J.; Wang, Y.; Tong, L. Microfiber optical sensors: A Review. *Sensors* **2014**, *14*, 5852.
15. Li, J.; Gai, L.; Li, H.; Hu, H. A high sensitivity temperature sensor based on packaged microfiber knot resonator. *Sens. Actuators B Chem.* **2017**, *263*, 369–372. [CrossRef]
16. Fan, R.; Yang, J.; Li, J.; Meng, F. Temperature measurement using a microfiber knot ring encapsulated in PDMS. *Phys. Scr.* **2019**, *94*, 125706. [CrossRef]
17. Li, J.; Yang, J.; Ma, J. Load sensing of a microfiber knot ring (MKR) encapsulated in polydimethylsiloxane (PDMS). *Instrum. Sci. Technol.* **2019**, *47*, 511–521. [CrossRef]
18. Dass, S.; Jah, R. Square knot resonator-based compact bending sensor. *IEEE Photon. Technol. Lett.* **2018**, *30*, 1649–1652. [CrossRef]
19. Ly, Y.; Pu, S.; Zhao, Y.; Yao, T. Fiber-optic magnetic field sensing based on microfiber knot resonator with magnetic fluid cladding. *Sensors* **2018**, *18*, 4358.

20. Liao, Y.; Wang, J.; Yang, H.; Wang, X.; Wang, S. Salinity sensing based on microfiber knot resonator. *Sens. Actuators A Phys.* **2015**, *233*, 22–25. [CrossRef]

21. Eryürek, M.; Karadag, Y.; Tasaltın, N.; Kilinc, N.; Kiraz, A. Optical sensor for hydrogen gas based on a palladium-coated polymer microresonator. *Sens. Actuators B Chem.* **2015**, *212*, 78–83. [CrossRef]

22. Wu, X.; Gu, F.; Zeng, H. Palladium-coated silica microfiber knots for enhanced hydrogen sensing. *IEEE Photon. Technol. Lett.* **2015**, *27*, 1228–1231.

23. Lim, K.S.; Harun, S.W.; Damanhuri, S.S.A.; Jasim, A.A.; Tio, C.K.; Ahmad, H. Current sensor based on microfiber knot resonator. *Sens. Actuators A Phys.* **2011**, *167*, 60–62. [CrossRef]

24. Available online: https://www.mordorintelligence.com/industry-reports/global-automotive-temperature-and-humidity-sensors-market-industry (accessed on 27 November 2019).

25. Leone, M.; Principe, S.; Consales, M.; Parente, R.; Laudati, A.; Caliro, S.; Cutolo, A.; Cusano, A. Fiber optic thermo-hygrometers for soil moisture monitoring. *Sensors* **2017**, *17*, 1451. [CrossRef] [PubMed]

26. Chavanne, X.; Frangi, J.P. Autonomous sensors for measuring continuously the moisture and salinity of a porous medium. *Sensors* **2017**, *17*, 1094. [CrossRef] [PubMed]

27. Yang, H.; Wang, S.; Wang, X.; Wang, J.; Liap, Y. Temperature sensing in seawater based on microfiber knot resonator. *Sensors* **2014**, *14*, 18515. [CrossRef]

28. Pang, Y.; Jian, J.; Tu, T.; Yang, Z.; Ling, J.; Li, Y.; Wang, X.; Qiao, Y.; Tian, H.; Yang, Y.; et al. Wearable humidity sensor based on porous graphene network for respiration monitoring. *Biosens. Bioelectron.* **2018**, *116*, 123–129. [CrossRef]

29. Le, A.D.D.; Han, Y.G. Relative humidity sensor based on a few-mode microfiber knot resonator by mitigating the group index difference of a few-mode microfiber. *J. Light. Technol.* **2018**, *36*, 904–909. [CrossRef]

30. Shin, J.C.; Yoon, M.S.; Han, Y.G. Relative humidity sensor based on an optical microfiber knot resonator with a polyvinyl alcohol overlay. *J. Light. Technol.* **2016**, *34*, 4511–4515. [CrossRef]

31. Xing, X.; Wang, Y.; Li, B. Nanofiber drawing and nanodevice assembly in poly (trimethylene terephthalate). *Opt. Express* **2008**, *16*, 10815–10822. [CrossRef]

32. Luo, H.; Sun, Q.; Li, X.; Yan, Z.; Li, Y.; Liu, D.; Zhang, Y. Refractive index sensitivity characteristics near the dispersion turning point of the multimode microfiber-based Mach-Zehnder interferometer. *Opt. Lett.* **2015**, *40*, 5042–5045. [CrossRef] [PubMed]

33. Xu, Z.; Sun, Q.; Li, B.; Luo, Y.; Lu, W.; Liu, D.; Shum, P.P.; Zhang, L. Highly sensitive refractive index sensor based on cascaded microfiber knots with Vernier effect MKR humidity sensor. *Opt. Express* **2015**, *23*, 6662–6672. [CrossRef] [PubMed]

34. Wu, Y.; Zhang, T.; Rao, Y.; Gong, Y. Miniature interferometric humidity sensors based on silica/polymer microfiber knot resonators. *Sens. Actuators B Chem.* **2011**, *155*, 258–263. [CrossRef]

35. Xu, K.; Li, H.; Liu, Y.; Wang, Y.; Tian, J.; Wang, L.; Du, J.; He, Z.; Song, A. Optical fiber humidity sensor based on water absorption peak near 2-μm waveband. *IEEE Photon. J.* **2019**, *11*, 7101308. [CrossRef]

36. Gouveia, M.A.; Pellegrini, P.E.S.; Santos, J.S.; Raimundo, I.M.; Cordeiro, C.M.B. Analysis of immersed silica optical microfiber knot resonator and its application as a moisture sensor. *Appl. Opt.* **2014**, *53*, 7454–7461. [CrossRef] [PubMed]

37. Sasikumar, G.; Ihm, J.W.; Ryu, H. Optimum Nafion content in PEM fuel cell electrodes. *Electrochim. Acta* **2004**, *50*, 601–605. [CrossRef]

38. Omosebi, A.; Besser, R.S. Electron beam assisted patterning and dry etching of Nafion membranes. *J. Electrochem. Soc.* **2011**, *158*, 603–610. [CrossRef]

39. Liu, S.Q.; Ji, Y.K.; Yang, J.; Sun, W.M.; Li, H.Y. Nafion film temperature/humidity sensing based on optical fiber Fabry-Perot interference. *Sens. Actuators A Phys.* **2018**, *269*, 313–321. [CrossRef]

40. Jin, X.L.; Li, W.; Sun, D.; Zhuang, Z.; Wang, X. Fabrication of relative humidity optical fiber sensor based on Nafion-crystal violet sensing film. *Spectrosc. Spectr. Anal.* **2005**, *25*, 1328–1331.

41. Santos, J.S.; Raimundo, I.M., Jr.; Cordeiro, C.M.B.; Biazolib, C.R.; Gouveia, C.A.J.; Jorge, P.A.S. Characterization of a Nafion film by optical fiber Fabry–Perot interferometry for humidity sensing. *Sens. Actuators B Chem.* **2014**, *196*, 99–105. [CrossRef]

42. Azzuhria, S.R.; Amiri, I.S.; Zulkhairi, A.S.; Salimd, M.A.M.; Razak, M.Z.A.; Khyasudeen, M.F.; Ahmad, H.; Zakariad, R.; Yupapin, P. Application of graphene oxide based microfiber-knot resonator for relative humidity sensing. *Results Phys.* **2018**, *9*, 1572–1577. [CrossRef]

Sensors **2019**, *19*, 5196

43. Yu, C.B.; Wu, Y.; Liu, X.L.; Yao, B.C.; Fu, F.; Gong, Y.; Rao, Y.J.; Chen, Y.F. Graphene oxide deposited microfiber knot resonator for gas sensing. *Opt. Mater. Express.* **2016**, *6*, 727–733. [CrossRef]

44. Dash, J.N.; Hegi, N.; Jha, R. Graphene oxide coated PCF interferometer for enhanced strain sensitivity. *J. Light. Technol.* **2017**, *35*, 5385–5390. [CrossRef]

45. Wang, M.; Li, D.; Wang, R.; Zhu, J.; Ren, Z. PDMS-assisted graphene microfiber ring resonator for temperature sensor. *Opt. Quantum Electron.* **2018**, *50*, 132. [CrossRef]

46. Fu, H.; Jiang, Y.; Ding, J.; Zhang, J.; Zhang, M.; Zhu, Y.; Li, H. Zinc oxide nanoparticle incorporated graphene oxide as sensing coating for interferometric optical microfiber for ammonia gas detection. *Sens. Actuators B Chem.* **2018**, *254*, 239–247. [CrossRef]

47. Xing, Z.; Zheng, Y.; Yan, Z.; Feng, Y.; Xiao, Y.; Yu, J.; Guan, H.; Luo, Y.; Wang, Z.; Zhong, Y.; et al. High-sensitivity humidity sensing of microfiber coated with three-dimensional graphene network. *Sens. Actuators B Chem.* **2019**, *281*, 953–959. [CrossRef]

48. Gastón, A.; Pérez, F.; Sevilla, J. Optical fiber relative-humidity sensor with polyvinyl alcohol film. *Appl Opt.* **2004**, *43*, 4127–4132. [CrossRef]

49. Wu, S.; Yan, G.; Lian, Z.; Chen, X.; Zhou, B.; He, S. An open-cavity Fabry-Perot interferometer with PVA coating for simultaneous measurement of relative humidity and temperature. *Sens. Actuators B Chem.* **2016**, *225*, 50–56. [CrossRef]

50. Wang, Y.; Liu, Y.; Zou, F.; Jiang, C.; Mou, C.; Wang, T. Humidity sensor based on a long-period fiber grating coated with polymer composite film. *Sensors* **2019**, *19*, 2263. [CrossRef]

51. Peng, Y.; Zhao, P.; Hu, X.G.; Chen, M.Q. Humidity sensor based on unsymmetrical U-shaped twisted microfiber coupler with wide detection range. *Sens. Actuators B Chem.* **2019**, *290*, 406–413. [CrossRef]

52. Zhao, Y.; Peng, Y.; Chen, M.-Q.; Xia, F.; Tong, R.-J. U-shaped microfiber coupler coated with polyvinyl alcohol film for highly sensitive humidity detection. *Sens. Actuators A Phys.* **2019**, *285*, 628–636. [CrossRef]

53. Liu, Y.; Deng, H.; Yuan, L. A novel polyvinyl alcohol and hypromellose gap-coated humidity sensor based on a Mach– Zehnder interferometer with off-axis spiral deformation. *Sens. Actuators B Chem.* **2019**, *284*, 323–329. [CrossRef]

54. Zhao, Y.; Tong, R.J.; Chen, M.Q.; Xia, F. Relative humidity sensor based on hollow core fiber filled with GQDs-PVA. *Sens. Actuators B Chem.* **2019**, *284*, 96–102. [CrossRef]

55. Faruki, M.J.; Ab Razak, M.Z.; Azzuhri, S.R.; Rahman, M.T.; Soltanian, M.R.K.; Brambilla, G.; Azizur Rahman, B.M.; Grattan, K.T.V.; Rue, R.D.L.; Ahmad, H. Effect of titanium dioxide (TiO_2) nanoparticle coating on the detection performance of microfiber knot resonator sensors for relative humidity measurement. *Mater. Express* **2016**, *6*, 501–508. [CrossRef]

56. Aneesh, R.; Khijwania, S.K. Titanium dioxide nanoparticle based optical fiber humidity sensor with linear response and enhanced sensitivity. *Appl. Opt.* **2012**, *51*, 2164–2171. [CrossRef] [PubMed]

57. Peng, Y.; Zhao, Y.; Chen, M.Q.; Xia, F. Research advances in microfiber humidity sensors. *Small* **2018**, *14*, 1800524. [CrossRef]

58. Yusoff, S.F.A.Z.; Lim, C.S.; Azzuhri, S.R.; Ahmad, H.; Zakaria, R. Studies of Ag/TiO_2 plasmonics structures integrated in side polished optical fiber used as humidity sensor. *Results Phys.* **2018**, *10*, 308–316. [CrossRef]

59. Chen, Z.; Lu, C. Humidity sensors: A review of materials and mechanisms. *Sensor Lett.* **2005**, *3*, 274–295. [CrossRef]

Review

Dual-Polarized Fiber Laser Sensor for Photoacoustic Microscopy

Xiangwei Lin [1,2], Yizhi Liang [3], Long Jin [3] and Lidai Wang [1,2,*]

[1] Department of Biomedical Engineering, City University of Hong Kong, 83 Tat Chee Ave,
Kowloon 999077, Hong Kong, China; lin_xiangwei@yeah.net
[2] City University of Hong Kong Shenzhen Research Institute, Yuexing Yi Dao, Nanshan District,
Shenzhen 518057, China
[3] Guangdong Provincial Key Laboratory of Optical Fiber Sensing and Communications, Institute of Photonics
Technology, Jinan University, Guangzhou 510632, China; liangyizhi88528@gmail.com (Y.L.);
iptjinlong@gmail.com (L.J.)
* Correspondence: lidawang@cityu.edu.hk

Received: 24 September 2019; Accepted: 19 October 2019; Published: 24 October 2019

Abstract: Optical resolution photoacoustic microscopy (OR-PAM) provides high-resolution, label-free and non-invasive functional imaging for broad biomedical applications. Dual-polarized fiber laser sensors have high sensitivity, low noise, a miniature size, and excellent stability; thus, they have been used in acoustic detection in OR-PAM. Here, we review recent progress in fiber-laser-based ultrasound sensors for photoacoustic microscopy, especially the dual-polarized fiber laser sensor with high sensitivity. The principle, characterization and sensitivity optimization of this type of sensor are presented. In vivo experiments demonstrate its excellent performance in the detection of photoacoustic (PA) signals in OR-PAM. This review summarizes representative applications of fiber laser sensors in OR-PAM and discusses their further improvements.

Keywords: fiber laser sensor; dual polarization; high sensitivity; high stability; optical resolution photoacoustic microscopy

1. Introduction

Optical resolution photoacoustic microscopy (OR-PAM) is a prosperous and growing biomedical imaging modality that provides high-resolution, non-invasive, and label-free functional imaging of healthy and diseased tissues [1–3]. Based on the photoacoustic (PA) effect, nanosecond laser pulses induce ultrasonic waves. Thus, an acoustic sensor should be able to detect the PA wave with high sensitivity, broad bandwidth, wide acceptance angle, and high stability [4–6]. Besides, the acoustic detection beam employed in OR-PAM needs to be aligned with the focused optical excitation beam to achieve high sensitivity and high resolution for in vivo imaging [7–12]. Most OR-PAM techniques use piezoelectric detectors to receive PA signals [13–16]. A piezoelectric detector may suffer from low sensitivity with reduced sensor size. Additionally, it is complicated to deliver a laser beam in a confined space such as an endoscope. Therefore, the development of a new photoacoustic detector is of urgent demand in photoacoustic microscopy.

Optical photoacoustic detection has been developed in recent years [13,17–26]. Compared with piezoelectric detectors, optical photoacoustic detectors usually possess high sensitivity. For example, a micro-ring resonator-based photoacoustic sensor has a 105 Pa noise-equivalent pressure (NEP) over a bandwidth of 280 MHz [18]. A planar Fabry–Perot polymer film has a 210 Pa NEP with 20 MHz bandwidth [19,20]. A two-wave mixing interferometer detects PA signals with a maximum bandwidth of 200 MHz [21,22]. Photoacoustic remote sensing microscopy with non-interferometric architecture achieves a measured signal to noise ratio (SNR) of 60 dB with 2.7 ± 0.5 μm spatial resolution [23,24].

A glass substrate-based gold nanostructure etalon reaches 40 MHz center frequency with a bandwidth of 57 MHz [25]. The wide bandwidth and high sensitivity of an optical ultrasound sensor dramatically enhances the performance of OR-PAM. However, the remaining challenge is that high sensitivity may cause poor resistance to external mechanical or thermal disturbances.

Besides the aforementioned sensors, optical fiber lasers are an emerging technique for PA detection in OR-PAM [27–31]. The sensitivity of the fiber laser sensor does not decrease as its size is reduced, making it perfectly match the miniaturization case. To the best of our knowledge, three types of optical fiber-based ultrasound sensors have been developed for photoacoustic imaging. The first one is a Fabry–Perot resonator on the fiber tip to detect the PA signal. The sensor can provide a NEP of 68.7 Pa over 80 MHz bandwidth. In vivo imaging of a mouse ear with a 10×10 mm^2 field of view was demonstrated in Ref. [29]. The second type is a pi-phase-shifted fiber Bragg grating-based sensor. Different from the resonator on the fiber tip, the Bragg grating sensor has an in-fiber cavity and thus can detect ultrasound in the radial direction of the fiber. A NEP of 440 Pa was achieved with 10 MHz bandwidth. Photoacoustic imaging was demonstrated via in vivo mouse ear imaging and ex vivo intravascular imaging [30,31]. The Fabry–Perot resonator sensor and the Bragg grating-based sensor have been systematically reviewed [13,19]. Here, we focus on reviewing a recently developed dual-polarization fiber-laser sensor for photoacoustic microscopy.

Considering the detection sensitivity, bandwidth, sensor size, and detection field of view [32–42], fiber laser sensors based on the dual-polarization principle have their unique advantages in OR-PAM. A fiber-laser-based ultrasound sensor can achieve a 40 Pa NEP over a 50 MHz bandwidth [43–46]. The smaller size makes it straightforward to combine the acoustic detection beam with the optical excitation beam. The cylindrical geometry and side-looking ability make it suitable for PA endoscopy or wearable devices. In addition, the differential detection between two polarizations ensures excellent stability while maintaining high detection sensitivity. Here, we review recent progress on the dual-polarization-based fiber laser sensor and its applications in OR-PAM. We first present the sensor principle, including sensor design, fabrication, signal demodulation, and noise analysis. We then discuss the characterization and optimization of sensitivity. The last part summarizes the in vivo application of the sensor in OR-PAM.

2. Principle of the Dual-Polarized Fiber Laser Sensor

2.1. Fabrication and Sensing Principle

A fiber laser with a short cavity and orthogonally polarized mode emits monochromatic light. Once an ultrasonic wave exerts pressure on the radial direction of the fiber laser cavity, the resonant frequencies of the two polarized modes change differently, and thus the beat frequency varies with ultrasonic pressure. A schematic of the dual-polarization fiber laser sensor is shown in Figure 1. The sensor was fabricated with an Er/Yd co-doped fiber (Er/Yb codoped fiber, EY305, CorActive, Canada). We photo-inscribed two wavelength-matched, highly reflective intracore Bragg gratings in the fiber core to form a resonant cavity. A 193 nm ArF excimer laser with a 1059 nm pitch phase mask was used to fabricate above gratings of length L_g and grating separation L_s. The two gratings and the gain medium between them formed the Fabry–Perot cavity of the fiber laser. To ensure the single longitude mode operation, the grating separation L_s was typically less than 1 cm. Fiber absorption of the pump wavelength (980 nm) was ~1337 dB/m, which offered high gain to the fiber laser. Each grating had a coupling strength of ~25 dB, providing strong optical reflectivity. The length of one grating L_g was 3.0 mm. After fabricating the gratings, the annealing process of the fiber laser was performed for 120 min at 120 °C to reduce the photon darkening effect induced by UV exposure.

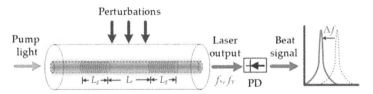

Figure 1. Principle of the dual-polarization fiber laser ultrasound sensor. PD, photodetector.

The lasing frequency is determined by the resonant cavity. The laser emits two linear polarization modes due to the weak birefringence of optical fiber. Each polarization mode can be expressed by [44,45]:

$$\frac{4\pi}{c} f_{x,y} \int_{-\infty}^{+\infty} n_{x,y}(z)|e(z)|^2 dz = 2M\pi \tag{1}$$

where c is the light speed in vacuum, $f_{x,y}$ is the lasing frequency for each polarization mode x and y, $n_{x,y}$ is the refractive index, and M denotes the resonant order. The term $|e(z)|^2$ represents the longitudinal profile of intracavity intensity, which is normalized as $\int_{-\infty}^{+\infty}|e(z)|^2 dz = 1$. When the intensity of the two polarization modes are detected by the photodetector, the subtle difference between lasing frequency yields a beat signal at radio-frequency (RF) range. The beat frequency can be expressed as:

$$\Delta f = \frac{c}{n_0\lambda} \int_{-\infty}^{+\infty} B(z)|e(z)|^2 dz \tag{2}$$

where $B(z) = (|n_x - n_y|)$ is the local birefringence. When the fiber is free from perturbation, the birefringence is mainly caused by imperfection. For example, the fiber cross section may deviate from a perfect circular geometry. Thermal stress and resultant non-uniform strain in the fiber core may induce additional birefringence. At the grating regions, the local birefringence may be affected by the UV side illumination in photo inscribing. Thus, each fiber laser has a unique beat frequency, even fabricated with the same fiber. The principle axis of the fiber is unknown during fabrication of the fiber laser. As heat can change birefringence, we used a CO_2 laser to irradiate the fabricated fiber laser, so that we could finely tune the beat frequency to the preassigned value.

The fiber laser sensing system is illustrated in Figure 2. A continuous-wave laser diode (980 nm) pumped the fiber laser through a wavelength division multiplexer (WDM). The pump power was dozens of milliwatt to maximize the laser output. An optical isolator was connected to the fiber laser output to prevent back reflections that makes the fiber laser unstable. The two polarization modes were orthogonal. To detect the beat signal, a linear polarizer was used to project the two orthogonal polarization modes into the same axis. A polarization controller adjusted the laser polarization states to maximize the beat signal, where its frequency and intensity could be stabilized by the polarization-maintaining fiber. The laser output power was about 0.5–1 mW. Then, we amplified the power to 28 mW using an erbium-doped fiber amplifier so that the SNR at the photodetector could be increased. The beat signal from the photodetector (DSC40S, Discovery, Ewing, NJ, USA) was measured with a vector signal analyzer to determine the frequency.

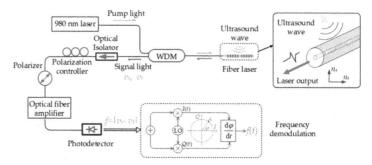

Figure 2. Ultrasound wave detection and signal demodulation based on the fiber laser sensor. Figure adapted with permission from Ref. [43].

As shown in Figure 3a, the two polarization modes can be visualized in the optical spectrum (BOSA 200 CL, Aragon Photonics Labs, Zaragoza, Spain), and these two modes output almost the same power. Considering the polarization-burning-hole effect, the mode competition in the fiber laser is negligible. Figure 3b illustrates the stably detected beat signal at the beat frequency of 2.74 GHz, where its intrinsic birefringence is 2.05×10^{-5}. From Equation (2), the change of fiber birefringence leads to beat frequency variation. Hence, the fiber laser sensor was a birefringence sensor. When the optical fiber engages the acoustic pressure, it is compressed along the ultrasound incident direction. The birefringence of the fiber laser is changed by the acoustic pressure in the radial direction, which subsequently induces the frequency shift in the beat signal [47]. As this frequency shift is proportional to the acoustic pressure, the PA signals can be recovered via a subsequent frequency demodulation procedure.

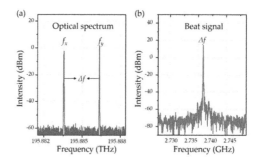

Figure 3. Output optical and radio-frequency spectrum of the fiber laser sensor. (**a**) Spectrum of two polarization modes measured with an optical spectrum analyzer. (**b**) Spectrum of beat signal measured by a radio-frequency (RF) analyzer after the photodetector. Figure adapted with permission from Ref. [43].

2.2. Signal Demodulation and Noise Analysis

The fiber laser sensor used in this work presented the frequency shift of the beat signal in response to incident acoustic waves. To recover the acoustic pressure, a frequency modulation and demodulation system based on I/Q frequency demodulation was required. The RF beat signal (carrier frequency $f_c = \sim 2.74$ GHz) from the photodetector was connected to a vector signal analyzer (Pxie-5646R, NI, Austin, TX, USA), where two low-noise quadrature signals were mixed with the modulated RF signal. The frequencies of the quadrature signals were close to the carrier frequency of the beat signal. After signal mixing and low-pass filtering, the I and Q quadrature signals were able to extract the phase φ of the beat signal. Then, the frequency was deduced as $f_b = d\varphi/dt$. The sampling rate for the I and Q quadrature signals was 100 MHz in the experiments, which allowed for acquiring PA signal with a 50 MHz bandwidth.

Our fiber laser sensor shared similar noise sources as a microwave photonics system [44,45]. The total noise n_0 mainly originates from the fiber laser n_{sen}, optical amplifier n_{edfa}, photodetector n_{pd} and data acquisition n_{acq}. The photodetector noise includes thermal noise n_{th} and shot noise n_{sh}. Each noise term depends differently on the optical power P_{opt}. The total noise can be expressed as $n_0 = n_{sen} + n_{edfa} + n_{th} + n_{sh} + n_{acq}$. Each noise term can be written as:

$$n_{sen} = k_1 P_{opt}^2, \; n_{edfa} = k_2 P_{opt}^2,$$
$$n_{sh} = k_3 P_{opt}, \; n_{th} = k_4 T, \; n_{acq} = k_5 \tag{3}$$

where T is the absolute temperature, and k_1, k_2, k_3, k_4, k_5 are constant coefficients.

The SNR of the demodulated signal can be written as:

$$\Gamma = \frac{1}{(\Delta f_{noise})} = \sqrt{\frac{3k_0 P_{opt}^2}{2nB^3}} = \frac{1}{B^{\frac{3}{2}}} \sqrt{\frac{3k_0 P_{opt}^2}{2(k_1 + k_2)P_{opt}^2 + k_3 P_{opt} + k_4 T + k_5}} \tag{4}$$

where B is the measurement bandwidth, P_{opt} is the RF signal power, and k_0 is the photon-to-electron conversion efficiency [48,49]. Using measurement data in Figure 4, the coefficients in Equations (3,4) are calculated as $k_1 = 5.8 \times 10^{-14}$, $k_2 = 7.8 \times 10^{-14}$, $k_3 = 1.6 \times 10^{-17}$, $k_4 = 1.6 \times 10^{-20}$, and $k_5 = 5.54 \times 10^{-19}$. Both thermal noise and shot noise at 1 mW are −177 dBc/Hz. From k_1, the noise of the fiber laser sensor is −145dBc/Hz, and the noise of the EDFA is ~6 dB. The fitted SNR curve is plotted as a dashed-dot line in Figure 4. In Zone 1, where we have low optical power ($P_{opt} < 1$ mW) on the photodetector, the SNR is mainly limited by thermal noise and is almost proportional to P_{opt}. In Zone 2 (1 mW < $P_{opt} < 3$ mW), shot noise is the dominant noise source. The SNR is approximately proportional to $P_{opt}^{1/2}$. When further increasing the optical power, both thermal and shot noises become less significant, and the SNR becomes stable (labeled as 'Zone 3'). When the input power exceeds the saturation power of the photodetector, the SNR may decrease due to reduced photodetector efficiency (k_0 in Equation (4)).

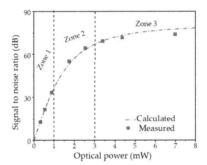

Figure 4. Measured and calculated signal to noise ratio (SNR) at indicated optical power at the photodetector.

3. Characterization and Optimization of Sensitivity

3.1. Frequency Response

The fiber sensor had a cylindrical shape. The pressure-induced deformation was able to be calculated with a vector acoustic scattering model. The scalar solution cannot describe the axially asymmetric modes, i.e., modes with nonzero azimuthal orders in the context, which are typically mixed with shear/longitudinal waves. Instead, a general model was applied to describe the fiber frequency response. We first calculated the plane-wave case, which can be simplified as a two-dimensional problem in the fiber cross-section plane. The solutions of the scalar φ and vector potentials $\vec{H}$ are:

$$\varphi = \sum_l A_n J_l(k_L r) \cos(l\theta) \tag{5}$$

$$H_z = \sum_l B_n J_l(k_S r) \sin(l\theta) \tag{6}$$

where A_n, B_n denote amplitudes of the potentials, J_l is the lth order Bessel function, $k_{S,L} = \omega/c_{S,L}$ is the wave number of the longitudinal or shear waves in the fiber. Here, only the frequency-domain response was considered, and the time-dependent factor $\exp(i\omega t)$ was ignored for simplification.

The displacement $\vec{u}(r, \theta)$ can be expressed as:

$$\vec{u} = -\nabla\varphi + \nabla \times \vec{H} \tag{7}$$

For free vibration, we have:

$$\begin{pmatrix} a_{11} & a_{12} \\ a_{21} & a_{22} \end{pmatrix} \begin{pmatrix} A_n \\ B_n \end{pmatrix} = 0 \tag{8}$$

where the matrix elements are $a_{11} = Z_L^2(q J_n(Z_L) - 2 J_n''(Z_L))/Z_S^2$, $a_{12} = 2n(J_n'(Z_S) - J_n(Z_S))/Z_S^2$, $a_{21} = 2n(J_n'(Z_L) - J_n(Z_L))$, and $a_{22} = -J_n''(Z_S) + J_n'(Z_S) - n^2 J_n(Z_S)$. $Z_L = k_L a$, $Z_S = k_S a$, $Z = k a$, $q = \lambda/\mu$, λ and μ are the Lamé elastic constants for compressibility and shear modulus, respectively.

The equations have nonzero solutions when its determinant $\begin{vmatrix} a_{11} & a_{12} \\ a_{21} & a_{22} \end{vmatrix} = 0$, yielding a discrete spectrum of acoustic resonance. Here, the $l = 2$ modes were investigated because only these modes induced differential stresses between the x and y directions, causing birefringence variation. Each eigen-mode was denoted as (l, n), where l and n are the azimuthal and radial mode index, respectively. The resonant frequencies of first and second radial order modes were $f_{(2, 1)} = 22.3$ MHz and $f_{(2, 2)} = 39.6$ MHz. Their displacement profiles are plotted in Figure 5. For $l = 0$ mode, i.e., the axial symmetrical one, $B_n = 0$, only the compressional waves exist. Equation (8) degenerates as $a_{11} A_n = 0$, which was used for mode calculations. When the fiber loses its axial symmetry via post-processing like side polishing, then mechanical modes with other azimuthal orders will be detected. Particularly, $l = 2$ modes with higher radial number are also simultaneously excited, but their resonant frequencies are beyond the detection bandwidth of frequency demodulation.

Considering these acoustic eigen modes of silica fiber damped by the surrounding medium, the fiber vibration can exert pressure waves. The waves can be depicted as outwards propagating cylindrical waves $C_n H_n^{(1)}(kr)$, where $H_n^{(1)}$ represents outwards propagating pressure and C_n denotes its amplitude. The interaction between the solid fiber and the surrounding medium can be expressed as:

$$a_{11} A_n + a_{12} B_n + a_{13} C_n = 0 \tag{9}$$

where a_{13} is $H(Z)/\rho_s \omega^2$, and b_1 equals zero. Though the shear waves are not supported in fluidic medium, the expression presenting zero shear stress at the boundary still holds, which can be rewritten as:

$$a_{21} A_n + a_{22} B_n = 0 \tag{10}$$

Also, the continuity of radial displacement demands:

$$a_{31} A_n + a_{32} B_n + a_{33} C_n = b_3 \tag{11}$$

where $a_{31} = -J_n'(Z_L)$, $a_{32} = n J_n(Z_S)$ and $a_{33} = -H_n^{(1)'}(Z)/\rho_s \omega^2$, and b_3 denotes the radial displacement created by the acoustic dipole source, which can be expanded as $b_3 =$

$\sum_l\big(b_{L,l}J_l(Z_L)\cos(l\theta) + b_{S,l}J_l(Z_S)\cos(l\theta)\big)$, where the subscript L and S denote the contributions from compressional and shear waves, respectively. Combining Equations (9)–(11), the coefficients A_n, B_n, and C_n can be solved.

The response in beat frequency shift Δf_b is proportional to the birefringence change, i.e., $\Delta f_b = \frac{c}{n_{eff}\lambda}\Delta B$. The birefringence change ΔB is determined by:

$$\Delta B = -p_{44}n_0^3\big(k_L^2 A_n - k_S^2 B_n\big)/2 = -p_{44}n_0^3 A_n\left(k_L^2 + k_S^2\frac{a_{31}}{a_{33}}\right)/2 \tag{12}$$

where n_{eff} means the effective index of the optical mode, p_{44} means the photoelastic coefficient. The model was experimentally verified, with the calculated frequency response found to be consistent with the measured results. As shown in Figure 5b, the original two peaks at 22.3 MHz and 39.6 MHz broadens because of acoustic interaction with the surrounding medium. The (2, 1) and (2, 2) modes present significantly different 3-dB bandwidth. The measured sensitivity at 39.6 MHz is lower than theoretical calculation, which may be caused by water absorption. The acoustic pressure-induced fiber cross-sectional deformations at different response frequency are illustrated in Figure 5c,d. The indexes of the in-plane vibration modes (azimuthal and radial) are denoted as (l, n), and thus the above two frequency response peaks correspond to (2, 1) and (2, 2), respectively. The (2, 1) mode in Figure 5c indicates the compression of the fiber along the ultrasound incident direction, and the (2, 2) mode in Figure 5d corresponds to the case of stretching the outer region while compressing the inner region of the fiber. From the above theoretical analysis, the fiber frequency response is dependent on the cross-sectional size of the fiber whereby can be adjusted by adjusting the fiber diameter. Here, we used HF-etching to reduce the fiber diameter to ~60 μm. The center frequency was tuned to ~42 MHz, and the bandwidth was extended to ~20 MHz.

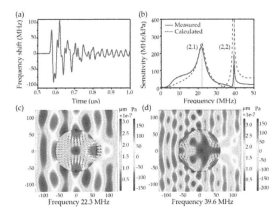

Figure 5. Frequency response of the fiber laser ultrasound sensor. (**a**) Transient response to a pulsed plane wave; (**b**) measured and calculated frequency responses; and (**c**) and (**d**) calculated displacement of the excited fiber vibration at the (2, 1) and (2, 2) modes.

3.2. Spatial Sensitivity

The fiber sensitivity to planer acoustic waves was calibrated using an unfocused ultrasound transducer (V358-SU, Panametrics, USA). Pulsed ultrasound waves propagated normally to the fiber laser sensor. The aperture of ultrasonic wave was ~6 mm, comparable with the fiber sensing region of ~5 mm in length. The acoustic wave induced perturbation uniformly over the entire sensor. The measured temporal and frequency responses are shown in Figure 6, where 198 MHz beat frequency shift occurs at 88 kPa acoustic pressure. The acoustic sensitivity for the 60 μm fiber is 2.25 kHz/Pa, with 40 Pa NEP over a 50 MHz bandwidth. The acoustic sensitivity for the 125 μm fiber is 1.7 kHz/Pa, and the NEP is ~45 Pa with frequency peaking at ~22 MHz and ~39 MHz.

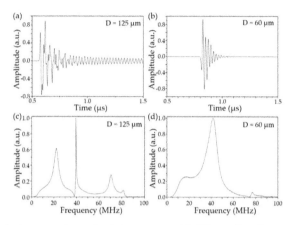

Figure 6. Transient response to a pulsed planar wave with different fiber diameter. Time domain responses of the 125 μm (**a**) and 60 μm (**b**) fiber sensor. Frequency responses of the 125 μm (**c**) and 60 μm (**d**) fiber sensor.

For photoacoustic microscopic imaging, the acoustic source is typically a point source. Thus, it was important to explore how the fiber sensor responded to a point source, i.e., the acoustic response at different positions (r, θ, z). Along the fiber direction, the fiber acts as an ideal line detector with cavity size L_c, and the lasing frequency depends on the resonant condition in Equation (1) at each polarization $f_{x,y}$. The intracavity optical intensity density $e(z)$, whose laser mode profile decides on the cavity length and the grating parameters, can weigh the sensitivity of the fiber laser sensor. The beat signal variation δf_b caused by the local birefringence change $\delta B(z)$ can be written as:

$$\delta f_b = \frac{c}{n_o \lambda} \int_{-L/2}^{L/2} \delta B\big(|p|, \omega, z\big)|e(z)|^2 \frac{exp(ik_a r)}{r} dz \tag{13}$$

For a spherical wave, the acoustic phase changes along the fiber due to different arrival times. The acoustic wave will be canceled out if the phase difference is beyond π. Thus, the sensitivity mainly originates from the perturbation accumulate over a confined region, where the phase is almost unchanged. As a result, Equation (13) can be approximated as:

$$\delta f_b = \frac{c}{n_o \lambda} L_{eq}(\omega) \delta B\big(|p|, \omega, 0\big)|e(z)|^2 \tag{14}$$

where $\delta B\big(|p|, \omega, 0\big)$ is the normal incident plane wave-induced birefringence change. We can see three determinative factors could contribute to the acoustic response: $\delta B\big(|p|, \omega, 0\big)$ depends on the fiber cross-sectional geometry and mechanical characters, $|e(z)|^2$ is the laser mode distribution, $L_{eq}(\omega)$ is the equivalent interaction length when considering phase cancellation. Figure 7 demonstrates that a point source S generates a spherical acoustic wave, which propagates along distance d to reach the line detector. The acoustic pressure is $p(\omega, r) = \frac{exp(ik_a r)}{r}$, where ka denotes the acoustic wave number, $r = (d^2 + z^2)_{1/2}$ is the propagation length, and z is the longitudinal position. Based on the relationship, the equivalent interaction length was calculated as $L_{eq} = 2.506d/ka$, which was much shorter than the effective sensing length, i.e,. 2–10 mm, of the fiber laser sensor.

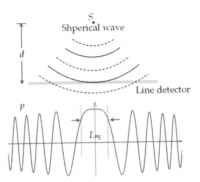

Figure 7. Schematic of the fiber laser sensor as an ideal line detector. A point source emits a spherical wave and the detected acoustic pressure distribution along the line detector. Figure adapted with permission from Ref. [45].

Figure 8 shows the measured frequency response of the fiber laser sensor. The acoustic source is located in the plane at a distance of $d = 1$ mm from the sensor, as shown in Figure 8a. Based on Equation (14), the frequency response decides on the product $L_{eq}(\omega)\delta B(|p|, \omega, 0)$. The effective length L_{eq} remains unchanged for the same distance but different lateral positions. In the x direction, the fiber laser sensor works as a point detector, which also shows a nearly flatten frequency respond with the acoustic source at different positions. As shown in Figure 8c, the measured frequency responses along the z axis are unchanged. In addition, the variation of the amplitude at the peak frequency of 39 MHz comes from the increased loss at further distance.

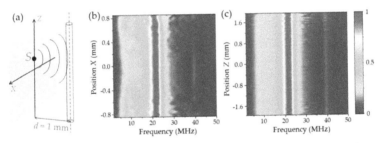

Figure 8. Measured frequency response of the fiber sensor. (**a**) Schematic of scanning acoustic source to measure the frequency response. (**b**) Measured frequency response with scanning source along the x axis (**c**) Measured frequency response with scanning source along the z axis. Figure adapted with permission from Ref. [45].

The cavity behavior of the laser mode was determined by the laser mode distribution, $|e(z)|^2$, which was affected by the grating separation, fiber gain, and grating coupling strength. Between the two gratings, the forward and backward lights experience amplifications and reach maximum intensity before arriving at the gratings. As a result, the Fabry–Perot fiber laser typically presents a profile with two peaks at the inner grating edges. Fibers with higher gain can create sharper peaks and those with lower gain lead to flat-top profiles; thus, higher coupling strengths enable higher slopes. Based on the coupled-mode theory [50,51], the intensities rapidly decrease over the gratings with a simple relation $T = 1 - \tanh^2(\kappa z)$, thus the normalized intensity profile can be expressed by [52]:

$$|e(z)|^2 = \kappa \cdot e^{-2\kappa|z|} \tag{15}$$

where κ is the coupling coefficient of the gratings, and z represents the penetration depth.

When the cavity laser is longer than 2 mm, it can be approximated as a Fabry–Perot laser. The corresponding rectangular mode profiles can be written as:

$$|e(z)|^2 = \begin{cases} \dfrac{1}{L_s} & -\dfrac{L_s}{2} < z < \dfrac{L_s}{2} \\ 0 & other\ regions \end{cases} \tag{16}$$

This model assumes that the intracavity light is evenly confined by the grating reflectors. The fiber lasers sensor can be generally characterized by the effective cavity length L_{eff}. The effective length is approximately equal to the grating separation for Fabry–Perot lasers, which can be expressed as $L_{eff} = 1/\kappa$ for the short structure.

The sensor responses are determined by the integration of laser mode distribution $|e(z)|^2$ over the interaction length L_{eq}. Because L_{eq} is much shorter than the laser cavity, the sensitivity curve is approximated as the laser intensity profile along the fiber cavity. We changed the laser mode distribution via different cavity lengths, L_s. Figure 9a plots the PA intensities along the fiber at different cavity lengths, where the source-to-fiber distance was 250 μm. The sensors showed flat-top profiles if the cavity lengths were 3 mm and 5 mm, whereas a Gaussian-like profile appeared in the 2 mm one. Also, the 2 mm one showed higher sensitivity due to its confined laser mode. The full width at half maximum of these sensors was calculated as 2.2 mm, 3.5 mm, and 4.6 mm, respectively.

When the acoustic pressure is along the principal axis, the induced birefringence variation can be maximized. At 45°, the acoustic pressure induces nearly equal phase changes for both polarization modes, thus the beat signal produces nearly zero frequency shift. Figure 9b shows the azimuthal transformation of the fiber sensitivity. It was measured by the sensor rotating at 10° per step, maintaining a fixed acoustic point source. The angular response exhibits a $|\cos(2\theta)|$ profile, which offers a 60° full angle at half maximum along this direction.

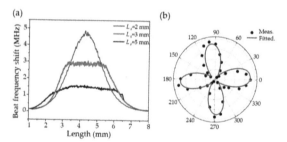

Figure 9. Measured acoustic responses along (a) longitudinal and (b) azimuthal direction. Figure adapted with permission from Ref. [45].

4. Photoacoustic Microscopic Imaging

The dual-polarized fiber laser sensor was used to develop OR-PAM, as shown in Figure 10. The optical beam from a 532 nm pulsed laser (VPFL-G-20, Spectral Physics, Santa Clara, CA, USA) with 1.8 ns pulse width and ~100 nJ pulse energy was collimated, reflected and then focused on the sample surface. The excited PA signals were detected by the fiber laser sensor, and the optical signal was measured by the photodetector and digitalized by a data acquisition card (see Figure 2 for details). To maximize the detection sensitivity, the optical focus and the fiber laser sensor were carefully aligned in the water tank. Meanwhile, the optical beam was carefully positioned to avoid being blocked by the optical fiber.

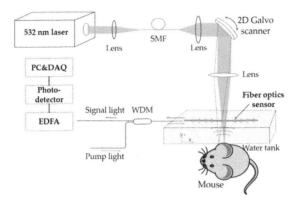

Figure 10. The optical resolution photoacoustic microscopy (OR-PAM) experimental setup, based on the fiber laser sensor. SMF: single mode fiber; WDM: wavelength-division multiplexer; EDFA: erbium-doped fiber amplifier; DAQ: data acquisition. Figure adapted with permission from Ref. [44].

To quantify the lateral resolution of the OR-PAM system, a sharp blade edge was linearly scanned with a step size of 0.18 μm. The lateral resolution was measured as 3.20 μm. To validate its imaging field of view (FOV), the Galvo mirror controlled laser beam was raster-scanned over a black tape with spatially uniform absorption. The fiber laser sensor was fixed 1.60 mm above it. The restored maximum intensity projection (MIP) image was over 3×3 mm^2, and the −6 dB FOV was calibrated as ~3×1.6 mm^2. The penetration depth of the photoacoustic microscopy (PAM) was estimated to be ~800 μm in the phantom experiment and ~200 μm for the in vivo imaging [43–45], as determined by the numerical aperture and laser wavelength. To test the sensor's stability, two human hairs were imaged in B-scan mode over 30 min. The peak to peak amplitude of the PA signal is shown in Figure 11. No noticeable noise variation of the sensor was observed. In a previous report [43], the fiber laser sensor remained stable while the sensor was scanned at ~10 mm/s in water due to the heterodyning detection.

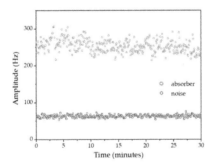

Figure 11. Photoacoustic (PA) amplitude extracted from the B-Scan maximum intensity projection (MIP) image of human hairs as absorbers for 30 min.

In vivo experiments on the mouse ear and brain were conducted to demonstrate the setup's imaging capability. Here, the laser pulse repetition rate was 100 kHz and the scanning rate along the slow and fast axis were 0.2 and 100 Hz, respectively. It took 5 seconds to capture a 3D image. The laser pulse energy on the tissue surface was 300 nJ and the imaging FOV was ~2×2 mm^2. Figure 12a,b exhibits the MIP and 3D images of the mouse ear, where both trunk vessels and capillaries can be resolved clearly in the PA microscopic image. The restored mouse brain images with and without skull are presented in Figure 12c,d, where certain amounts of capillaries become much clearer without the skull. Therefore, the fiber laser sensor exhibited excellent performance in OR-PAM, being able to detect PA signals with great stability and high sensitivity.

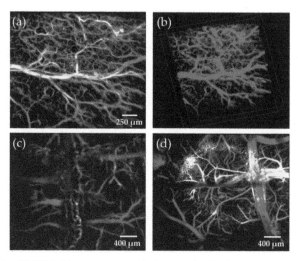

Figure 12. In vivo OR-PAM of the mouse ear and brain using the fiber laser sensor. (**a**) 2D MIP imaging of the mouse ear. (**b**) 3D volumetric image of the mouse ear, the white box is 2.2 × 2.2 × 0.52 mm. (**c**) In vivo imaging of the mouse brain with the skull. (**d**) In vivo imaging of the mouse brain without the skull. (a) and (b) adapted with permission from Ref. [44].

5. Conclusions

In this work, recent developments in dual-polarized fiber laser ultrasound sensors for application in optical-resolution photoacoustic microscopy were reviewed. The fiber laser sensor presented herein demonstrated excellent characteristics, such as high sensitivity, broad bandwidth, minimized size, and great stability. While the photoacoustic waves exerted pressure and induced harmonic vibration of the fiber, the frequency shift of the beating signal between the two orthogonal polarization modes could be captured efficiently. Specifically, a 60 μm fiber laser could achieve an NEP of 40 Pa over a 50 MHz bandwidth. Note that the NEP can be further improved separately via using either a high-power photodetector or averaging multiple duplicated optical signals to suppress the noise from the light source, optical amplifier, photodetector, and data acquisition card. Meanwhile, the frequency shift of the beat signal coming from the dual-polarization mode resisted external perturbations without any frequency-locking techniques. As a result, OR-PAM based on the fiber laser sensor was developed and calibrated, achieving a lateral resolution of 3.2 μm and a FOV of 3 × 1.6 mm². Moreover, excellent in vivo results of photoacoustic imaging in the mouse ear and brain were presented, wherein microvasculature can be clearly visualized. Therefore, the fiber laser ultrasound sensor offers a new tool for all-optical photoacoustic imaging. Moreover, the miniature size and side-looking manner give it the potential for photoacoustic endoscopy.

Author Contributions: All authors conceived the proposed subject. Y.L. and L.J. performed the experiments and provided the data; X.L. and L.W. analyzed the data and results; X.L. and Y.L. wrote the original draft; all authors reviewed and revised the paper.

Funding: This work was partially supported by the National Natural Science Foundation of China (NSFC) (81627805, 61805102); Shenzhen Basic Research Project (JCYJ20160329150236426, JCYJ20170413140519030); Research Grants Council of the Hong Kong Special Administrative Region (21205016, 11215817, 11101618); Fundamental Research Funds for the Central Universities (21618319).

Conflicts of Interest: The authors declare no conflict of interest.

References

1. Wang, L.V.; Yao, J. A practical guide to photoacoustic tomography in the life sciences. *Nat. Methods* **2016**, *13*, 627–638. [CrossRef] [PubMed]
2. Hai, P.; Yao, J.; Maslov, K.I.; Zhou, Y.; Wang, L.V. Near-infrared optical-resolution photoacoustic microscopy. *Opt. Lett.* **2014**, *39*, 5192–5195. [CrossRef] [PubMed]
3. Simandoux, O.; Stasio, N.; Gateau, J.; Huignard, J.P.; Moser, C.; Psaltis, D.; Bossy, E. Optical-resolution photoacoustic imaging through thick tissue with a thin capillary as a dual optical-in acoustic-out waveguide. *Appl. Phys. Lett.* **2015**, *106*, 094102. [CrossRef]
4. Upputuri, P.K.; Pramanik, M. Fast photoacoustic imaging systems using pulsed laser diodes: A review. *Biomed. Eng. Lett.* **2018**, *8*, 167–181. [CrossRef] [PubMed]
5. Shu, X.; Li, H.; Dong, B.; Sun, C.; Zhang, H.F. Quantifying melanin concentration in retinal pigment epithelium using broadband photoacoustic microscopy. *Biomed. Opt. Express* **2017**, *8*, 2851–2865. [CrossRef] [PubMed]
6. Liu, W.; Zhang, H.F. Photoacoustic imaging of the eye: A mini review. *Photoacoustics* **2016**, *4*, 112–123. [CrossRef] [PubMed]
7. Cao, R.; Li, J.; Ning, B.; Sun, N.; Wang, T.; Zuo, Z.; Hu, S. Functional and oxygen-metabolic photoacoustic microscopy of the awake mouse brain. *NeuroImage* **2017**, *150*, 77–87. [CrossRef]
8. Maslov, K.; Zhang, H.F.; Hu, S.; Wang, L.V. Optical-resolution photoacoustic microscopy for in vivo imaging of single capillaries. *Opt. Lett.* **2008**, *33*, 929–931. [CrossRef]
9. Hu, S.; Maslov, K.; Wang, L.V. Second-generation optical-resolution photoacoustic microscopy with improved sensitivity and speed. *Opt. Lett.* **2011**, *36*, 1134–1136. [CrossRef]
10. Wang, L.; Maslov, K.; Yao, J.; Rao, B.; Wang, L.V. Fast voice-coil scanning optical-resolution photoacoustic microscopy. *Opt. Lett.* **2011**, *36*, 139–141. [CrossRef]
11. Yao, J.; Wang, L.; Yang, J.M.; Maslov, K.; Wong, T.T.; Li, L.; Huang, C.H.; Zou, J.; Wang, L.V. High-speed label-free functional photoacoustic microscopy of mouse brain in action. *Nat. Methods* **2015**, *12*, 407–410. [CrossRef] [PubMed]
12. Kim, J.Y.; Lee, C.; Park, K.; Lim, G.; Kim, C. Fast optical-resolution photoacoustic microscopy using a 2-axis water-proofing MEMS scanner. *Sci. Rep.* **2015**, *5*, 7932. [CrossRef] [PubMed]
13. Wissmeyer, G.; Pleitez, M.A.; Rosenthal, A.; Ntziachristos, V. Looking at sound: Optoacoustics with all-optical ultrasound detection. *Light Sci. Appl.* **2018**, *7*, 53. [CrossRef] [PubMed]
14. Wang, L.; Zhang, C.; Wang, L.V. Grueneisen relaxation photoacoustic microscopy. *Phys. Rev. Lett.* **2014**, *113*, 174301. [CrossRef] [PubMed]
15. Xie, Z.; Chen, S.L.; Ling, T.; Guo, L.J.; Wang, X. Pure optical photoacoustic microscopy. *Opt. Express* **2011**, *19*, 9027–9034. [CrossRef]
16. Lutzweiler, C.; Razansky, D. Optoacoustic imaging and tomography: Reconstruction approaches and outstanding challenges in image performance and quantification. *Sensors* **2013**, *13*, 7345–7384. [CrossRef]
17. Hou, Y.; Kim, J.S.; Ashkenazi, S.; Huang, S.W.; Donnell, M.O. Broadband all-optical ultrasound transducer. *Appl. Phys. Lett.* **2007**, *91*, 73507–73509. [CrossRef]
18. Dong, B.; Hao, L.; Zhang, Z.; Zhang, K.; Chen, S.; Sun, C.; Zhang, H.F. Isometric multimodal photoacoustic microscopy based on optically transparent micro-ring ultrasonic detection. *Optica* **2015**, *2*, 169–176. [CrossRef]
19. Guggenheim, J.A.; Li, J.; Allen, T.J.; Colchester, R.J.; Noimark, S.; Ogunlade, O.; Parkin, I.P.; Papakonstantinou, I.; Desjardins, A.E.; Zhang, E.Z.; et al. Ultrasensitive plano-concave optical microresonators for ultrasound sensing. *Nat. Photonics* **2017**, *11*, 714–719. [CrossRef]
20. Zhang, E.; Laufer, J.; Beard, P.C. Backward-mode multiwavelength photoacoustic scanner using a planar Fabry Perot polymer film ultrasound sensor for high-resolution three dimensional imaging of biological tissues. *Appl. Opt.* **2008**, *47*, 561–577. [CrossRef]
21. Chitnis, P.V.; Lloyd, H.; Silverman, R.H. An adaptive interferometric sensor for all-optical photoacoustic microscopy. In Proceedings of the 2014 IEEE International Ultrasonics Symposium (IUS), Chicago, IL, USA, 3–6 September 2014.
22. Shnaiderman, R.; Wissmeyer, G.; Seeger, M.; Soliman, D.; Estrada, H.; Razansky, D.; Rosenthal, A.; Ntziachristos, V. Fiber interferometer for hybrid optical and optoacoustic intravital microscopy. *Optica* **2017**, *4*, 1180–1186. [CrossRef]

23. Hajireza, P.; Shi, W.; Bell, K.; Paproski, R.J.; Zemp, R.J. Non-interferometric photoacoustic remote sensing microscopy. *Light Sci. Appl.* **2017**, *6*, e16278. [CrossRef] [PubMed]

24. Hajireza, P.; Shi, W.; Zemp, R.J. Label-free in vivo fiber-based optical-resolution photoacoustic microscopy. *Opt. Lett.* **2011**, *36*, 4107–4109. [CrossRef] [PubMed]

25. Hou, Y.; Kim, J.S.; Huang, S.W.; Ashkenazi, S.; Guo, L.J.; Donnell, M.O. Characterization of a broadband all-optical ultrasound transducer-from optical and acoustical properties to imaging. *IEEE Trans. Ultrason. Ferroelectr. Freq. Control* **2008**, *55*, 1867–1877. [PubMed]

26. Wissmeyer, G.; Soliman, D.; Shnaiderman, R.; Rosenthal, A.; Ntziachristos, V. All-optical optoacoustic microscope based on wideband pulse interferometry. *Opt. Lett.* **2016**, *41*, 1953–1956. [CrossRef] [PubMed]

27. Wang, Y.; Li, C.; Wang, R.K. Noncontact photoacoustic imaging achieved by using a low-coherence interferometer as the acoustic detector. *Opt. Lett.* **2011**, *36*, 3975–3977. [CrossRef]

28. Hochreiner, A.; Marschallinger, J.B.; Burgholzer, P.; Jakoby, B.; Berer, T. Non-contact photoacoustic imaging using a fiber-based interferometer with optical amplification. *Biomed. Opt. Express* **2013**, *4*, 2322–2331. [CrossRef]

29. Allen, T.J.; Ogunlade, O.; Zhang, E.; Beard, P.C. Large area laser scanning optical resolution photoacoustic microscopy using a fibre optic sensor. *Biomed. Opt. Express* **2018**, *9*, 650–660. [CrossRef]

30. Rosenthal, A.; Razansky, D.; Ntziachristos, V. High-sensitivity compact ultrasonic detector based on a pi-phase-shifted fiber Bragg grating. *Opt. Lett.* **2011**, *36*, 1833–1835. [CrossRef]

31. Rosenthal, A.; Kellnberger, S.; Bozhko, D.; Chekkoury, A.; Omar, M.; Razansky, D.; Ntziachristos, V. Sensitive interferometric detection of ultrasound for minimally invasive clinical imaging applications. *Laser Photonics Rev.* **2014**, *8*, 450–457. [CrossRef]

32. Yao, J.; Wang, L.V. Sensitivity of photoacoustic microscopy. *Photoacoustics* **2014**, *2*, 87–101. [CrossRef] [PubMed]

33. Esenaliev, R.O.; Karabutov, A.A.; Oraevsky, A.A. Sensitivity of laser opto-acoustic imaging in detection of small deeply embedded tumors. *IEEE J. Sel. Top. Quantum Electron.* **1999**, *5*, 981–988. [CrossRef]

34. Winkler, A.M.; Maslov, K.; Wang, L.V. Noise-equivalent sensitivity of photoacoustics. *J. Biomed. Opt.* **2013**, *18*, 097003. [CrossRef] [PubMed]

35. Razansky, D.; Baeten, J.; Ntziachristos, V. Sensitivity of molecular target detection by multispectral optoacoustic tomography (MSOT). *Med. Phys.* **2009**, *36*, 939–945. [CrossRef] [PubMed]

36. Strohm, E.M.; Berndl, E.S.L.; Kolios, M.C. High frequency label-free photoacoustic microscopy of single cells. *Photoacoustics* **2013**, *1*, 49–53. [CrossRef] [PubMed]

37. Guo, Z.; Favazza, C.P.; Uribe, A.G.; Wang, L.V. Quantitative photoacoustic microscopy of optical absorption coefficients from acoustic spectra in the optical diffusive regime. *J. Biomed. Opt.* **2012**, *17*, 066011. [CrossRef] [PubMed]

38. Shelton, R.L.; Mattison, S.P.; Applegate, B.E. Volumetric imaging of erythrocytes using label-free multiphoton photoacoustic microscopy. *J. Biophotonics* **2014**, *7*, 834–840. [CrossRef]

39. Zeng, L.; Liu, G.; Yang, D.; Ji, X.; Huang, Z.; Dong, J. Compact optical-resolution photoacoustic microscopy system based on a pulsed laser diode. *Chin. J. Lasers* **2014**, *10*, 131–136.

40. Chen, S.L.; Xie, Z.; Ling, T.; Guo, L.J.; Wei, X.; Wang, X. Miniaturized all-optical photoacoustic microscopy based on microelectromechanical systems mirror scanning. *Opt. Lett.* **2012**, *37*, 4263–4265. [CrossRef]

41. Qin, W.; Jin, T.; Guo, H.; Xi, L. Large-field-of-view optical resolution photoacoustic microscopy. *Opt. Express* **2018**, *26*, 4271–4278. [CrossRef]

42. Xia, J.; Li, G.; Wang, L.; Nasiriavanaki, M.; Maslov, K.; Engelbach, J.A.; Garbow, J.R.; Garbow, L.V. Wide-field two-dimensional multifocal optical-resolution photoacoustic-computed microscopy. *Opt. Lett.* **2013**, *38*, 5236–5239. [CrossRef] [PubMed]

43. Liang, Y.; Jin, L.; Wang, L.; Bai, X.; Cheng, L.; Guan, B.O. Fiber-laser-based ultrasound sensor for photoacoustic imaging. *Sci. Rep.* **2017**, *7*, 40849. [CrossRef] [PubMed]

44. Liang, Y.; Liu, J.W.; Jin, L.; Guan, B.O.; Wang, L. Fast-scanning photoacoustic microscopy with a side-looking fiber optic ultrasound sensor. *Biomed. Opt. Express* **2018**, *9*, 5809–5816. [CrossRef] [PubMed]

45. Bai, X.; Liang, Y.; Sun, H.; Jin, L.; Ma, J.; Guan, B.O.; Wang, L. Sensitivity characteristics of broadband fiber-laser-based ultrasound sensors for photoacoustic microscopy. *Opt. Express* **2017**, *25*, 17616–17626. [CrossRef] [PubMed]

46. Liang, Y.; Liu, J.W.; Wang, L.; Jin, L.; Guan, B.O. Noise-reduced optical ultrasound sensor via signal duplication for photoacoustic microscopy. *Opt. Lett.* **2019**, *44*, 2665–2668. [CrossRef]

47. Zheng, J. Single-mode birefringent fiber frequency-modulated continuous- wave interferometric strain sensor. *IEEE Sens. J.* **2010**, *10*, 281–285. [CrossRef]

48. Jiang, H.; Taylor, J.; Quinlan, F.; Fortier, T.; Diddams, S.A. Noise floor reduction of an Er: Fiber laser-based photonic microwave generator. *IEEE Photonics J.* **2011**, *3*, 1004–1012. [CrossRef]

49. Devgan, P.S.; Urick, V.J.; Mckinney, J.D.; Williams, K.J. Cascaded noise penalty for amplified long-haul analog fiber-optic links. *IEEE. Trans. Microw. Theory Tech.* **2007**, *55*, 1973–1977. [CrossRef]

50. Ronnekleiv, E.; Ibsen, M.; Cowle, G.J. Polarization characteristics of fiber DFB lasers related to sensing applications. *IEEE J. Quantum Electron.* **2000**, *36*, 656–664. [CrossRef]

51. Hernández-Cordero, J.; Kozlov, V.A.; Carter, A.L.G.; Morse, T.F. Polarization effects in a high-birefringence elliptical fiber laser with a Bragg grating in a low-birefringence fiber. *Appl. Opt.* **2000**, *39*, 972–977. [CrossRef]

52. Foster, S. Spatial mode structure of the distributed feedback fiber laser. *IEEE J. Quantum Electron.* **2004**, *40*, 884–892. [CrossRef]

MDPI
St. Alban-Anlage 66
4052 Basel
Switzerland
Tel. +41 61 683 77 34
Fax +41 61 302 89 18
www.mdpi.com

Sensors Editorial Office
E-mail: sensors@mdpi.com
www.mdpi.com/journal/sensors

CPSIA information can be obtained
at www.ICGtesting.com
Printed in the USA
LVHW021632250820
664158LV00009B/280